Kasthurirangan Gopalakrishnan, Halil Ceylan, and Nii O. Attoh-Okine (Eds.)

Intelligent and Soft Computing in Infrastructure Systems Engineering

Studies in Computational Intelligence, Volume 259

Editor-in-Chief
Prof. Janusz Kacprzyk
Systems Research Institute
Polish Academy of Sciences
ul. Newelska 6
01-447 Warsaw
Poland
E-mail: kacprzyk@ibspan.waw.pl

Further volumes of this series can be found on our homepage: springer.com

Vol. 237. George A. Papadopoulos and Costin Badica (Eds.)
Intelligent Distributed Computing III, 2009
ISBN 978-3-642-03213-4

Vol. 238. Li Niu, Jie Lu, and Guangquan Zhang
Cognition-Driven Decision Support for Business Intelligence, 2009
ISBN 978-3-642-03207-3

Vol. 239. Zong Woo Geem (Ed.)
Harmony Search Algorithms for Structural Design Optimization, 2009
ISBN 978-3-642-03449-7

Vol. 240. Dimitri Plemenos and Georgios Miaoulis (Eds.)
Intelligent Computer Graphics 2009, 2009
ISBN 978-3-642-03451-0

Vol. 241. János Fodor and Janusz Kacprzyk (Eds.)
Aspects of Soft Computing, Intelligent Robotics and Control, 2009
ISBN 978-3-642-03632-3

Vol. 242. Carlos Artemio Coello Coello, Satchidananda Dehuri, and Susmita Ghosh (Eds.)
Swarm Intelligence for Multi-objective Problems in Data Mining, 2009
ISBN 978-3-642-03624-8

Vol. 243. Imre J. Rudas, János Fodor, and Janusz Kacprzyk (Eds.)
Towards Intelligent Engineering and Information Technology, 2009
ISBN 978-3-642-03736-8

Vol. 244. Ngoc Thanh Nguyen, Rados law Piotr Katarzyniak, and Adam Janiak (Eds.)
New Challenges in Computational Collective Intelligence, 2009
ISBN 978-3-642-03957-7

Vol. 245. Oleg Okun and Giorgio Valentini (Eds.)
Applications of Supervised and Unsupervised Ensemble Methods, 2009
ISBN 978-3-642-03998-0

Vol. 246. Thanasis Daradoumis, Santi Caballé, Joan Manuel Marquès, and Fatos Xhafa (Eds.)
Intelligent Collaborative e-Learning Systems and Applications, 2009
ISBN 978-3-642-04000-9

Vol. 247. Monica Bianchini, Marco Maggini, Franco Scarselli, and Lakhmi C. Jain (Eds.)
Innovations in Neural Information Paradigms and Applications, 2009
ISBN 978-3-642-04002-3

Vol. 248. Chee Peng Lim, Lakhmi C. Jain, and Satchidananda Dehuri (Eds.)
Innovations in Swarm Intelligence, 2009
ISBN 978-3-642-04224-9

Vol. 249. Wesam Ashour Barbakh, Ying Wu, and Colin Fyfe
Non-Standard Parameter Adaptation for Exploratory Data Analysis, 2009
ISBN 978-3-642-04004-7

Vol. 250. Raymond Chiong and Sandeep Dhakal (Eds.)
Natural Intelligence for Scheduling, Planning and Packing Problems, 2009
ISBN 978-3-642-04038-2

Vol. 251. Zbigniew W. Ras and William Ribarsky (Eds.)
Advances in Information and Intelligent Systems, 2009
ISBN 978-3-642-04140-2

Vol. 252. Ngoc Thanh Nguyen and Edward Szczerbicki (Eds.)
Intelligent Systems for Knowledge Management, 2009
ISBN 978-3-642-04169-3

Vol. 253. Roger Lee and Naohiro Ishii (Eds.)
Software Engineering Research, Management and Applications 2009, 2009
ISBN 978-3-642-05440-2

Vol. 254. Kyandoghere Kyamakya, Wolfgang A. Halang, Herwig Unger, Jean Chamberlain Chedjou, Nikolai F. Rulkov, and Zhong Li (Eds.)
Recent Advances in Nonlinear Dynamics and Synchronization, 2009
ISBN 978-3-642-04226-3

Vol. 255. Catarina Silva and Bernardete Ribeiro
Inductive Inference for Large Scale Text Classification, 2009
ISBN 978-3-642-04532-5

Vol. 256. Patricia Melin, Janusz Kacprzyk, and Witold Pedrycz (Eds.)
Bio-inspired Hybrid Intelligent Systems for Image Analysis and Pattern Recognition, 2009
ISBN 978-3-642-04515-8

Vol. 257. Oscar Castillo, Witold Pedrycz, and Janusz Kacprzyk (Eds.)
Evolutionary Design of Intelligent Systems in Modeling, Simulation and Control, 2009
ISBN 978-3-642-04513-4

Vol. 258. Leonardo Franco, David A. Elizondo, and José M. Jerez (Eds.)
Constructive Neural Networks, 2009
ISBN 978-3-642-04511-0

Vol. 259. Kasthurirangan Gopalakrishnan, Halil Ceylan, and Nii O. Attoh-Okine (Eds.)
Intelligent and Soft Computing in Infrastructure Systems Engineering, 2009
ISBN 978-3-642-04585-1

Kasthurirangan Gopalakrishnan, Halil Ceylan,
and Nii O. Attoh-Okine (Eds.)

Intelligent and Soft Computing in Infrastructure Systems Engineering

Recent Advances

Springer

Prof. Kasthurirangan Gopalakrishnan
354 Town Engineering Bldg.
Dept. of Civil, Constr. & Env. Engg.
Iowa State University
Ames, IA 50011-3232
USA
E-mail: rangan@iastate.edu

Dr. Nii O. Attoh-Okine
343B DuPont Hall
Dept. of Civil and Environmental Engineering
University of Delaware
Newark, DE 19716
USA

Prof. Halil Ceylan
482B Town Engineering Bldg.
Dept. of Civil, Constr. & Env. Engg.
Iowa State University
Ames, IA 50011-3232
USA

ISBN 978-3-642-04585-1 e-ISBN 978-3-642-04586-8

DOI 10.1007/978-3-642-04586-8

Studies in Computational Intelligence ISSN 1860-949X

Library of Congress Control Number: 2009937151

Typeset & Cover Design: Scientific Publishing Services Pvt. Ltd., Chennai, India.

Printed in acid-free paper

9 8 7 6 5 4 3 2 1

springer.com

Preface

The term "soft computing" applies to variants of and combinations under the four broad categories of evolutionary computing, neural networks, fuzzy logic, and Bayesian statistics. Although each one has its separate strengths, the complementary nature of these techniques when used in combination (hybrid) makes them a powerful alternative for solving complex problems where conventional mathematical methods fail.

The use of intelligent and soft computing techniques in the field of geomechanical and pavement engineering has steadily increased over the past decade owing to their ability to admit approximate reasoning, imprecision, uncertainty and partial truth. Since real-life infrastructure engineering decisions are made in ambiguous environments that require human expertise, the application of soft computing techniques has been an attractive option in pavement and geomechanical modeling.

The objective of this carefully edited book is to highlight key recent advances made in the application of soft computing techniques in pavement and geomechanical systems. Soft computing techniques discussed in this book include, but are not limited to: neural networks, evolutionary computing, swarm intelligence, probabilistic modeling, kernel machines, knowledge discovery and data mining, neuro-fuzzy systems and hybrid approaches. Highlighted application areas include infrastructure materials modeling, pavement analysis and design, rapid interpretation of nondestructive testing results, porous asphalt concrete distress modeling, model parameter identification, pavement engineering inversion problems, subgrade soils characterization, and backcalculation of pavement layer thickness and moduli.

This book belongs to the "Studies in Computational Intelligence (SCI)" series published by Springer Verlag. Each chapter contained in this book has been peer-reviewed by at least two anonymous referees to assure the highest quality. The valuable contributions of the following individuals in assisting with the review process are greatly appreciated: Sunghwan Kim (Iowa State University), Roger W. Meier (The University of Memphis), Fwa Tien Fang (National University of Singapore), P. Chris Marshall (Golder Associates Inc.), Abhisek Mudgal (Iowa State University), and Amit Pande (Iowa State University).

Researchers and practitioners engaged in developing and applying soft computing and intelligent systems principles to solving real-world infrastructure engineering problems will find this book very useful. This book will also serve as an excellent state-of-the-art reference material for graduate and postgraduate students in transportation infrastructure engineering.

August 13, 2009

Kasthurirangan (Rangan) Gopalakrishnan
Ames, Iowa

About This Book

The use of intelligent and soft computing techniques in the field of geomechanical and pavement engineering has steadily increased over the past decade owing to their ability to admit approximate reasoning, imprecision, uncertainty and partial truth. Since real-life infrastructure engineering decisions are made in ambiguous environments that require human expertise, the application of soft computing techniques has been an attractive option in pavement and geomechanical modeling. The objective of this carefully edited book is to highlight key recent advances made in the application of soft computing techniques in pavement and geomechanical systems. Soft computing techniques discussed in this book include, but are not limited to: neural networks, evolutionary computing, swarm intelligence, probabilistic modeling, kernel machines, knowledge discovery and data mining, neuro-fuzzy systems and hybrid approaches. Highlighted application areas include infrastructure materials modeling, pavement analysis and design, rapid interpretation of nondestructive testing results, porous asphalt concrete distress modeling, model parameter identification, pavement engineering inversion problems, subgrade soils characterization, and backcalculation of pavement layer thickness and moduli. Researchers and practitioners engaged in developing and applying soft computing and intelligent systems principles to solving real-world infrastructure engineering problems will find this book very useful. This book will also serve as an excellent state-of-the-art reference material for graduate and postgraduate students in transportation infrastructure engineering.

Written for

Researchers and practitioners engaged in developing and applying soft computing and intelligent systems principles to solving real-world geomechanical and pavement engineering problems.

Keywords

Pavement engineering; artificial intelligence; artificial neural networks; evolutionary computing; genetic algorithms; particle swarm optimization; shuffled complex evolution; support vector machines; data mining; rough set; neuro-fuzzy; decision trees; genetic polynomial; relief ranking filter; extended Kalman filter.

Contents

Rapid Interpretation of Nondestructive Testing Results Using Neural Networks

Imad N. Abdallah[1] and Soheil Nazarian[2]

[1] Center for Transportation Infrastructure Systems,
University of Texas at El Paso, El Paso, Texas
emadn@utep.edu

[2] Center for Transportation Infrastructure Systems,
University of Texas at El Paso, El Paso, Texas
nazarian@utep.edu

Abstract. Artificial neural network tools for structural pavement evaluation have been developed to facilitate the determination of the integrity of existing flexible pavements. With the onset of the movement toward more mechanistic pavement design, such as Mechanistic Empirical Pavement Design Guide, nondestructive testing techniques play a major role to determine properties of pavement structures. Conventional methods such as backcalculating the layer properties are complex and either require a significant computational effort and/or frequent operator intervention. Studies are presented that show the power of artificial neural networks to estimate pavement layer properties and allow for capabilities in developing pavement performance curves and for estimating and monitoring remaining life.

1 Introduction

Many highways agencies are attempting to incorporate this new design process in the state-of-practice. Currently, nondestructive testing (NDT) devices such as the Falling Weight Deflectometer (FWD) and the Seismic Pavement Analyzer (SPA) are available for collecting field data. Each of these technologies provides support to the design process. These tools have significantly contributed to pavement maintenance and rehabilitation strategies.

In this chapter, a discussion of the conventional use of NDT data is described, followed by an overview of the use of the artificial neural networks to supplant the conventional methods. Finally, two studies are presented to demonstrate the power of incorporating ANN in pavement evaluation.

2 Conventional Analysis of NDE Programs Using FWD and SPA

2.1 Estimating Modulus of Pavement Layers Using Falling Weight Deflectometer

The Falling Weight Deflectometer is the most popular NDT device. As shown in Figure 1a, the FWD applies an impulse load to the pavement and seven or more

K. Gopalakrishnan et al. (Eds.): Intel. & Soft Comp. in Infra. Sys. Eng., SCI 259, pp. 1–19.
springerlink.com

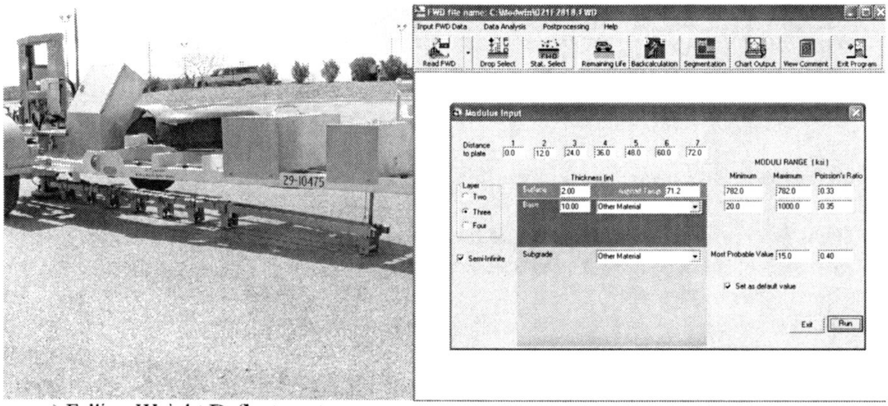

a) Falling Weight Deflectometer b) MODULUS Program

Fig. 1. Falling Weight Deflectometer and MODULUS backcalculation program.

sensors measure the deflections of the pavement. The deflections obtained from the sensors are analyzed to determine the layer moduli of the pavement using a backcalculation algorithm.

Several tools are available in the market that use the FWD data to backcalculate the pavement layer moduli. The determination of pavement moduli using the static layer elastic backcalculation method is, by far, the most widely used procedure (Bush, 1980; Lytton, et al., 1985; Uzan, et al., 1988) [1-3]. The application of layered theory for in-situ material characterization requires the estimation of only one unknown parameter, the modulus of each layer. MODULUS (see Figure 1b, Liu and Scullion, 2001 [4]) is an example of a backcalculation tool used by several agencies including TXDOT.

2.2 Estimating Modulus of Pavement Layers Using Seismic Methods

The Seismic Pavement Analyzer is a trailer-mounted, nondestructive testing device, as shown in Figure 2a. The SPA is similar in size to the FWD. However, the SPA uses more transducers with higher frequencies and more sophisticated interpretation techniques. The measurement is rapid. A complete testing cycle at one point takes less than one minute (lowering sources and receivers, making measurements, and withdrawing the equipment). A detailed discussion on the background of the device can be found in Nazarian et al. (1995) [5]. Its operating principle is based on generating and detecting stress waves in a layered medium. Several seismic testing techniques are combined.

The SPA is mainly designed to determine the variation in modulus with depth and to diagnose the structural condition of pavements. The SPA records the pavement response produced by high- and low-frequency pneumatic hammers on five accelerometers and three geophones. The equipment has been used in several

a) Seismic Pavement Analyzer

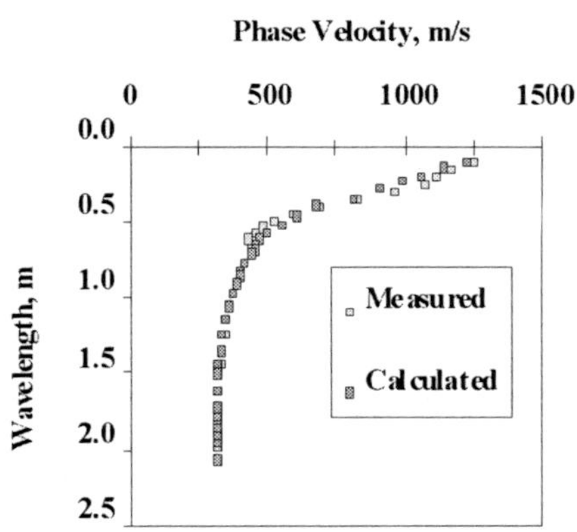

b) Dispersion Curve

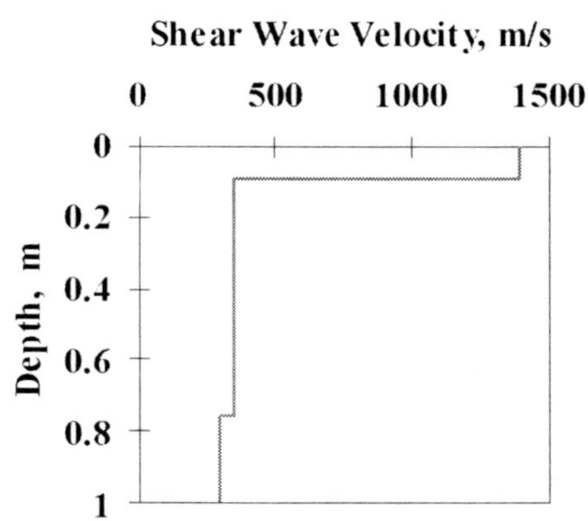

c) Inversion Profile

Fig. 2. Seismic Pavement Analyzer

applications such as analyzing pavement conditions in project-level surveys, diagnosing specific distress precursors to aid in selecting a maintenance treatment, and monitoring pavement conditions after construction as a quality control tool.

The Spectral-Analysis-of-Surface-Waves (SASW) method is used in the SPA to determine the modulus profiles of pavement sections by measuring the dispersive nature of surface waves. The procedure includes collecting data, determining a experimental dispersion curve, and obtaining the stiffness profile (see Figure 2b and 2c). The SASW result, namely the linear-elastic modulus profile, is further incorporated into an algorithm that accounts for the nonlinear behavior of the pavement materials under actual truck traffic. The algorithm is referred to as an equivalent-linear analysis method. Further information on this process can be found in Abdallah et al. (2003) [6].

3 Limitation of Conventional Methods

In its simplest definition, backcalculation is an iterative process that requires varying a set of moduli until a best match between the measured and calculated quantities is obtained. The problem with the backcalculation process is the non-uniqueness of the results. A good match between the measured and estimated values for any of the methods above does not guarantee that the backcalculated moduli are reasonable for that section and, as a consequence, the remaining life of the section could be grossly under or over estimated. This process requires an experienced analyst.

4 Artificial Neural Network

The use of ANN is not new in pavement engineering. Several applications have already been published in the literature. For example, Gucunski et al. (1998) [7], Kim and Kim (1998) [8] and Meier and Rix (1998) [9] implemented of neural networks in pavement engineering. Some of the early applications include: 1) parameter determination, such as the pavement section moduli; 2) assessment of the condition of the pavement and 3) selection of maintenance strategies. Developing a successful ANN model requires several steps as discussed below.

4.1 Generating a Database

To generate a database for any system or model, the input and output parameters to be used in the development process and the tools for generating them need to be identified. ANN models should be ideally developed based on actual field data. As pertained to pavement evaluation, a comprehensive database where the data has passed through a rigorous quality assurance process is difficult to obtain. The cost associated with carrying out extensive testing for this purpose is tremendous and the process is rather difficult. A synthetic database, where the output and possibly input parameters are generated from a numerical model is the next ideal process. A synthetic database allows for flexibility in considering a wide possible range for pavement profiles to cover all the possible conditions.

Figure 3 presents the process of generating a synthetic database that can be used in training, testing, or validating the ANN models. Assume the ANN model to be developed requires deflection data as input and pavement performance as output. The possible ranges of the input parameters have to be defined first. A Monte Carlo simulation (Ang and Tang, 1984) [10] is then conducted using the following assumptions: 1) the variables (e.g., thickness and modulus) are not correlated and 2) these variables are uniformly distributed.

After defining the pavement variables, the input and output parameters are simulated. For example, the deflection profiles can be determined using any layered elastic program for each pavement section. The output parameters (e.g. critical stresses and strains under typical traffic loads can be determined using the same layered elastic program. Based on the critical strains, the remaining life can then

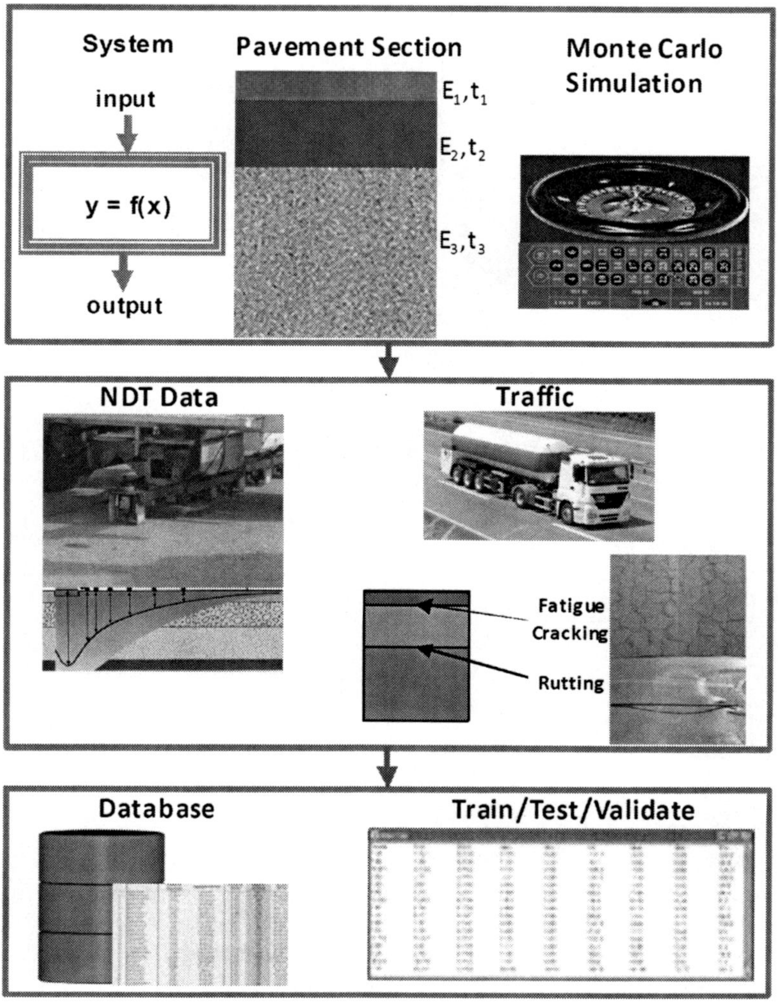

Fig. 3. Process of Developing a Synthetic Database

be calculated from well-established pavement performance models. The main performance parameters are typically fatigue cracking and rutting. Once both the input and output parameters are available for each pavement section, a database is ready to be preprocessed to develop ANN models.

4.2 Data Mining

Data processing or data mining has been recognized as one of the most important aspect of developing ANN models. Data mining can consume up to 70% of the time used to develop a neural network model. Figure 4 presents the results of two

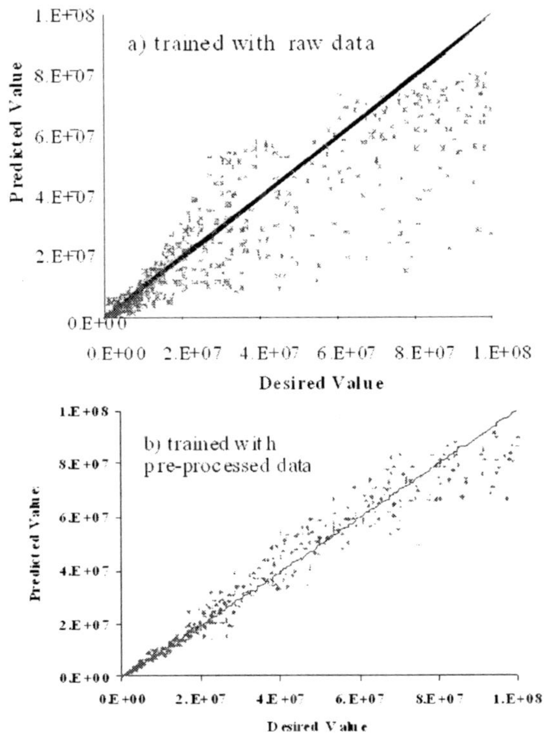

Fig. 4. Effect of Data Preprocessing on ANN Model.

ANN models; one trained with the raw data and the other trained with pre-processed data. The accuracy gained by pre-processing of the data is evident.

Even though ANN modeling is very powerful and is able to map highly nonlinear processes, data mining activities are prudent before training to ensure robust and realistic models. The three general data mining approaches used are: a) correlation analysis, b) evaluation of data sparseness, and c) data transformation (see Figure 5). If some of the input variables are highly correlated, then the model may lack enough independent information. On the other hand, if the input parameters have no correlation to the output a priori, the chance of developing a robust model is small. Correlation analysis can be used to optimize the number of independent input parameters (especially when a large number of input parameters are used) and ensure that the input and output parameters are reasonably correlated.

Data sparseness is defined as uneven and fractionated distribution of the input and output parameters. As shown in Figure 5, the distribution of the deflections is not even and clustered. In many cases, the data should be mined to ensure that the range of data is presented evenly in training the ANN model. In this example, since the deflection data is calculated, there is no control to populate the entire range of

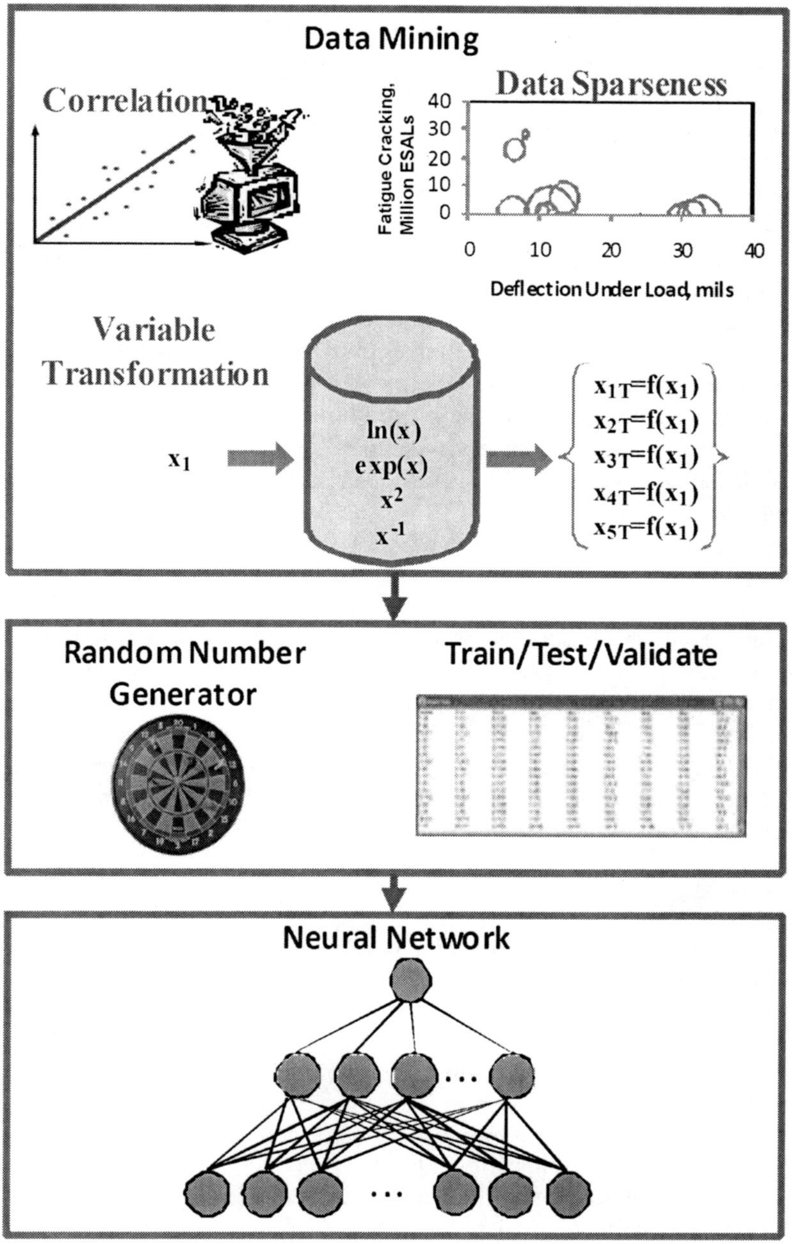

Fig. 5. Process for Data Preprocessing.

deflections. However, for the ranges where the data is denser than others, the number of exemplars can be reduced so that the variable has a more even distribution. This will provide for a better training set when developing the ANN model.

Data transformation also provides a means to improve the data quality of the input and output variables to ensure that the ANN learning process is not inhibited. In data transformation, the variables are subjected to mathematical transformations before being used in the training of ANN models. A combinatorial analysis can be conducted to select a suitable set of transformations for each of the input and output variables. The transformation algorithm generates hybrid or offspring variables based on the original input and output variables. These offspring variables are intended to exhibit more linear and smoother characteristics than the original variables. The selection of candidate variables from the pool of variable transformation can be done in many activities. A genetic algorithm can be implemented to choose the best set of transformations. The criterion used to select the transformations is the minimization of the root mean square (RMS) error. Gucumski et al. (2000) [11] and Abdallah et al. (2000a) [12] provide more information regarding the transformation process.

The next process is to generate the databases to be used for training testing and validating the ANN model. With a synthetic database, the number of examples is usually not an issue. However, the selection of examples for each database is important. To inhibit bias in the data, a random number generator can be used to scatter the examples and ensure that the entire range of pavement section is equally represented in each file. In most instances the database is divided into three sets with usually 7-3-1 ratio for training, testing and validating, respectively. Generally, the larger the database is, the more comprehensive the trained models will be. One concern with a large training database is the possibility of the so-called "overtraining" the model, which deprives it from generalization.

4.3 ANN Models

Figure 6 graphically shows a model for an ANN and its main components. In general, an ANN consists of at least three layers of interconnected processing elements (PEs): (1) input, (2) hidden, and (3) output layers. The number of PEs in the input layer is the same as the number of input variables defined to predict the desired output. The PEs in the output layer represent the output parameters to be predicted. One or several layers of PEs are incorporated in hidden layers. The number of hidden PEs within these layers is decided by trial and error depending on the complexity and nonlinearity of the problem.

In most types of ANN, the PEs between two adjacent layers are usually interconnected. The strength of each connection is expressed by a numerical value called a weight. The weights are determined through a "training" process by presenting input and output examples to the network. The ANN is supposed to learn the relationship between the input and the output by adapting the weights of the connections.

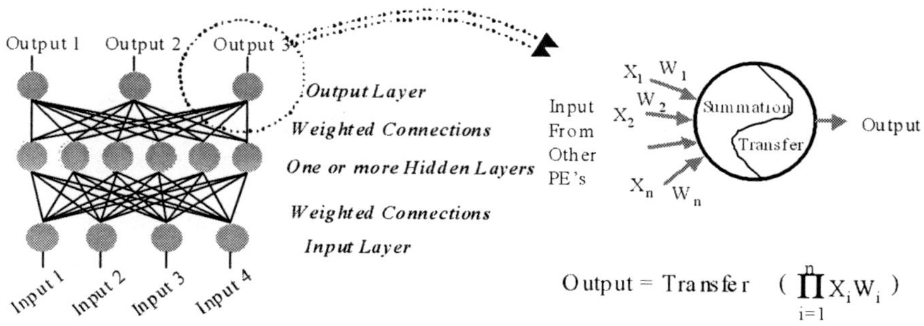

Fig. 6. Components of an Artificial Neural Network.

The popular backpropagation model was implemented in all applications presented in this chapter. Training of a backpropagation neural network involves the forward and backward transfer of information as detailed by Smith (1993) [13] amongst others.

The main interactions between the developer and the software include introducing noise (uncertainty) in the data to simulate measurement uncertainties, selecting the evaluation function used in training, characterizing the closeness between the desired and actual outputs, and sometimes selecting the reasonable variable transformations of the inputs and outputs.

The appropriateness of the model is carried out usually subjectively to either a maximum acceptable root mean square (RMS) errors or a minimum acceptable coefficient of correlation between the calculated and desired output of the ANN model. In cases where several ANN models exhibited similar RMS errors and coefficients of correlation, the number of hidden nodes can be a deciding factor. The fewer the nodes used in the hidden layer, the more general and stable the model will be. Using too many nodes increases the complexity of the model and may increase the risk of over-training the model. Once the network is trained, the model is then tested and validated. Once it has been validated, the development process is completed.

4.4 Limitation

The two main limitations of the ANN models are that (1) the model does not have the ability to extrapolate beyond the range of parameters used in training and (2) the development of robust models is time consuming. The first limitation can be overcome by carefully selecting the training set. The second limitation only affects the model developers and not the users.

5 Case Studies

Two case studies are presented that show the power of ANN to replace or improve the conventional analyses.

5.1 Use of ANN to Improve the Seismic Inversion Process

An algorithm for the rapid reduction of the SASW data was developed where both thickness and modulus of each pavement layer in a three-layer pavement system were estimated. Five individual ANN models were developed, three models to estimate moduli of the three layers and two to estimate the thickness of the asphalt concrete (AC) and base layers. Based on experience, this process ensures that the model accuracy and architecture are only influenced by only one output variable. To evaluate the accuracy of each model and to ensure that no over-training was carried out, a validation database of 250 cases (that were not used in training and testing the model) was used. Depending on the validation results, the developed models can be used either directly to substitute backcalculation or they can be used as a first approximation in traditional backcalculation process.

As an example, Figure 7a presents the quality of the model for estimating the thickness of the AC layer. An upper-lower bound of 10% of the desired value is plotted to help visualize the accuracy of the model's predictions. Only a few data points fall outside those bounds. The histogram of the errors, as shown in Figure 7b,

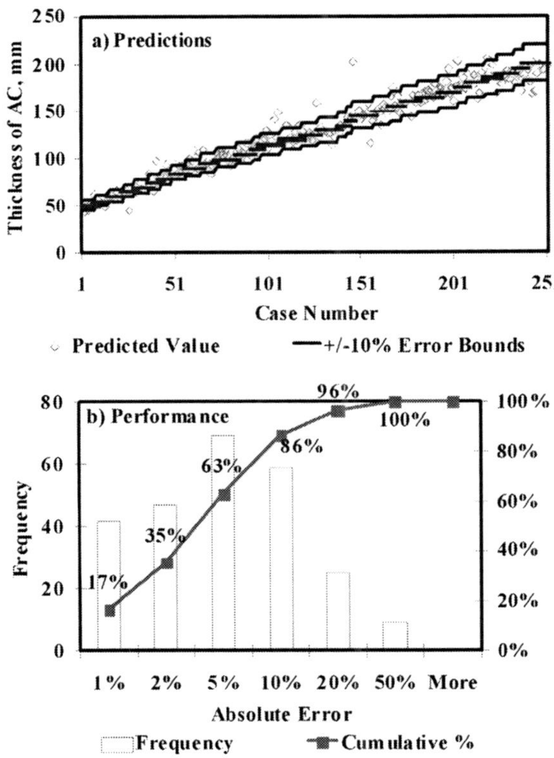

Fig. 7. Results of ANN Model for Predicting the Thickness of AC.

Table 1. Results of ANN Models Based on the Percentage of Data

ANN Model	Layer	Architecture (I/H/O)[a]	Cumulative % of Results Based on Absolute Error					
			1%	**2%**	**5%**	**10%**	**20%**	**50%**
Thickness	AC	(15/30/1)	17%	35%	63%	86%	96%	100%
	Base	(19/14/1)	3%	6%	17%	32%	64%	91%
Modulus	AC	(10/28/1)	45%	74%	96%	100%	-	-
	Base	(13/25/1)	12%	22%	50%	75%	91%	99%
	Subgrade	(14/28/1)	9%	16%	37%	66%	88%	99%

a – (I/H/O) are the processing elements in input, hidden layer and output respectively.

indicates that 86% of the data is predicted with an error of less than 10%. The results of this model are encouraging enough that perhaps this ANN model can replace the backcalculation process. Table 1 contains the information about all five models. The architecture of the models changes, even though the original input parameters to all models is the same. Four models can be considered satisfactory since they estimate their respective output with an accuracy of better than 20% in 88% of the cases. The base thickness model is not satisfactory because of the limited resolving power of dispersion data. Therefore, additional training of this model would not improve performance by much. Other information would be necessary for developing a better ANN model for base thickness.

To demonstrate the use of these neural networks in analysis of SASW data, a pavement section in El Paso, Texas was tested. Fifteen points were collected using the SPA. The data from the SPA was analyzed using the following three approaches: a) Traditional backcalculation, b) ANN models and c) ANN models as a first estimate to the traditional back-calculation. In the third process, the outcome of the ANN was considered as the a priori information (a.k.a. seed value) for backcalulation process.

The estimated moduli of the AC base and subgrade layers from different processes are compared in Figure 8. The ANN results show more consistency along the pavement section. Overall, the moduli from the ANN seem slightly greater for the AC layer and subgrade layers and smaller for the base layer.

Another observation is that the results from the ANN model showed overlap with results to the traditional method when used as a first estimate to the traditional inversion process.

In Figure 9 the results of the thickness of the AC layer are presented. The results of the ANN model are consistent throughout the pavement section. Since the results of the ANN model for predicting base thickness showed only 64% of the data having less than 20% error, no model was selected for the base layer thickness.

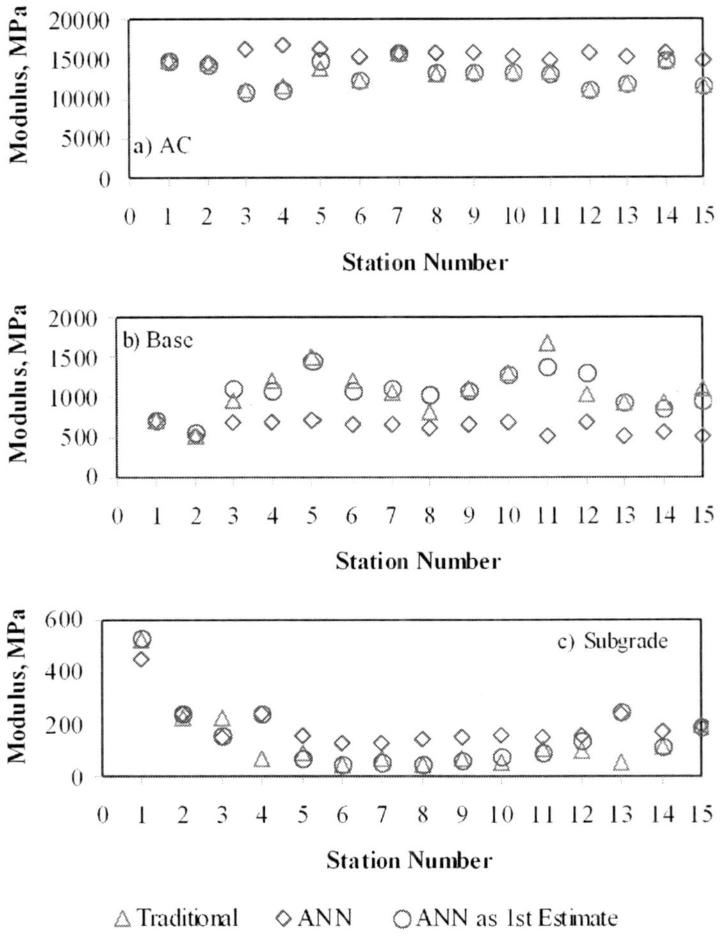

Fig. 8. Comparison of ANN Predicted Modulus Results with Traditional Method.

These encouraging results are significant and are an indication that the ANN modeling tool shows a lot of promise in the analysis of SASW data and could perhaps be a tool that would reduce expert intervention.

Overall, the developed ANNs represent a significant step to moving the advancement of the SASW inversion process one step closer to fully automating the process. This study is also a good illustration of capabilities of artificial neural networks in building tools for a real time evaluation of pavements. Nazarian et al. (2004) [14] provides more detail on this study showing how the use of ANN can become more and more important to quality control and evaluation of pavements.

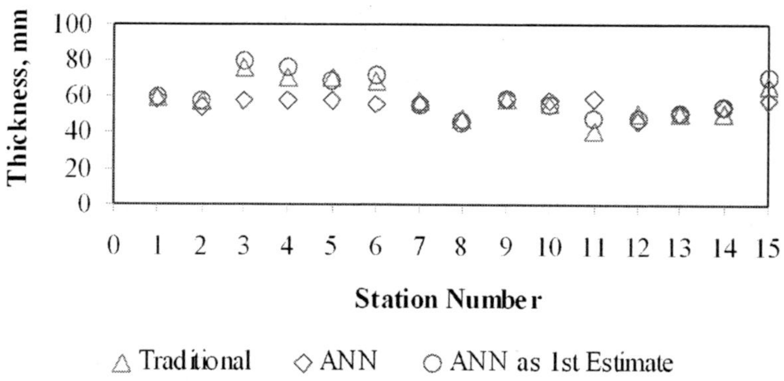

Fig. 9. Comparison of ANN Predicted Thickness Results with Traditional Method.

5.2 Use of ANN in Pavement Performance

As part of a research study conducted for the Texas Department of Transportation, work was carried out to evaluate the structural integrity of the pavement using the Falling Weight Deflectometer. The standard process consists of backcalculating the moduli of the layers, estimating the critical strains from the backcalculated moduli, estimating the remaining life of the pavement from these strains and finally predicting the performance of the pavement with time. The major shortcoming with the traditional backcalculation process is the non-uniqueness of the estimated moduli that may yield vastly different critical strains from the same deflection basin. A series of ANN models were developed to directly predict the critical strains for more reliable estimation of the pavement performance as shown in Figure 10. The inputs to these ANN models where only the best estimates of the thickness of each layer and the surface deflections obtained from a FWD. As such, the backcalculation process is eliminated.

The results of the ANN models are presented in Table 2 for three and four layer flexible pavement systems. The models were separated into thin and thick layers of asphalt-concrete. An additional model to predict the depth to bedrock was developed. The results in Table 2 clearly show the goodness of fit of all the models based on their absolute errors.

The failure limits shown in the figure represents the maximum damage level that can be tolerated before the pavement is repaired. In this study, the failure was defined as 0.5 in. of rutting, or 45% area of the wheel path for fatigue cracking, as recommended by the Asphalt Institute. The performance curve can be developed with at least two points. The first point is the results of the ANN model and is set from the time the FWD testing was conducted. The second is assumed at the beginning of the pavement development, no damage to the pavement. In addition, if condition survey is conducted at the time of the FWD testing, a third point can be incorporated into the function. Furthermore, the instantaneous estimation of an ANN model allows for the flexibility of incorporating a probabilistic algorithm

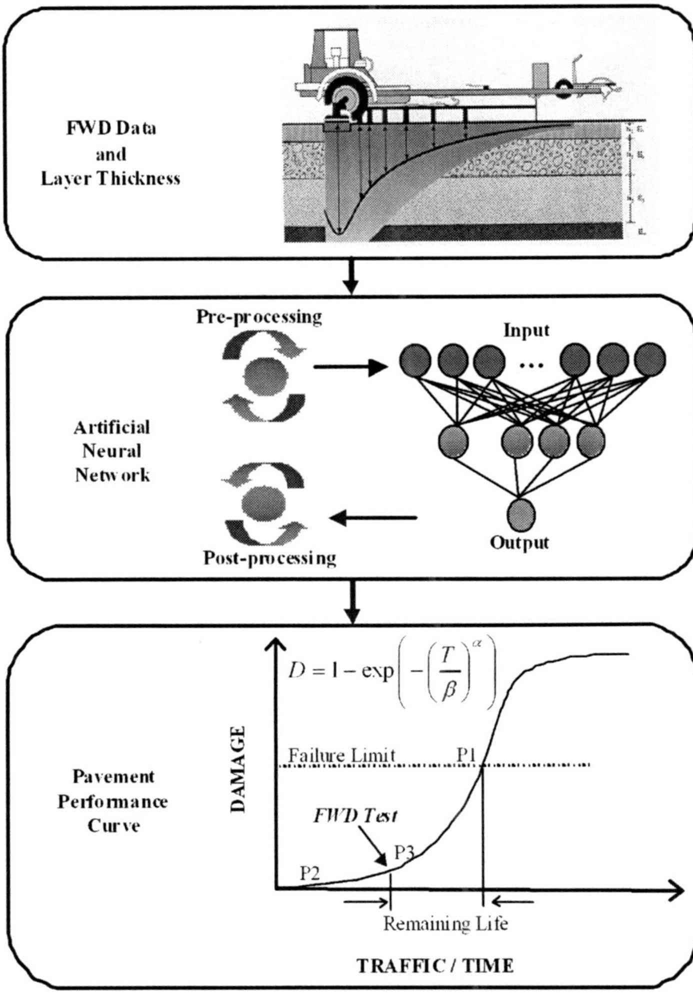

Fig. 10. Summary of Methodology Used to Describe Service Life of a Pavement

into the process. The process allows for thickness and deflection variability to be incorporated in the input for a more realistic estimation of the pavement performance. Further detail on the development of the performance curve is documented in Abdallah et al. (2000b) [16].

This algorithm was validated with real data based on a site located in Texas, which was trafficked to failure using a Mobile Load Simulator (MLS). The site was also tested with FWD and condition survey collected at predetermined load applications. The advantage of this site was that this site was trafficked to failure so that all the pavement performance history is documented. The validation process was carried out in two steps. The first step was to verify that the theoretical

Table 2. Summary and Performance of ANN Models for Flexible Pavement Systems

ANN Models		Parameter	Architecture (I/H/O)[a]	Cumulative % of Results Based on Absolute Error			
				1%	5%	10%	20%
3-Layers	Thin AC	Tangential Strain	40/25/1	58	92	97	99
		Compressive Strain	40/24/1	49	89	95	98
	Thick AC	Tangential Strain	42/50/1	66	99	-	-
		Compressive Strain	42/42/1	67	98	-	-
		AC Modulus	42/30/1	28	82	95	99
	Depth to Bedrock		39/31/1	39	87	96	99
4-Layers	Thin AC	Tangential Strain	45/12/1	23	83	95	99
		Compressive Strain	41/7/1	25	66	81	91
	Thick AC	Tangential Strain	40/30/1	66	95	98	-
		Compressive Strain	43/10/1	62	99	-	-
		AC Modulus	45/30/1	25	79	93	99
	Depth to Bedrock		43/30/1	21	68	89	97

a – (I/H/O) are the processing elements in input, hidden layer and output respectively.

remaining lives reported by the ANN models and those calculated by the conventional method were reasonably similar. The second step consisted of comparing the predicted remaining lives and performance curve with those observed at the site. In the second stage, the impacts of site-related and material-related variability were also considered.

5.3 Comparison of ANN Results with Conventional Methods

The results after the application of predetermined number of axles are shown in Figure 11 along four transverse lines of the testing area. The two methods yield fairly close results. The remaining lives along the centerline (see Figure 11b) are independent of the MLS loading. This is a desirable outcome, indicating that the constant decrease in the remaining lives along the two wheel-paths is primarily due to the damage of the pavement during loading and is not related much to the environmental condition during the MLS loading.

In general, results of this study indicate that in most cases the ANN models can estimate the remaining lives at least as well as a trained design engineer. Since the ANN models yield the results almost instantaneously and without requiring any engineering judgment, it seems reasonable to use the ANN models.

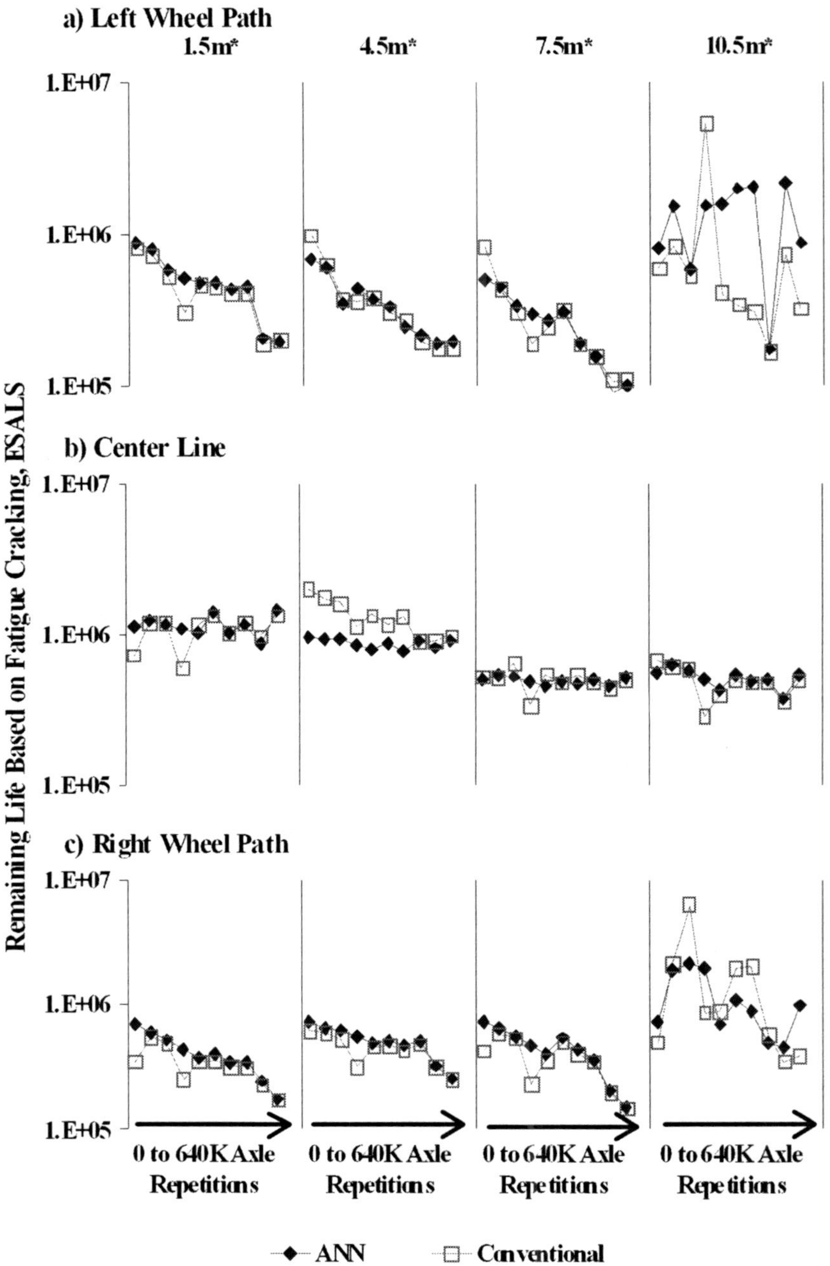

Fig. 11. Comparison of Fatigue Cracking Remaining Lives from ANN Models and Conventional Method

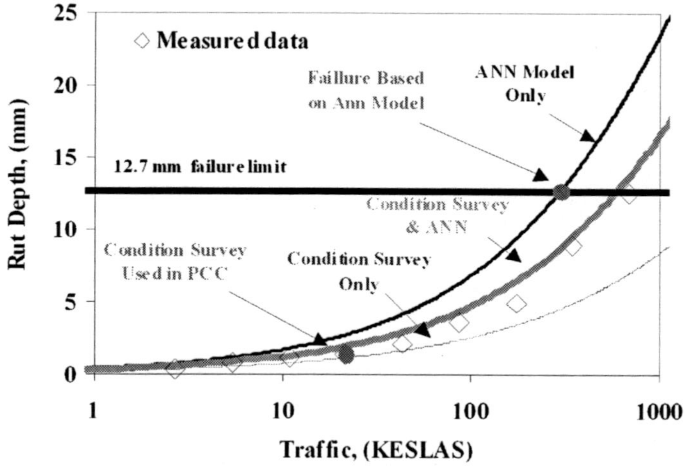

Fig. 12. Comparison of Actual Rutting Performance Curve with Calculated Ones at 20,000 Repetitions Using Several Strategies

The progression of rutting from the ANN is compared with the actual performance data in Figure 12. The PPC when the FWD deflections collected at 20,000 axle repetitions was used alone under-predicts the performance of the pavement. On the other hand, when the condition survey alone was used alone, the performance of the pavement is significantly over predicted. However, when the deflections and the condition survey are combined, the PPC follows the actual performance of the pavement fairly well.

Summary

The artificial neural network technology has proved to be a feasible and practical for developing models to assess the integrity of pavements using data that is readily available to pavement engineers. This is particularly advantageous because other approaches require information from laboratory tests, making the assessment more tedious and time-consuming. The ANN technology represents a significant step toward automating many processes in pavement design and analysis since it allows for building tools for a real time evaluation of pavements. One advantage of the ANN models over traditional approaches is that the remaining life can be calculated without having to backcalculate the elastic moduli of each pavement layer. Another advantage of the ANN models is its inherent capabilities for incorporating uncertainty without sacrificing processing time.

References

[1] Bush, A.J.: Development of the Nondestructive Testing for Light Aircraft Pavements. Phase I, Evaluation of NDT Device Report No. FAA-RD-80-9, Washington, D.C (1980)

[2] Lytton, R.L., Roberts, R.L., Stoffels, S.: Determination of Asphaltic Concrete Pavement Structural Properties by Nondestructive Testing. NCHRP Report No. 10-27, Texas A & M University, College Station, Texas (1985)

[3] Uzan, J., Scullion, T., Michalek, C.H., Parades, M., Lytton, R.L.: A Microcomputer Based Procedure for Backcalculating Layer Moduli from FWD Data. Texas Transportation Institute: Research Report No. 1123-1, Texas A&M University, College Station, Texas (1988)

[4] Liu, W., Scullion, T.: Flexible Pavement Design System FPS 19W: Users's Manual. Research Report 1869-2, Texas Transportation Institute, Texas A&M University, College Station, Texas (2001)

[5] Nazarian, S., Yuan, D., Baker, M.R.: Rapid Determination of Pavement Moduli with Spectral-Analysis-of-Surface-Waves Method. Research Report 1243-1F, Center for Geotechnical and Highway Materials Research, The University of Texas at El Paso, El Paso, TX, 76 p. (1995)

[6] Abdallah, I., Yuan, D., Nazarian, S.: Validation of software Developed for Determining Design Modulus from Seismic Testing. Research Report 1780-5, Center for Highway Materials Research, the University of Texas at El Paso, TX (2003)

[7] Gucunski, N., Krstic, V., Maher, M.H.: Backcalculation of Pavement Profiles from the SASW Test by Neural Networks. In: Flood, I., Kartam, N. (eds.) Artificial Neural Networks for Civil Engineers: Advanced Features and Applications, ASCE, ch. 8, pp. 191–222 (1998)

[8] Kim, Y., Kim, Y.R.: Prediction of Layer Moduli from Falling Weight Deflectometer And Surface Wave Measurements Using Artificial Neural Network. Transportation Research Record 1639, Washington, DC, pp. 53–61 (1998)

[9] Meier, R.W., Rix, G.J.: Backcalculation of Flexible Pavement Moduli from Falling Weight Deflectometer Data using Artificial Neural Networks. In: Flood, I., Kartam, N. (eds.) Artificial Neural Networks for Civil Engineers: Advanced Features and Applications, ASCE, ch. 7, pp. 162–191 (1998)

[10] Ang, A.H.-S., Tang, W.H.: Probability Concepts in Engineering, Planning and Design, vol. 2. John Wiley & Sons, Inc., New York (1984)

[11] Gucumski, N., Abdallah, I., Nazarian, S.: Backcalculation of Pavement Profiles from the SASW Test Using Artificial Neural Network – Individual Layer Approach. In: Proceedings Geo-Denver 2000, Specialty Conference on Pavement Subgrade, Unbound Materials, and Nondestructive Testing, Denver, CO (2000)

[12] Abdallah, I., Melchor-Lucero, O., Ferregut, C., Nazarian, S.: Artificial Neural Network Models for Assessing Remaining Life of Flexible Pavements. Research Report 1711-2, Center for Highway Materials Research, University of Texas El Paso (2000a)

[13] Smith, M.: Neural Networks for Statistical Modeling. Van Nostrand Reinhold, 115 Fifth Ave., New York, NY, 10003 (1993)

[14] Nazarian, S., Abdallah, I., Yuan, D.: Rapid-Reduction to Interpretation of SASW Results Using Neural Networks. Journal of Transportation Research Board No. 1868, Washington, DC, 150–155 (2004)
[15] Garcia-Diaz, A., Riggins, M., Liu, S.J.: Development of Performance Equations and Survivor Curves for Flexible Pavements. Research Report 284-5, Texas Transportation Institute, Texas A&M University, pp. 15-47 (1984)
[16] Abdallah, I., Nazarian, S., Melchor-Lucero, O., Ferregut, C.: Calibration and Validation of Remaining Life Models Using Texas Mobile Load Simulator. In: Proceedings International Conference on Accelerated Pavement Testing, Reno, NV (2000b)

Probabilistic Inversion: A New Approach to Inversion Problems in Pavement and Geomechanical Engineering

Rambod Hadidi[1] and Nenad Gucunski[2]

[1] Senior Engineer, MACTEC Engineering and Consulting, Inc., 28 Second Street, Suite 700, San Francisco, CA, 94105
rhadidi@mactec.com

[2] Professor, Department of Civil and Environmental Engineering, Rutgers University, 623 Bowser Road, Piscataway, NJ, 08854
gucunski@rci.rutgers.edu

"Far better an approximate answer to the right question, which is often vague, than the exact answer to the wrong question, which can always be made precise." John Tukey, Statistician (1915-2000)

Abstract. A wide range of important problems in pavement and geomechnical engineering can be classified as inverse problems. In such problems, the observational data related to the performance of a system is known, and the characteristics of the system that generated the observed data are sought. There are two general approaches to the solution of inverse problems: deterministic and probabilistic. Traditionally, inverse problems in pavement and geomechanical engineering have been solved using a deterministic approach, where the objective is to find a model of the system for which its theoretical response best fits the observed data. In this approach, it is implicitly assumed that the uncertainties in the problem, such as data and modeling uncertainties, are negligible, and the "best fit" model is the solution of the problem. However, this assumption is not valid in some applications, and these uncertainties can have significant effects on the obtained results. In this chapter, a general probabilistic approach to the solution of the inverse problems is introduced. The approach offers the framework required to obtain uncertainty measures for the solution. To provide the necessary background of the approach, few essential concepts are introduced and then the probabilistic solution is formulated in general terms using these concepts. Monte Carlo Markov Chains (MCMC) and its integration with Neighborhood Algorithm (NA), a recently developed global optimization and approximation algorithm, are introduced as computational tools for evaluation of the probabilistic solution. Finally, the presented concepts and computational tools are used to solve inverse problems in Falling Weight Deflectometer (FWD) backcalculation and seismic waveform inversion for shallow subsurface characterization. For each application, the probabilistic formulation is presented, solutions defined, and advantages of the probabilistic approach illustrated and discussed.

1 Introduction

A wide range of problems in pavement and geomechnical engineering involves solution of inverse problems. In such problems, the observational data regarding

K. Gopalakrishnan et al. (Eds.): Intel. & Soft Comp. in Infra. Sys. Eng., SCI 259, pp. 21–45.
springerlink.com

the performance of a system is known and the information about the system is sought. Examples of inverse problems are interpretation of nondestructive testing data, determination of material constitutive parameters from laboratory or field tests, model calibration, etc.

There are two general approaches to the solution of inverse problems: deterministic and probabilistic. In the deterministic approach, which has been historically used for applications in civil engineering (Santamarina and Fratta 1998), the objective is to find the model of a system for which its theoretical response best fits the observed data. The obtained best fit model is then generally considered to be the solution of the inverse problem. This approach provides a single model as the solution of the problem, implying that uncertainties in the inputs to the problem (i.e. the observed data and the theoretically calculated model predictions) are not considered. However, uncertainties are always present, and their effects on the obtained results need to be considered. The probabilistic approach to the solution of inverse problems, a new approach in pavement and geomechnical engineering, provides the framework and mathematical techniques required for obtaining the solution of the inverse problem and evaluating uncertainty measures (Tarantola, 2005).

Introduction and application of the probabilistic approach to problems in pavement and geomechnical engineering is the main focus of this chapter. To provide the necessary background of the approach, few essential concepts are initially introduced and afterwards the probabilistic solution is formulated in general terms using these concepts. Monte Carlo Markov Chains (MCMC) and its integration with Neighborhood Algorithm (NA), a recently developed global optimization and approximation algorithm, are introduced as computational techniques for evaluation of the probabilistic solution. Finally, the presented concepts and computational techniques are used to solve example inverse problems of Falling Weigh Deflectometer (FWD) backcalculation and seismic waveform inversion for shallow subsurface characterization. For each application, the probabilistic formulation is presented and solutions obtained, and advantages of the probabilistic approach illustrated and discussed.

2 Probabilistic Solution of the Inverse Problem

An inverse problem is a mathematical problem where the objective is to obtain information about a parameterized system from observational data, theoretical relationships between system parameters and data, and any available a priori information. To be able to mathematically formulate the probabilistic solution of an inverse problem, there are a few essential concepts that need to be introduced. A detailed description of the concepts presented herein can be also found in other references (Menke 1984, Parker 1994, Tarantola 2005).

2.1 Uncertainty

The concept of uncertainty, as it is related to "simple measurements", is a very familiar and accepted concept in engineering, and represents the effect of uncertainties in a measurement process. For example, a weight measurement has an uncertainty based on the type of the scale, measurement environment, and the care exercised in performing the measurement. Simple uncertainty measures in engineering are often shown with an interval around a measurement value (e.g. 100 +/-1 gram). These uncertainties often accompany results of measurements. The concept of uncertainty should not be confused with the concept of error. Error refers to the difference between the true value of a quantity subject to measurement, called measurand, and measurement results. A measurement can unknowingly be very close to the unknown value of the measurand, thus having a negligible error; however, it may have a large uncertainty. Since the exact value of a measurand can never be evaluated, error is an abstract concept, which cannot be quantified. However, uncertainty is a measure that can be quantified for every measurement.

Any inverse problem can be viewed as a "complex measurement" (Tarantola 2005), where the parameters of interest in the problem are evaluated indirectly by measurement of another set of parameters and establishing the theoretical relationship between these two sets of parameters. In principle, there is no fundamental difference between a "simple measurement" and a "complex measurement". In fact, the measurements that are considered simple are simple forms of inverse problems. For example, a measurement of the weight by a scale is a simple inverse problem. In this problem, the parameter of interest, the weight of an object, is evaluated indirectly by measuring the displacement of a spring. The theoretical relationship linking the observed parameter (deflection) to the parameter of interest (weight) is a simple linear relationship, which is often solved using a calibrated gauge.

Generally speaking, there are uncertainties associated with any measurement, including the complex measurements, which should be presented with the measurement results. These uncertainties are the result of the uncertainties in observed data as well as the uncertainties in the theoretical relationship linking the observed data and the parameters of interest. In simple measurements, the theoretical relationship between the observed data and parameters of interest is usually simple and often do not contribute significantly to the uncertainty of a measurement. Therefore, characterization of uncertainties in the observed data is enough to understand the uncertainties in the measurement process. However, in complex measurements, in addition to data uncertainties, the uncertainties in the theoretical relationship can be substantial. Additionally, depending on the problem, the mapping of the data uncertainty to the parameters of interest can produce significant uncertainties, which should be quantified. The probabilistic approach to inverse problems provides the structure necessary to evaluate the uncertainties in complex measurements. This approach can be viewed as a generalization of the familiar concept of quantifying uncertainties in simple measurements.

2.2 State of Information

In the probabilistic approach, any information about the problem, including the solution of the problem, is expressed by probability distributions that are interpreted using the concept of the state of information. The state of information is an intuitive concept associated with the concept of probability. In addition to the statistical interpretation of probability, a probability distribution can also be interpreted as the subjective degree of knowledge of the true value of a given parameter. The subjective interpretation of the probability theory is usually named Bayesian, in honor of British mathematician Thomas Bayes (1702-1761). The Bayesian interpretation of probability is a very common concept in everyday life, which is used in many situations, such as in weather forecast reports. For example, the forecast predicting a certain probability of having precipitation presents the subjective knowledge of the meteorologist based on all the available information. In engineering problems, the subjective knowledge about any parameter may also be represented by a probability distribution. If there is no specific information about the value of a parameter, this lack of information can also be represented by a homogeneous probability distribution, where all the possible values have the same probability. The other extreme situation is when the exact value of the parameter is known. This precise information can be also represented by a Dirac delta probability distribution. In general, the spread of the probability distribution is an indication of how precise the knowledge about the underlying parameter is; the narrower the spread of the distribution, the more precise the information.

2.3 Probabilistic Formulation

For any given inverse problem, it is possible to select a set of parameters (i.e. parameterization) that adequately describes the system under investigation. Generally, the choice and number of system parameters is not unique. However, once a particular parameterization is chosen (i.e. model), it is possible to introduce a space of values that contains all the possible values of the system parameters. This abstract space is termed the model space, and is denoted by M. Individual sets of parameters describing a specific model, $\boldsymbol{m}=\{m_1,m_2,\ldots\}$, are basically points in the model space.

In an inverse problem, the values of parameters $\boldsymbol{m}$ are of the main interest; however, they are not directly measurable. The goal of the inverse problem is to obtain information on the values of $\boldsymbol{m}$ by making a direct observation on another set of parameters denoted as $\boldsymbol{d_{obs}}$. Similar to the concept of a model space, an abstract idea of a data space D can be introduced, which is the space of all conceivable observed responses, such as $\boldsymbol{d}$. The actual observed response is in fact a point in this space represented by $\boldsymbol{d_{obs}}=\{d_{obs1},d_{obs2},\ldots\}$.

In the approach presented here, the density function of the probability distribution, representing the state of information on system parameters prior to the solution, is denoted by $\rho_M(\boldsymbol{m})$ and is termed the model a priori probability. This information can be based on prior experience and judgment, specific measurement, or theoretical considerations. Similarly, the a priori information on the observed data prior to the solution can be expressed as data a priori probability distribution, with a density function denoted by $\rho_D(\boldsymbol{d})$. This probability describes the uncertainties in the measurement of observed data (i.e. uncertainties in simple measurements).

The a priori information on system parameters is independent of the a priori information on data. This notion of independence can be used to define a joint probability over the joint model and data space as a product of the two marginal probability densities. The density function of this probability distribution is denoted by:

$$\rho(\boldsymbol{m},\boldsymbol{d})=k\ \rho_M(\boldsymbol{m})\ \rho_D(\boldsymbol{d}) \tag{1}$$

where $\rho(\boldsymbol{m},\boldsymbol{d})$ is referred to as joint a priori probability and k is normalization constant. This distribution can be graphically depicted as a "cloud" of probability centered on the observed data and a priori model, as shown in Figure 1.

In the probabilistic approach, information contained in the theoretical relationship between model parameters and data is also represented by a probability distribution, $\theta(\boldsymbol{m},\boldsymbol{d})$. If this relationship is exact and without any uncertainty, a single response will be predicted for a given model. This can be represented using a forward operator defined as $\boldsymbol{d}=g(\boldsymbol{m})$. However, the predictions of a theoretical relationship are rarely exact and there are modeling approximations involved. The distribution $\theta(\boldsymbol{m},\boldsymbol{d})$ represents these uncertainties, where for a given model parameter, a probability in the data space is predicted representing the modeling uncertainties. The distribution $\theta(\boldsymbol{m},\boldsymbol{d})$ is also graphically depicted in Figure 1.

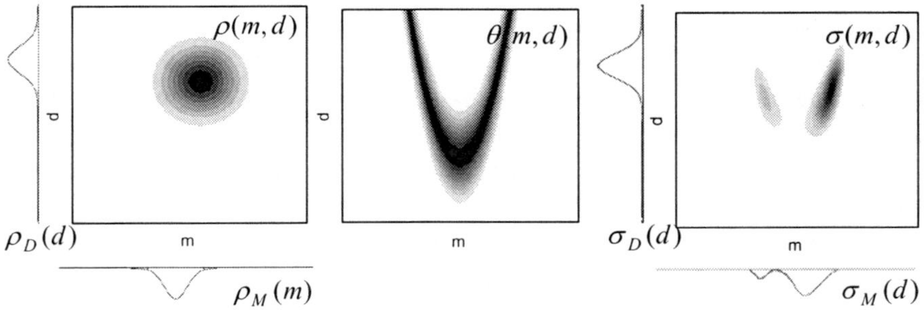

Fig. 1. Conceptual presentation of the general probabilistic solution of an inverse problem, $\sigma(\boldsymbol{m},\boldsymbol{d})$, as a combination of the a priori information, $\rho(\boldsymbol{m},\boldsymbol{d})$, and information obtained from the forward model, $\theta(\boldsymbol{m},\boldsymbol{d})$. Darker shades indicate higher probabilities.

The solution of an inverse problem can be defined as a probability distribution combining the a priori information, $\rho(\boldsymbol{m},\boldsymbol{d})$, with the information obtained from the theoretical relationship, $\theta(\boldsymbol{m},\boldsymbol{d})$. This combination can be accomplished by multiplication of the probability distributions (Tarantola 2005).

The resulting probability distribution, $\sigma(\boldsymbol{m},\boldsymbol{d})$, is termed a posteriori probability distribution and is denoted by:

$$\sigma(\boldsymbol{m},\boldsymbol{d}) = k\ \theta(\boldsymbol{m},\boldsymbol{d})\ \rho(\boldsymbol{m},\boldsymbol{d}) \tag{2}$$

where k is a normalization constant. This multiplication and obtained distribution are conceptually depicted in Figure 1. As shown in the figure, the marginal probability of $\sigma(\boldsymbol{m},\boldsymbol{d})$ in the model space, denoted by $\sigma_M(\boldsymbol{m})$, is the main solution of the inverse problem.

In most applications, including the applications considered in the chapter, data a priori and modeling uncertainties can be modeled by Gaussian probabilities. Assuming Gaussian uncertainties, it can be shown that the problem solution can be reduced to (Tarantola 2005):

$$\sigma_M(\boldsymbol{m}) = k\ \rho_M(\boldsymbol{m})\ \lambda(\boldsymbol{m}) \tag{3}$$

where k is a normalization constant and $\lambda(\boldsymbol{m})$ is the likelihood function represented by:

$$\lambda(\boldsymbol{m}) = k\ exp(-0.5(\boldsymbol{d}_{obs} - g(\boldsymbol{m}))^T (\boldsymbol{C}_T^{-1} + \boldsymbol{C}_D^{-1})(\boldsymbol{d}_{obs} - g(\boldsymbol{m}))) \tag{4}$$

where $\boldsymbol{d}_{obs}$ is the observed data, $\boldsymbol{C}_D$ is the covariance matrix representing observational uncertainties, $\boldsymbol{C}_T$ is the covariance matrix representing uncertainties in theoretical relationship, k is a normalization constant, and $g(\boldsymbol{m})$ is the forward model operator defined earlier. This equation is adopted for numerical implementation in the presented examples.

It is worthy to mention that, by inspection of these equations, it can be observed that the deterministic solution is a very special case of the probabilistic solution. If there is no a priori information (i.e. homogenous probability distribution), the maximum of the a posteriori probability occurs at the maximum of the likelihood function. Additionally, if the covariance matrices are multiplications of the identity matrix, it can be shown that the maximum of the likelihood function is the L_2 norm best fit to the observed data. This is the same solution obtained in the deterministic approach. Therefore, it can be concluded that the deterministic solution is a very special case of the probabilistic solution.

3 Evaluation of Probabilistic Solution

3.1 Analytical Evaluation

If the a priori and forward model probability distributions are simple, it might be feasible to obtain a closed form analytical solution of the problem directly using the presented equations (e.g. Hadidi and Gucunski 2008). However, in practice, this is rarely the case and a numerical solution of the problem might be the only option.

3.2 Direct Sampling Evaluation

In cases where the a priori and/or forward model probability distributions are not simple, the a posteriori probability can be numerically evaluated by direct sampling of the probability distribution. The direct sampling is a numerical technique that has as an ultimate objective randomly generating a large representative collection of models according to the a posteriori probability, $\sigma_M(\boldsymbol{m})$, and then analyzing the sampled models to derive information about the underlying probability distribution. Generation of random samples according to $\sigma_M(\boldsymbol{m})$ can be accomplished using Markov chains.

A Markov chain, named after the Russian mathematician Andrey Markov (1856-1922), is a sequence of random samples, x_n, such that the next value or state of the sequence depends only on the previous state (Bhat and Miller 2002). This dependency is mathematically described by a transition kernel, $\psi(x_{n+1}|x_n)$, which is the probability distribution representing the conditional probability of transition to x_{n+1} from x_n. It can be shown that, given certain conditions (Tierney 1994), the distribution of random samples will always converge to a stationary or target distribution, denoted by $\Psi(x)$. Hence, by an appropriate definition of the transition kernel, a set of samples according to $\Psi(x)$ can be generated.

Given a desired target probability, $\Psi(x)$, one approach in the generation of samples according to this distribution is through application of Metropolis acceptance rules (Metropolis et al 1953; Hastings 1970). To define these rules, a Markov chain with a given state x_n is considered. To move to the next state, a candidate state, y, is randomly generated from a uniform distribution. If the value of $\Psi(y) > \Psi(x_n)$, the candidate point is accepted and $x_{n+1}=y$. If $\Psi(y) < \Psi(x_n)$, the candidate point is only accepted with a probability of $\Psi(y)/\Psi(x_n)$. Otherwise, the candidate point is rejected and $x_{n+1} = x_n$. If this sequence is repeated, it can be shown that the Markov chain generated in this manner will converge to $\Psi(x)$. Since random numbers are involved in generation of a Markov Chain, the chain is often further described as a Monte Carlo Markov Chain (MCMC). A collection of sampled points for a two dimensional target probability distribution generated by a MCMC is graphically presented in Figure 2. MCMC is a very useful tool and it is

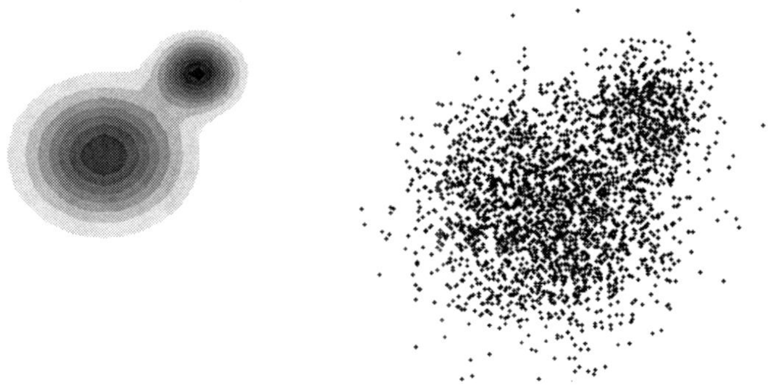

Fig. 2. A contour plot of two dimensional probability (left) and collected samples based on that probability (right) are depicted. Darker shades indicate higher probabilities.

a very active area of statistical research. A detailed treatment of MCMC, including the issues of convergence and stability, is beyond the scope of this discussion and can be found in other references (Gelman et al. 2004).

To obtain a numerical solution of an inverse problem, the Metropolis acceptance rule in a cascade form can be used to generate samples of the model space at a rate proportional to a posteriori probability, $\sigma_M(\boldsymbol{m})$. In the approach based on Equation (3), samples of the model space, according to the a priori probabilities $\rho_M(\boldsymbol{m})$, are initially generated using the Metropolis acceptance rules. The generation of these samples can be simply accomplished by starting with a random set of values and proceeding with generation of another random candidate, which should be accepted or rejected according to Metropolis rules with a target probability of $\rho_M(\boldsymbol{m})$. If this process is repeated, a Markov chain with the target probability of $\rho_M(\boldsymbol{m})$ will be generated. However, if the samples generated based on the target probability of $\rho_M(\boldsymbol{m})$ are then treated as candidate samples for another set of Metropolis rules with the target distribution of $\lambda(\boldsymbol{m})$, it can be shown that the resulting Markov chain from samples accepted using both Metropolis rules will have a target probability of $\rho_M(\boldsymbol{m})\lambda(\boldsymbol{m})$ (Mosegaard and Tarantola 1995). Such a target distribution, according to Equation (3), is the a posteriori probability, $\sigma_M(\boldsymbol{m})$. The described sampling algorithm is a cascade implementation of Metropolis acceptance rules with the target probabilities $\rho_M(\boldsymbol{m})$ and $\lambda(\boldsymbol{m})$. In this algorithm, system parameters that are consistent with a priori information, as well as observations, are sampled most often, whereas others that are incompatible with either a priori information or observational data are sampled rarely. This procedure generates a collection of samples for which their frequency becomes asymptotically proportional to a posteriori probability distribution in the model space. A more comprehensive description of this procedure can be found in other references (Mosegaard and Tarantola 1995).

3.3 Direct Sampling Evaluation with Neighborhood Approximation

To determine the acceptance or rejection of a sample in a direct sampling technique, the likelihood function and in turn the forward model have to be evaluated. Because the number of samples generated in the direct sampling approach is large, if the forward model is complex (such as a Finite Element Model), the technique becomes inefficient. In such cases, one desirable approach is to initially build a "good" approximation to the likelihood function (See Equation 4) using limited evaluations of the forward model. Such an approximation, instead of the actual likelihood function, can then be used to directly sample and appraise the a posteriori probability. This approach is implemented in this chapter using a recently developed search and approximation algorithm, referred to as Neighborhood algorithm (NA) (Sambridge 1999a,b). As hinted, there are two distinct stages in this numerical implementation: the search/approximation stage and appraisal stage.

In the search/approximation stage, the model space is searched and an approximation to the likelihood function is constructed throughout the model space. This stage is implemented using the NA. The Neighborhood algorithm falls in the same class of global optimization methods, such as simulated annealing and genetic algorithm, and can be directly applied to optimization problems. However, this algorithm, rather than seeking a single optimal point, provides an approximation of the likelihood function, which is preferentially sampled more at the "good" regions of the space (i.e. maxima or minima points).

The basic premise of the NA algorithm is to use previously evaluated samples to construct an approximation to the likelihood function throughout the model space and use this approximation to guide further evaluations for further refining the approximation. The generalized algorithm can be presented as follows:

- Construct the approximate likelihood function surface from the previous set of inputs for which the forward model has been solved;
- Use this approximation to generate the next set of inputs and find the likelihood function value for them;
- Repeat the above steps until an approximation with desired accuracy is reached.

To construct an approximation using a limited number of forward model evaluations, NA uses a mathematical construct known as Voronoi diagram, named after Georgy Voronoi (1868-1908). Given n points in the space, Voronoi diagram is a unique way of dividing the n-dimensional space into n unique regions called cells. Each cell is simply the nearest neighborhood region about one of the points, as measured by a particular distance measure, most often L_2 norm. In NA, the approximate value of the likelihood function within each Voronoi cell is assumed to be constant and equal to the value of the likelihood function evaluated for the point inside the cell. Once the NA approximation is constructed, NA uses this approximation to guide selection of the new samples until the desired degree of accuracy is achieved. A typical NA approximation of a test function $f(x,y)=x\exp(-x^2-y^2)$ during the progress of NA is presented in Figure 3. Details of the implementation of this algorithm are presented in two companion papers by Sambridge (1999a, b).

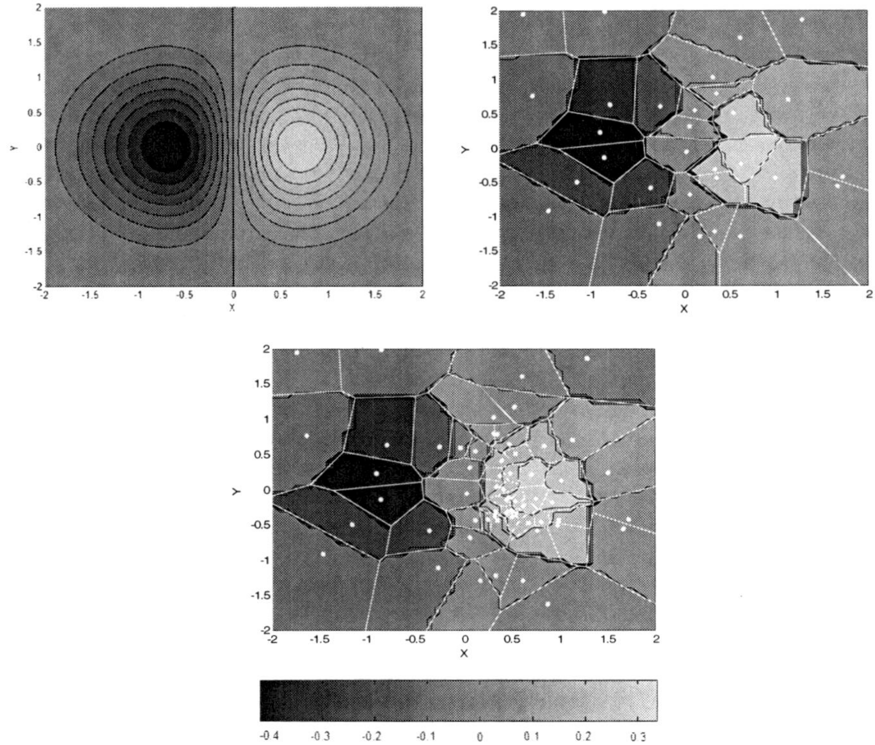

Fig. 3. Contour plot of $f(x,y)=x exp(-x^2-y^2)$ (top left) and NA approximation of the same function using Voronoi cells with 30 (top right) and 60 (bottom) random samples during the progress of NA.

Once an adequate NA approximation to the likelihood function is obtained, the a posteriori probability distribution is evaluated in the appraisal stage using the direct sampling approach presented earlier.

To illustrate the application of the presented probabilistic approach, two examples of inverse problems in pavement and geomechancial engineering are presented. Those include:

- Backcalculation of FWD data, and
- Inversion of seismic waveforms for shallow subsurface characterization.

4 Probabilistic FWD Backcalculation

4.1 Background

The Falling Weight Deflectometer (FWD) test, is the most widely accepted, used and studied technique for in-situ non-destructive evaluation of pavements. The

FWD test is routinely used by pavement engineers to evaluate in-situ pavement layer moduli. The objective of the test is to excite the pavement by an impact of a falling weight equivalent to those applied on by traffic and measure its response under those load levels (see Figure 4). During the test, the pavement deflection response is measured by transducers at different offsets from the load. The maximum pavement displacements at transducer locations (collectively referred to as the deflection bowl) or the displacement time histories at each receiver location are then reported as pavement response. With pavement layer thicknesses as a given input, the measured pavement response is then analyzed or backcalculated to obtain the in-situ pavement layer elastic moduli. The backcalculated pavement moduli are then used to design overlays, estimate remaining life of pavement sections, evaluate the load transfer efficiency of joints in rigid pavements, identify weak areas in the pavement structure, and perform network level monitoring.

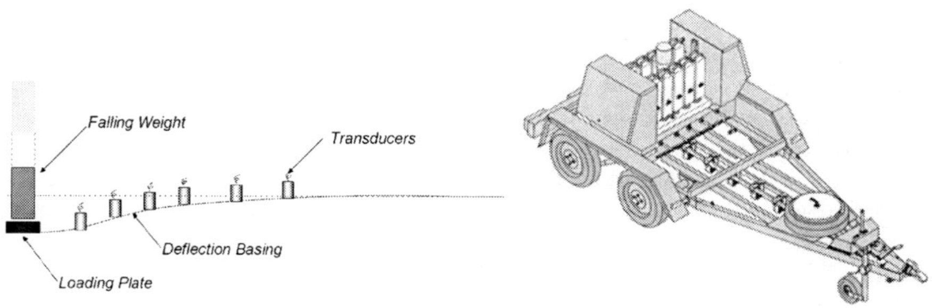

Fig. 4. Schematics of FWD test setup (left) and Dynatest® model 8000 FWD trailer (right)

FWD backcalculation is mathematically an inverse problem, with in-situ pavement layer moduli as its solution. Over the past few decades, numerous researchers have investigated different aspects of interpretation and backcalculation of FWD test results. Many of the important findings of these studies were summarized in several volumes of American Society for Testing and Material (ASTM) special technical publications (ASTM 1989; ASTM 1994; ASTM 2000). The review of currently available backcalculation procedures indicates that the available procedures follow a deterministic approach. These procedures can be broadly categorized as either a static or a dynamic procedure based on the type of the forward model used to evaluate the theoretical response of the pavement. The probabilistic approach to FWD backcalculation is a new approach introduced in this section. Results of synthetic FWD test are utilized to illustrate the use of the probabilistic approach in FWD backcalculation.

4.2 Probabilistic Formulation

Using the notion of the presented generalized measurement, the FWD backcalculation can be considered as the measurement of pavement layer moduli using surface deflection measurements. The generalized notion of the measurement combined with the tools developed in the previous section is used herein to backcalculate pavement layer moduli from a set of synthetic test data and to obtain uncertainty measures.

4.2.1 Forward Model and Synthetic Pavement Response

A linear two dimensional axisymmetric finite element forward model with absorbing boundary elements was developed using ABAQUS® program for the analysis of the FWD test and use in the probabilistic backcalculation procedure (both static and dynamic procedures). Even though the pavement response to an FWD impulse is nonlinear, a linear model is considered in this work since linear model procedures are currently the dominant form of FWD backcalculation. However, having introduced the general probabilistic approach, any other modeling approach can be easily incorporated using a different pavement model.

The developed axisymmetric finite element model of a pavement is presented in Figure 5. The FWD load pulse was modeled as a time varying, spatially uniform load applied on a 0.15 m radius plate (Al-Khoury et al. 2001). The model boundaries were placed 10 m away from the impact location. To minimize numerical inaccuracies, the size of the elements was selected not to be larger than one half of the shortest wavelength of the generated waves within the frequency range of interest, which was 0-100 Hz. The absorbing boundary elements were used at the lower and right boundaries of the model to reduce spurious wave reflections in the dynamic analysis. Extra care was taken in the selection of a sufficiently small time step to ensure correct modeling of the dynamic pavement response. The material properties presented in Table 1 were assigned to the pavement layers.

The response of the developed pavement model to loading, described by an idealized loading wavelet, was evaluated. The idealized loading wavelet and the pavement response, described in terms of the deflection time histories and deflection bowl (obtained from the deflection histories), are presented in Figure 6. To verify the accuracy of the finite element analysis, the material properties of the layers were selected to be equal to those considered in other studies (Al-Khoury et al. 2001).

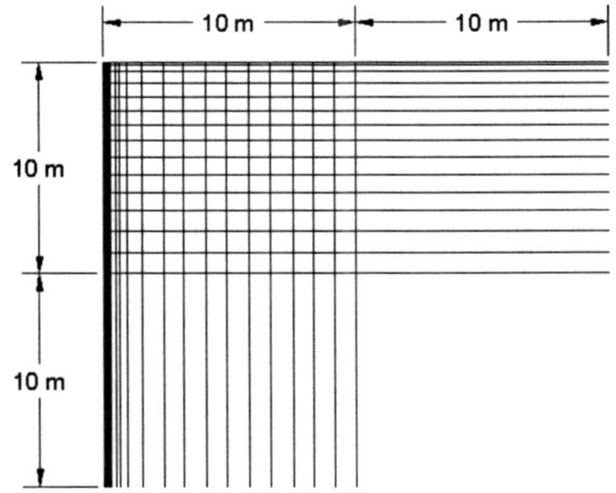

Fig. 5. Axisymmetric ABAQUS® Finite Element meshes with absorbing boundary elements for theoretical modeling of FWD analysis.

Table 1. Geometrical and material properties used for numerical evaluation and verification of finite element results.

Material Type	Thickness (m)	Material Model	Elastic Modulus (MPa)	Rayleigh Damping Ratio	Poisson's Ratio	Mass Density (kg/m^3)
Asphalt Concrete	0.15	Linear Elastic	1000	0.001	0.35	2300
Aggregate Base Course	0.25	Linear Elastic	200	0.001	0.35	2000
Subgrade	infinity	Linear Elastic	100	0.001	0.35	1500

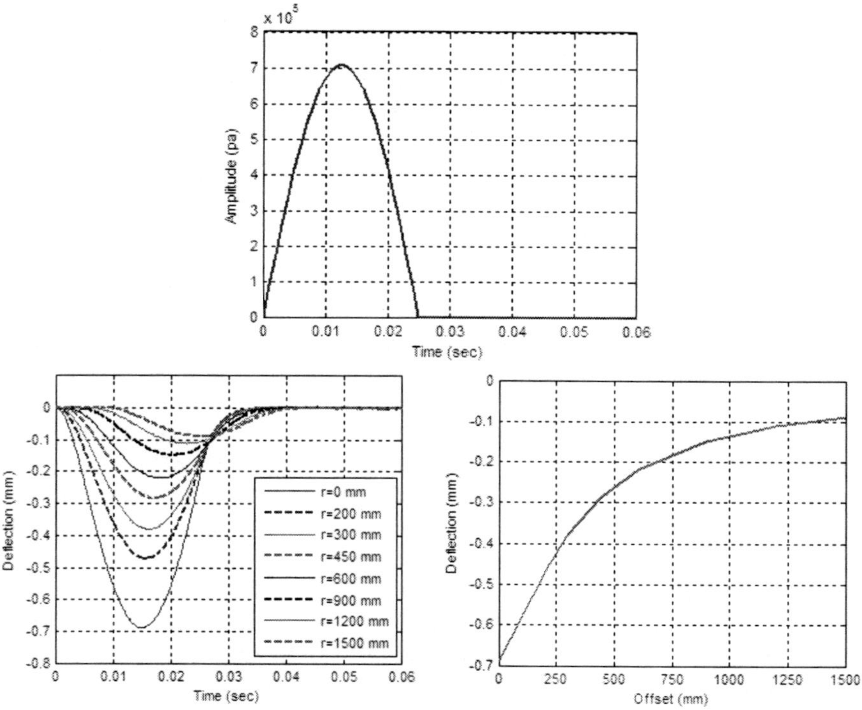

Fig. 6. (top) Time history of an idealized loading wavelet, (bottom left) predicted pavement surface deflection time histories from the finite element model and (bottom right) the corresponding deflection bowl.

4.2.2 Model and Data a Priori Information

The deflection time histories and deflection bowl presented in Figure 6 were used as synthetic data in the backcalculation. Because the pavement system has only three layers, three model parameters (i.e. modulus of each layer) were included in the backcalculation.

For the pavement layer moduli, most often, there is no prior information available other than possible limits of moduli based on experience and engineering judgment.

This information, or lack of more specific information for that matter, can be presented by a homogenous probability distribution. Homogeneous a priori probability densities considered for all pavement layer properties in this example can be presented in the form of:

$$\begin{aligned} \rho_M(m_i) &= 1/(m_{max} - m_{min}) \qquad && \text{if } m_{min} < m < m_{max} \\ \rho_M(m_i) &= 0 && \text{otherwise} \end{aligned} \tag{5}$$

where m_{max} and m_{min} are respectively the maximum and minimum limits of the value of the parameter of interest, m_i. The choice of the limits of $\mathbf{m}$ generally depends on the problem and the experience and judgment of the analyst. However, when in doubt, a larger interval can be selected. The data uncertainty was considered to follow a Gaussian distribution with the mean equal to the observed value (i.e. deflection of the pavement) and a coefficient of variation of two percent (Bentsen et al. 1989). Forward modeling uncertainties are discussed below.

4.2.3 Backcalculation Results

Backcalculation of Modulus Based on Deflection Bowl Measurements Using Linear Static Forward Model: The first set of backcalculation analyses results are the results of backcalculation of layer moduli based on the deflection bowl measurements. Estimates of marginal a posteriori probability densities of the layer moduli from the backcalculation are presented in Figure 7. The forward model in these backcalculation analyses is the static linear finite element model. Although present, no modeling uncertainty was considered for the first set of analyses.

A comparison between the observed deflection bowl and the defection bowl corresponding to the most probable modulus values (i.e. values representing the peaks of each probability, which produce the best fit) is presented in the same figure. It can be observed that, although there is a reasonably good match between the observed and backcalculated deflection bowls, the backcalculated layer moduli are very different from the values used in the generation of the synthetic data (i.e. the correct solution), shown with arrows on the figure. It is interesting to note that the values used in the generation of the data are not even among probable solutions. This discrepancy is basically due to modeling uncertainties that are not included in the backcalculation. The synthetic observed deflection bowl, as presented in Figure 6, was generated using a linear dynamic finite element model. But, the backcalculation routine uses a static model to backcalculate layer moduli, which does not necessarily produce the same deflection bowl. Unless this discrepancy in modeling is explicitly considered in the backcalculation, the final results will not be close to the values used in the generation of the deflection bowl. It should be mentioned that this approach to the FWD backcalculation is similar to the current dominant practice, where the results of the FWD test (a truly dynamic test) are summarized in terms of the deflection bowl and the backcalculation is

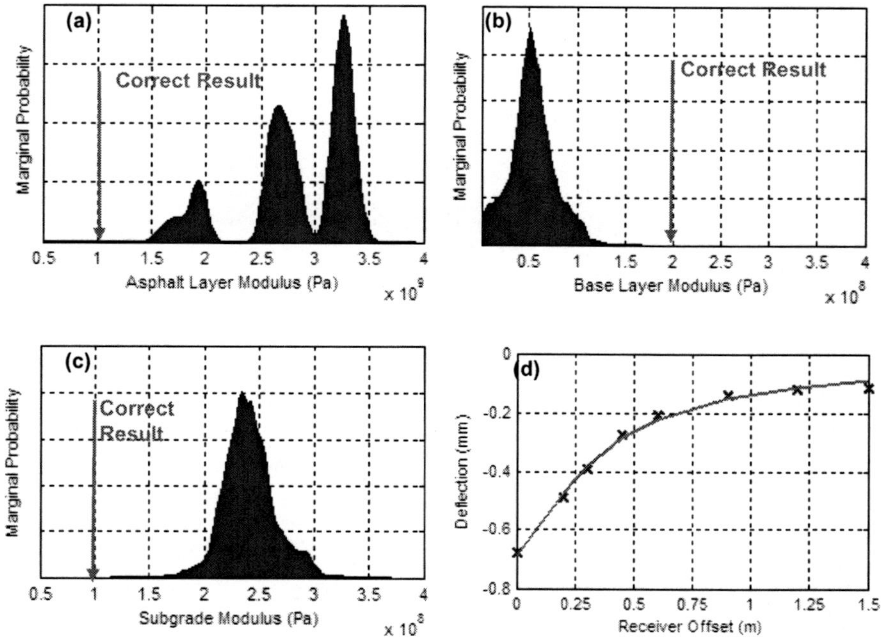

Fig. 7. (a, b and c) Estimates of marginal a posteriori probability densities for the layer moduli obtained from the probabilistic backcalculation of the deflection bowl with the static forward model without considering modeling uncertainties and (d) comparison of the backcalculated deflection bowls for the pavement section with the most probable layer moduli and the observed data.

conducted using a static forward model. It is clear that in the commonly used deterministic approach where only the best fit model is sought (i.e. values corresponding to the peaks of each probability), the solution will be very different than the actual values used to generate synthetic data. Based on this analysis, it can be observed that the static backcalculation, without consideration of the modeling uncertainty, may result in incorrect backcalculated values.

As explained, in the probabilistic backcalculation, the modeling uncertainties, if evaluated, can be explicitly considered in the analysis. To illustrate such an approach, the modeling uncertainty for this analysis has been evaluated by comparing the deflection bowls from static and dynamic analyses of the same pavement model for a range of layer moduli values. For this study the paving layer modulus was varied from 500 to 4000 MPa in 250 MPa intervals. The variation in the base and subgrade moduli was from 50 to 400 MPa in 50 MPa intervals. This resulted in about 960 different pavement models, which were evaluated using both static and dynamic models. Typical histograms and density estimates of the difference between calculated deflections from dynamic and static analyses for the first

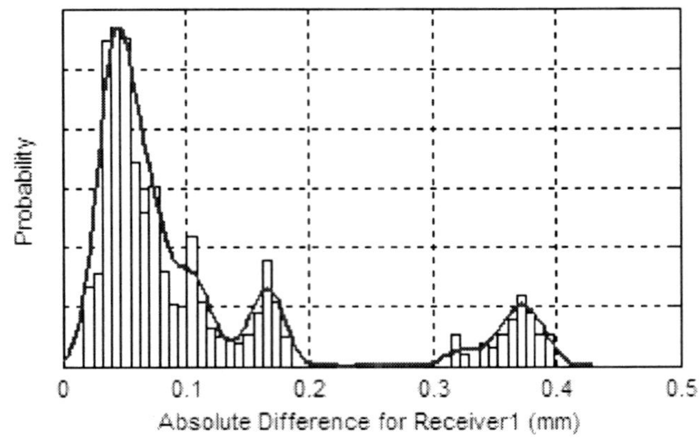

Fig. 8. Typical histogram and density estimate of the absolute difference between the calculated maximum deflections from the dynamic and static analyses for receiver 1.

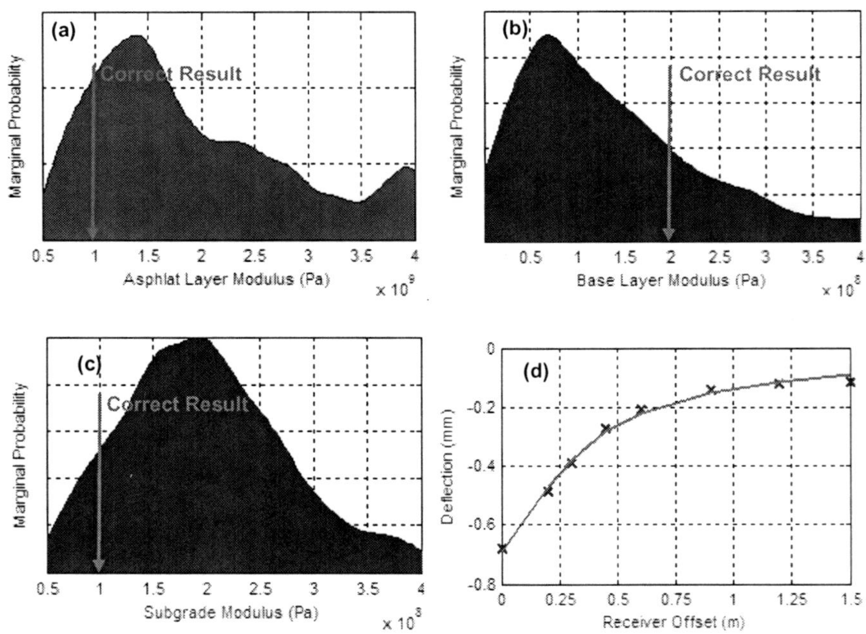

Fig. 9. (a, b and c) Estimates of marginal a posteriori probability densities for the layer moduli obtained from probabilistic backcalculation of the deflection bowl with the static forward model that includes the modeling uncertainties and (d) comparison of the backcalculated deflection bowls for the pavement section with the most probable layer moduli and the observed data.

transducer from all 960 analyses are depicted in Figure 8. Similar histograms were also generated for other receivers. The modeling uncertainty at each receiver location to be used as an input in the backcalculation was then evaluated by calculating the covariance of the difference between calculated deflections from dynamic and static analyses for each receiver. Based on this analysis, as the first estimate, a Gaussian distribution with a standard deviation of 0.1 mm was used to represent the modeling uncertainty for all receivers.

Including the evaluated modeling uncertainties, a new set of analyses was performed, which included the combined effect of data and modeling uncertainties. One-dimensional marginal probability densities of the layer moduli from the backcalculation analysis are presented in Figure 9. Based on these results, it can be observed that by including modeling uncertainties, the correct results are among the probable results of the analysis. However, due to high modeling uncertainties, there is a high uncertainty in the backcalculation results represented by the spread of the probabilities. These results highlight the unreliability of static backcalculation procedures.

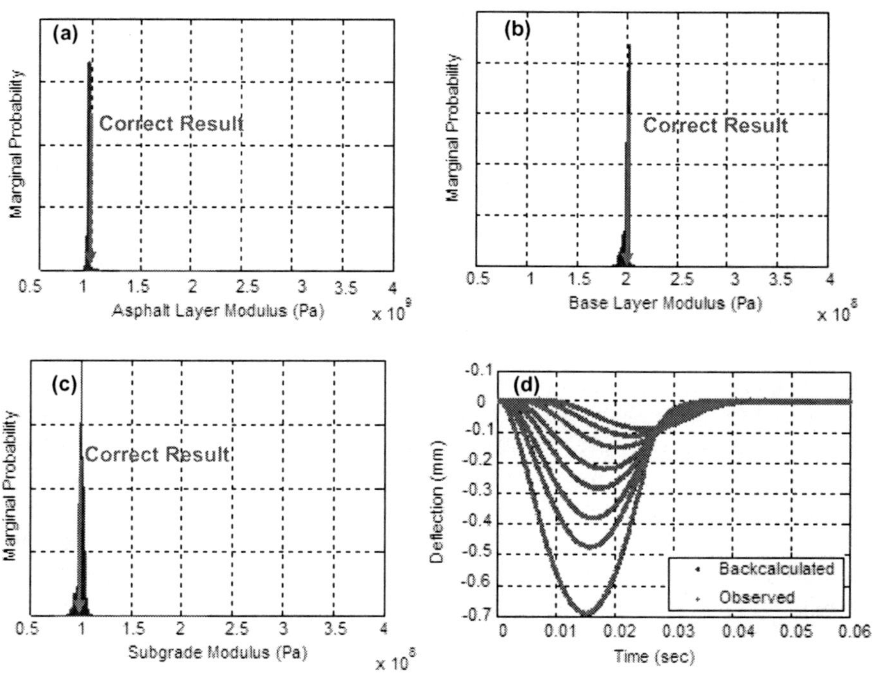

Fig. 10. (a, b and c) Estimates of marginal a posteriori probability densities for the layer moduli obtained from the probabilistic backcalculation of the deflection time history with the dynamic forward model and (d) comparison of the backcalculated deflection time histories for the pavement section with the most probable layer moduli and the observed data.

Backcalculation of Modulus Based on Deflection Time History Measurements Using Linear Dynamic Forward Model: Intuitively, FWD test deflection time histories carry significantly more information regarding the pavement than the discrete deflection bowl data. Therefore, it is expected that the backcalculation procedures based on the use of time histories will provide more reliable results. This in fact can be observed from the backcalculation of synthetic test results using the dynamic forward model. For this backcalculation, the data uncertainty coefficient of variation was selected to be two percent and no modeling uncertainty was considered, because the same model was used in generation of the synthetic results and theoretical data. Estimates of marginal a posteriori probability densities for the layer moduli obtained from the probabilistic backcalculation are presented in Figure 10. A comparison of the backcalculated deflection time histories for the pavement section with the most probable layer moduli and the observed data are also presented in the same figure. As shown, these results have much sharper peaks at the correct moduli values. Therefore, it can be concluded that using the complete deflection time histories as the data is a much more reliable approach in the FWD backcalculation.

5 Probabilistic Seismic Waveform Inversion

5.1 Background

Elastic waves carry substantial information about the characteristics of the media they propagate in. The seismic evaluation techniques use the information carried by elastic waves to infer information about the properties of the media. These techniques are generally nondestructive and are being used increasingly in engineering applications, such as in determination of stiffness and integrity of structural elements, evaluation of elastic moduli of soil deposits and pavement systems, void detection and sizing in geotechnical engineering, crack detection, etc. There are several techniques that are routinely used for shallow subsurface investigations, namely; spectral analysis of surface waves (SASW) (Nazarian, 1984), impulse response (IR) (Reddy, 1992), impact echo (IE) (Sansalone, 1997) and multi-channel analysis of surface waves (MASW) (Park, Miller and Xia, 1999). However, these standard methods usually use a limited portion of the information carried by elastic waves, such as travel time, peak return frequency, or wave velocity dispersion. Seismic waveform inversion seeks to use the full information content of the seismic waveform. Its objective is to find a reasonable model, consistent with a given a priori information, for which its predicted waveforms match reasonably well the observed waveforms.

This section presents the probabilistic formulation of the seismic waveform inversion problem for evaluation of subsurface properties. Results of a synthetic seismic experiment are used to illustrate the application of the probabilistic approach in waveform inversion. It should be mentioned that the waveform inversion considered here is a relatively new technique for applications in pavement and geomechanical engineering. The technique is a much more versatile and powerful technique in comparison to the traditional techniques, such as SASW, MASW, IR, IE, for which their inherent assumptions limit their applicability. The traditional techniques generally assume that subsurface layers are horizontal, whereas the waveform inversion can invert virtually any type of geometry. The only theoretical limitation in its use would arise from the limitation in the parameterization and modeling of the test setup.

5.2 Synthetic Seismic Experiment

A hypothetical geological soil profile, as depicted in Figure 11, will be used in the illustration of the probabilistic inversion approach. The profile consists of horizontal and inclined layers underlain by a half space or bedrock. Each layer is parameterized in terms of its thickness, T, compressional wave velocity, V_P, density, γ, and poisson's ratio ν. For the experiment presented in this section, the objective of the seismic test is to quantify material properties of the subsurface layers, in terms of their compressional wave velocities, from the waveforms recorded at the surface. It is assumed that the thickness and dipping angle of the layers are available from other information collected at the site, such as boring logs. To achieve the objective, a seismic test is performed by generating an elastic wave field using a

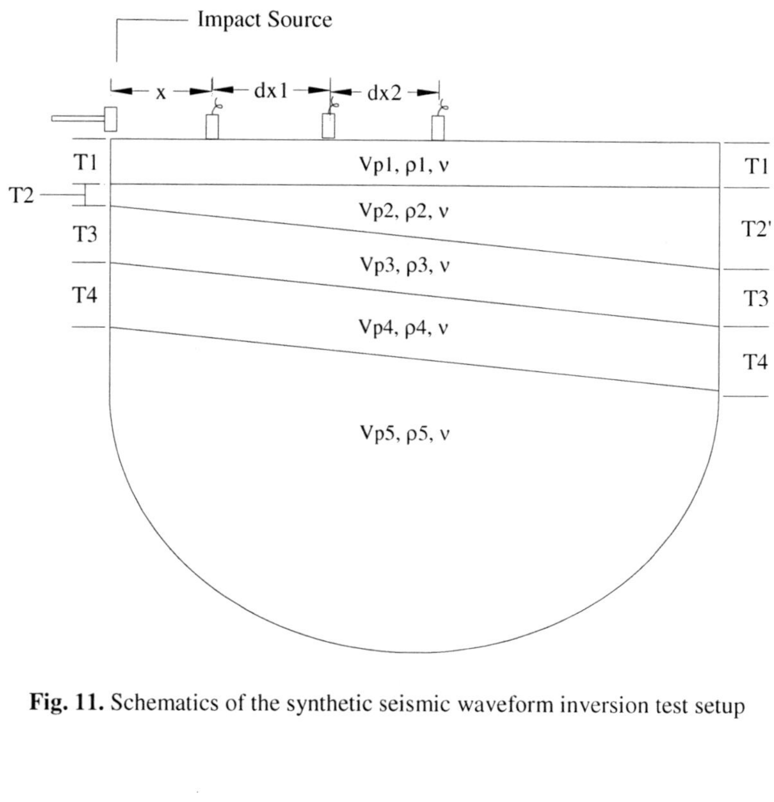

Fig. 11. Schematics of the synthetic seismic waveform inversion test setup

known impact source and seismic waveforms are recorded at several locations along the surface. The recorded waveforms are then analyzed to invert the unknown parameters of the model.

5.3 Probabilistic Formulation

Using the notion of the presented generalized measurement, the waveform inversion can be considered as a measurement of the subsurface properties, such as compressional wave velocities, from surface measurements. The generalized notion of measurement combined with the tools developed in the previous section is used here to calculate layer properties and obtain uncertainty measures.

5.3.1 Forward Model and Synthetic Waveform

In general, determination of the surface response to impact loads mathematically falls into the area of wave propagation theory. Numerical solutions are required to obtain the solutions in general. However, closed form solutions and/or simplified techniques exist when the problem boundaries and geometry are simple. These simplifications, if possible, present considerable savings in terms of the computational effort.

For the analysis in this section, a finite element forward model is used to model the wave propagation forward problem. A finite element model is selected as the forward model in the inversion of synthetic data to conserve the generality of the presented procedure and provide a framework for considering other classes of problems.

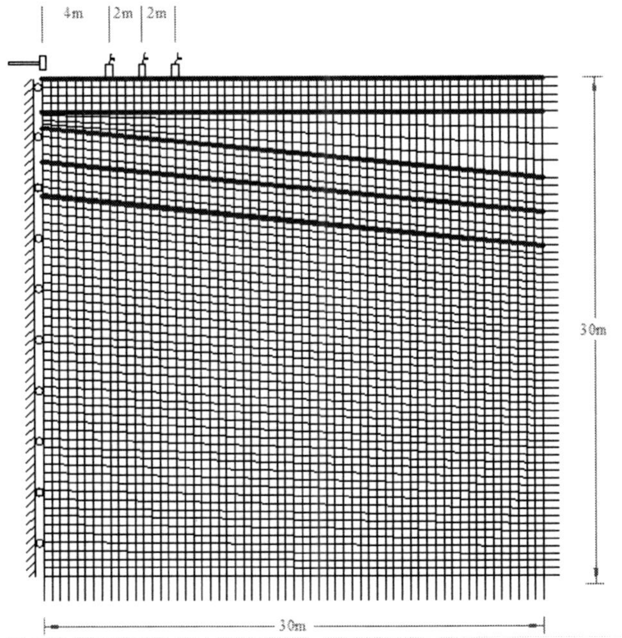

Fig. 12. Finite element model with the test setup superimposed on the mesh

Table 2. Subsurface profile parameters in the generation of synthetic waveform inversion example.

Parameter	Value	Parameter	Value
V_p1	80 m/s	T3	2 m
V_p2	50 m/s	T4	2 m
V_p3	70 m/s	ρ1 thru ρ5	1900
V_p4	100 m/s	v	0.30
V_p5	60 m/s	x	4
T1	2 m	dx1	2 m
T2	1 m	dx2	2 m
T2'	4 m		

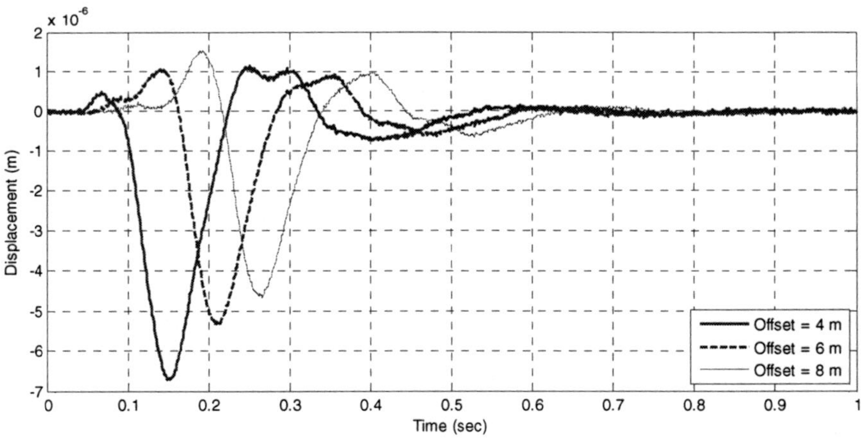

Fig. 13. Synthetic waveforms used in the inversion.

To simulate the test and generate a set of synthetic data, the presented seismic test is modeled by finite elements using ABAQUS® software. The test setup can be described by an axi-symmetric model with an impact loading at the center. Explicit time integration of the equation of motion is used to obtain the solution. Because the strain levels during the seismic tests are small, linear elastic material models were considered for all layers. To accurately model the wave propagation, several criteria were imposed during the modeling to ensure accuracy of the simulation. The element size was selected relatively small to capture short wavelengths, while the overall model was relatively large to allow propagation of large wavelengths and reduce the boundary effects. Infinite absorbing elements were used at the boundaries to further reduce the reflections from the boundaries.

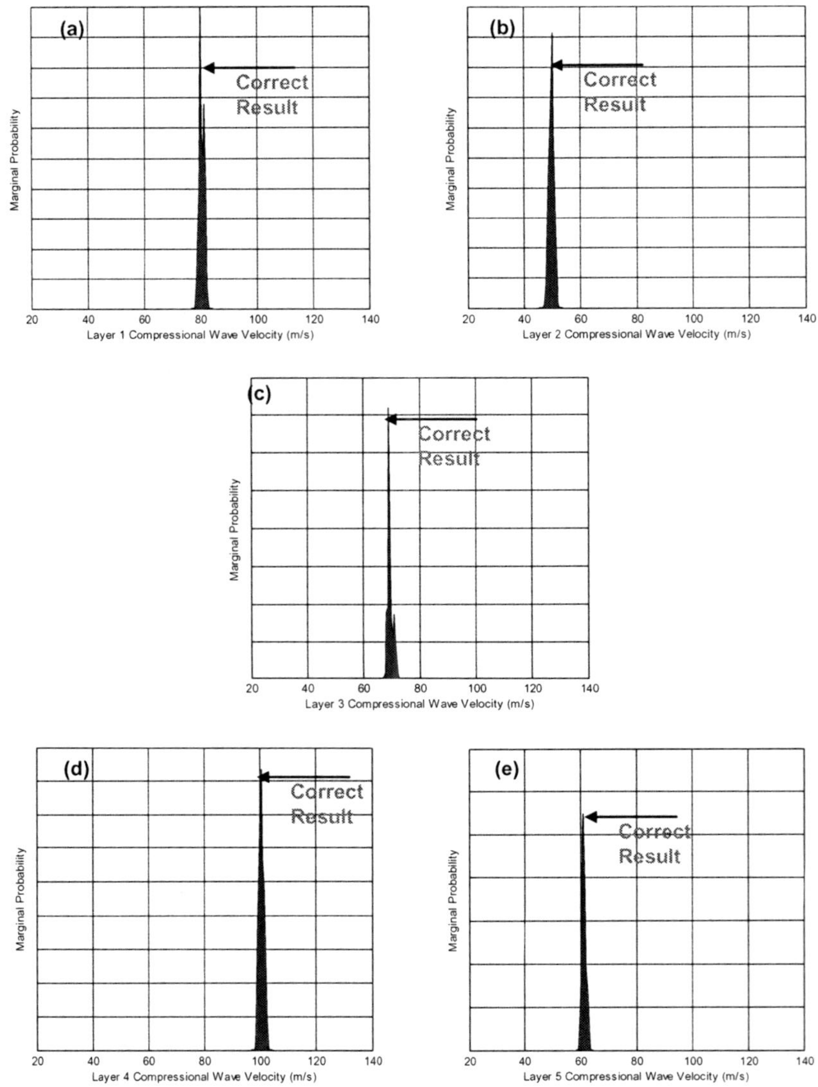

Fig. 14. (a, b, c, d and e) Kernel density estimates of marginal a posteriori probabilities for the layer compressional wave velocities.

The developed finite element model is presented in Figure 12. Receiver locations, as well as boundaries of the layers, for a typical test setup are superimposed on the finite element model in this figure.

Using the developed model, a set of synthetic waveforms was generated at three receiver locations shown in Figure 12. The geometric and material properties used in the generation of synthetic data are presented in Table 2. To simulate field conditions, artificial Gaussian random noise was added to the calculated waveforms. The waveform at each receiver is one second long and is sampled at 0.001 second intervals. These waveforms are depicted in Figure 13 and are used as the synthetic data for inversion.

5.3.2 Model and Data a Priori Information

Uniform probability density, as presented in Equation 5, was considered for a priori probability density of all model parameters. This probability density represents the information on the limits of the parameters. The data uncertainty was considered by adding a Gaussian noise to the synthetic records.

5.3.3 Results of Inversion of Synthetic Seismic Test

The presented synthetic data was analyzed to obtain the layer compressional wave velocities. Kernel density estimates of one-dimensional marginal a posteriori probability densities for this example are presented in Figure 14. As presented, the inversion analysis results are very close to the target compressional wave velocities presented in Table 2. It can be observed that the inversion process has effectively inverted the profile and has clearly resolved the target compressional wave velocities. The calculated marginal a posteriori probability densities have clearly the peaks at the target compressional wave velocity values, which represent very low uncertainties of the wave velocities. It can be observed that the waveform inversion can solve problems where the assumptions of traditional seismic analysis techniques limit their application.

6 Summary and Conclusions

In this chapter, a general probabilistic approach to the solution of the inverse problems was introduced as a new approach to the solution of inverse problems. The mathematical framework required for implementation of this approach was presented in detail. Following the mathematical formulation of the approach, techniques for evaluation the probabilistic solution using Monte Carlo Markov Chains (MCMC), with and without Neighborhood Algorithm (NA) approximation, were introduced and explained. The application of the presented concepts and techniques was then illustrated by solving two important problems in pavement and geomechanical engineering: FWD backcalculation and seismic waveform inversion.

The probabilistic FWD backcalculation was introduced and formulated, and the results of the backcalculation of synthetic test data presented. The probabilistic backcalculation was then used as a tool to compare different backcalculation procedures, such as static and dynamic backcalculations. Based on the presented results, it was shown that the static backcalculation procedures fail to capture the essential dynamic nature of the test and consequently can not be relied upon for accurate backcalculation. The dynamic deflection time history backcalculation offers the best and most reliable approach in the FWD backcalculation.

The probabilistic formulation of a seismic waveform inversion problem for evaluation of shallow subsurface properties was presented. Using a set of synthetic data, the potential of the waveform inversion for evaluation of the subsurface profiles was illustrated. It was shown that the waveform inversion is a powerful technique for evaluation of shallow subsurface properties, which is not limited by many of the assumption of other seismic techniques, such as Spectral Analysis of Surface Waves (SASW).

References

Bush III, A.J., Baladi, G.Y. (eds.) American Society for Testing and Material (ASTM), Nondestructive testing of pavement and Backcalculation of moduli., ASTM Special Technical Publication 1026, ASTM, Philadelphia (1989)

Von Quintus, H.L., Bush III, A.J., Baladi, G.Y. (eds.): ASTM, Nondestructive testing of pavement and Backcalculation of moduli, vol. 2. Special Technical Publication 1198, ASTM, Philadelphia (1994)

Tayabji, S.D., Lukanen, E.O. (eds.): ASTM, Nondestructive testing of pavement and Backcalculation of moduli, vol. 3. ASTM Special Technical Publication 1375, West Conshohocken (2000)

Al-Khoury, R., Scarpas, A., Kasbergen, C., Blaauwendraad, J.: Spectral Element Technique for Efficient Parameter Identification of Layered Media: Part I: Forward Model. International Journal of Solids and Structures 38(9), 1605–1623 (2001)

Bentsen, R.A., Nazarian, S., Harrison, A.: Reliability of Seven Nondestructive Pavement Testing Devices. In: Bush III, A.J., Baladi, G.Y. (eds.) Nondestructive testing of pavement and Backcalculation of moduli. ASTM Special Technical Publication 1198, ASTM, Philadelphia (1989)

Bhat, U.N., Miller, G.K.: Elements of Applies Stochastic Processes. John Willey & Sons Inc., Hoboken (2002)

Gelman, A., Carlin, J.B., Stern, H.S., Rubin, D.B.: Bayesian Data Analysis. Chapman & Hall/CRC, Boca Raton (2004)

Hadidi, R., Gucunski, N.: A Probabilistic Approach to the Solution of Inverse Problems in Civil Engineering. ASCE Journal of Computing in Civil Engineering 22(6), 338–347 (2008)

Hastings, W.K.: Monte Carlo sampling methods using Markov chains and their applications. Biometrika 57, 97–109 (1970)

Menke, W.: Geophysical Data Analysis: Discrete Inverse Theory. Academic Press Inc., Orlando (1984)

Metropolis, N., Rosenbluth, A.W., Rosenbluth, M.N., Teller, A.H., Teller, E.: Equation of state calculations by last computing machines. J. Chem. Phys. 21, 1087–1092 (1953)

Mosegaard, K., Tarantola, A.: Monte Carlo Sampling of Solution to Inverse Problems. Journal of Geophysical Research 100(B7), 12431–12447 (1995)

Nazarian, S.: In situ determination of elastic moduli of soil deposits and pavement systems by spectral analysis of surface waves method. PhD thesis, Univ. of Texas at Austin, Texas (1984)

Park, C.B., Miller, R.D., Xia, J.: Multichannel Analysis of Surface Waves. Geophysics 64(3), 800–808 (1999)

Parker, R.L.: Geophysical Inverse Theory. Princeton University Press, Princeton (1994)

Reddy, S.: Improved Impulse Response Testing - Theoretical and Practical Validations. M.S. Thesis, The University of Texas at El Paso (1992)

Sambridge, M.: Geophysical Inversion with Neighborhood Algorithm – I. Searching the Parameter Space. Geophysical Journal International 138, 479–494 (1999a)
Sambridge, M.: Geophysical Inversion with Neighborhood Algorithm – II. Appraising the Ensemble. Geophysical Journal International 138, 727–746 (1999b)
Santamarina, J.C., Fratta, D.: Introduction to Discrete Signals and Inverse Problems in Civil Engineering. ASCE Press, Reston (1998)
Sansalone, M.: Impact-Echo: The Complete Story. ACI Structural Journal 94(6), 777–786 (1997)
Tarantola, A.: Inverse Problem Theory. SIAM, Philadelphia (2005)

Neural Networks Application in Pavement Infrastructure Materials

Sunghwan Kim[1], Kasthurirangan Gopalakrishnan[2], and Halil Ceylan[3]

[1] Iowa State University, Ames, IA 50011, USA
sunghwan@iastate.edu
[2] Iowa State University, Ames, IA 50011, USA
rangan@iastate.edu
[3] Iowa State University, Ames, IA 50011, USA
hceylan@iastate.edu

Abstract Interest on artificial neural networks (ANN) in infrastructure materials research and practice has increased in recent years. This chapter presents a review of ANN applications in characterization of infrastructure materials focusing on portland cement concrete (PCC) and asphalt concrete (AC) materials. The principles of ANN are briefly introduced and summarized. The strengths and limitations of ANN for modeling behavior of infrastructure materials are discussed. Various applications of the ANN approach in infrastructure materials testing, analysis and design problems are discussed.

1 Introduction

Over the last few years or so, development of computer hardware and software has inspired new approaches of data processing and analysis. From among these approaches soft computing has been recognized as low cost and complete solution yielding analysis tool to solve complex problems in many areas of engineering. In recent years artificial neural networks (ANN) among soft computing have been applied to complex engineering problems in various relevant civil engineering areas such as pavement and geotechnical engineering, structure engineering, water sources and environmental engineering [Adeli 2001].

It is well recognized that the engineering properties of materials are critical in the infrastructure design, construction and performance. The engineering properties of infrastructure materials have been characterized through laboratory or field experimental testing, which can require cost and time of testing. A number of numerical modeling or physical models have been developed to estimate the engineering properties of infrastructure materials as an alternative to laboratory or field testing. However, these models could not be able to simulate perfectly since the properties of infrastructure materials are influenced by a multitude of tributary factors and the uncertainty involved in the experimental tests. Based on data processing advantage, ANN has been used to model the properties and behaviors of infrastructure materials. This has demonstrated some degree of success.

This paper highlights key advances of ANN applications in testing, analysis and design problems related to infrastructure materials focusing on Portland Cement

K. Gopalakrishnan et al. (Eds.): Intel. & Soft Comp. in Infra. Sys. Eng., SCI 259, pp. 47–66.
springerlink.com

Concrete (PCC) and Asphalt Concrete (AC). The objective of this review is to promote more considerations of using ANN in infrastructure material property prediction and design. The attempt has made to include those individuals who have made major contributions, but also the acknowledgement is provided to the efforts of many others who have been unintentionally overlooked.

2 Artificial Neural Network Approach

ANN is the computational intelligence system that simulates the behavior of the human brain and nervous system. The basic element in the ANN is a processing element, called as artificial neuron or node. Each neuron contains a very limited amount of local memory and performs basic mathematical operations on data passing through them. These neurons are highly interconnected in layers such as an input layer, an output layer and one or more hidden layers. The computational power of ANN comes from this interconnection which makes input data concurrently processed in artificial neurons [TRB Circular 1999].

An artificial neuron receives information (signal) from other neurons, processes it, and then relays the filtered signal to other neurons [Tsoukalas and Uhrig 1997]. The receiving end of the neuron has incoming signals (x_1, x_2, x_3....and x_n). Each of them is assigned a weight (w_{ji}) that is based on experience and likely to change during the training process. The summation of all the weighted signal amounts yields the combined input quantity (I_j) which is sent to a preselected transfer function (f), sometimes called an activation function. A filtered output (y_j) is generated in the outgoing end of the artificial neuron (j) through the mapping of the transfer function. The parameters can be written as per the following equations:

$$I_j = \sum_{i=1}^{n} w_{ji} x_i \tag{1}$$

$$y_j = f(I_j) \tag{2}$$

There are several types of transfer functions that can be used, including sigmoid, threshold, and Gaussian functions. The transfer function most often used is the sigmoid function because of its differentiability. The sigmoid function can be represented by the following equation:

$$f(I_j) = \frac{1}{1+\exp(-\varphi I_j)} \tag{3}$$

Where φ = positive scaling constant, which controls the steepness between the two asymptotic values 0 and 1 [Tsoukalas and Uhrig 1997].

The ANN performs two major functions: learning (training) and testing. A training data set and an independent testing data set are prepared for these functions. Inputs from a training data set are presented to the input layer to start the propagation of data. Inside the network, weights are adjusted when data pass between artificial neurons along the connections. Since interconnected neurons have the flexibility to adjust the weights, ANN has the ability to analyze complex problems. It uses a learning rule to find a set of weights such that the error is minimum. This process is called "learning" or "training" [Shahin et al. 2001]. The following are the three broad types of learning in neural network technology [TRB Circular 1999]

- Supervised learning: system/weight is adjusted by comparing the network output with a given or desired output
- Unsupervised training: the network is trained to form categories based on similarity among the data and identify irregularities in data
- Reinforcement learning: the network attempts to learn the input-output vectors by trial and error through maximizing a performance function. The system can identify whether a given output is correct or not but cannot estimate the exact output

Once the training phase of the model has been successfully accomplished, the network performance is verified by presenting independent testing datasets to the ANN. This process is called "testing." Details regarding the theory and mathematics behind the ANN is available in several sources [Aleksander and Morton 1990, Fausett 1994, Haykin 1998, Bishop 1995, Swingler 1996].

There are nearly as many different types of ANN used in many areas of engineering. These differ in the arrangement and degree of connectivity of their neurons, the types of calculations performed within each neuron, the degree of supervision they receive during training, the determinism of the learning process, and the overall learning theory under which they operate [Mehra and Wah 1992]. However, certain types of ANN are more repeatedly used, either because they are broadly applicable to a wide variety of problems or ideally suited for a narrow range of problems [TRB Circular 1999]. These include hopfield nets, adaptive resonance theory (ART) networks, self-organized feature maps (SOFM), backpropagation neural networks (BPNN), feedback (sequential) neural networks (FBNN), counter propagation networks, radial basis function network (RBF), and generalized regression neural networks (GRNN).

BPNN is one of most ANN referred in the civil engineering application of ANN because of its powerfulness, versatility, and simplicity [TRB Circular 1999, Adeli 2001]. BPNN can be taught a mapping from one data space to another using a representative set of patterns/examples to be learned. The term "backpropagation network" actually refers to a multi-layered, feed-forward neural network trained using an error backpropagation algorithm [TRB Circular 1999]. The learning process performed by this algorithm is called "backpropagation learning" which is mainly an error minimization technique between the correct responses and the predicted outputs [Haykin 1998]. The details of BPNN have been described in many sources [Hegazy et al 1994, Adeli and Hung 1995, Mehrotra et al. 1997, Topping and Bahreininejad 1997, Haykin 1998].

3 Strengths and Limitations of Artificial Neural Networks

For modeling infrastructure materials behaviors, ANN provide an analytical alternative to classical mathematics and traditional techniques which are often limited by strict assumptions of normality, linearity, variable independence, etc. Since ANN can learn and generalize many kinds of relationships between variables and responses to provide meaningful solutions to problems even when input data contain errors and are incomplete [Rafiq et al 2001], it allows the user to quickly and relatively easily model a phenomenon which otherwise is very difficult. ANN offers a number of advantages, including requiring less formal statistical training, ability to implicitly detect complex nonlinear relationships between dependent and independent variables, ability to detect all possible interactions between predictor variables, and the availability of multiple training algorithms [Tu 1996]. In modeling of infrastructure martial behaviors, ANN is developed from experimental data on which it has been trained without a priori assumptions about the material behavior. As more experimental data are acquired about the behavior of a material, this ANN-based material model can be updated through a training session that includes the new data [Wu et al 1990].

A major drawback of ANN is its "black box" nature which lack in the ability to present explicit rules between the input variables and the output. In problems where explaining rules may be crucial, ANN is not the choice. Thus, it is the tool of choice when results are more important than understanding how they are obtained. However, ANN can explain which inputs are more important than others through sensitivity analysis performed either inside the network by using the errors generated from back propagation or externally by comparing the accuracy of the network using specific inputs. ANN may suffer from overfitting and overtraining, which takes away the ability to use the model for data not in the training set. These problems may be avoided by selecting a suitable architecture and by using a training set/control set protocol [Livingstone et al. 1997]. The development of ANN model requires the amount of data to provide answers close to the observed behaviors [Ghaboussi et al. 1991]. The ANN is not the tool of choice when there is scarcity or appropriate data.

Although ANN has some limitations, these are superseded by its advantages. On a practical level, ANN has been considered as a promising tool for predicting various engineering properties of infrastructure materials.

4 Artificial Neural Networks for Infrastructure Materials

The application of ANN in infrastructure materials has focused on prediction of certain properties of materials and analyzing the relationships between a number of variable parameters and response properties of materials. ANN can be developed to predict certain properties of materials. Using ANN, sensitivity of certain properties of materials can be evaluated by changing values of input parameters.

4.1 Portland Cement Concrete (PCC)

PCC is one of the most widely used infrastructure materials. PCC is a composite material composed of aggregates and paste. The paste consists of cement and water. The paste glues aggregates into a rocklike mass as it hardens resulting from the chemical reaction of the cement and water [Kosmatka et al 2002]. The properties of concrete changes from plastic (semi-fluid) at freshly mixed concrete to elastic and gains strength as it hardens with age. Fresh concrete properties, especially workability, significantly affect handling concrete for fabrication and the hardened concrete properties. Hardened concrete properties including strength and durability should be within acceptable standard.

It is well recognized that the characterization of concrete properties is important in the quality and cost of concrete construction. However, the characterization of concrete properties is more complex and uncertain with advancement of concrete materials such as high performance concrete (HPC) and special types of concrete over the past decade. This enabled some researchers to apply the ANN technique not only to predict properties of concrete but also of mix design.

4.1.1 Properties of Fresh Concrete

The properties of fresh concrete mainly considered for construction are workability and setting. Workability of concrete is the property of freshly mixed concrete that determines the ease with which it can be mixed, placed, consolidated, and finished to a homogenous condition [Kosmatka et al 2002]. Workability can be measured by slump, compacting factor, Vebe time and rheological parameters. Setting is the degree to which fresh concrete has lost its plasticity and hardened by the continuing hydration of cement [Kosmatka et al 2002]. ANN was used in studies to characterize the workability and setting of concrete.

Bai et al [2003] developed ANN models for workability, measured by slump, compacting factor and Vebe time for concrete incorporating fly ash and metakaolin. Three independent ANN models were developed for each of the two water/binder (w/b) ratios. Each of the networks had two inputs (fly ash and metakaolin) and one output (slump, compacting factor and Vebe time). Using developed ANN models, the effects of fly ash and metakaolin on workability were investigated. The results indicate that the ANN models developed are usable in practice to predict the workability of cement–fly ash–metakaolin. El-Chabib et al [2003] explored ANN to predict rheological and mechanical properties of underwater concrete (UWC). The input vectors included the quantity values of the mixture variables influencing the behavior of UWC mixtures (that is, cement, silica fume, fly ash, slag, water, coarse and fine aggregates, and chemical admixtures) and corresponding output vectors consisted of slump, slump-flow, washout resistance, and compressive strength.

Stegemann and Buenfeld [2001] used ANN to construct models for setting time as a function of mix composition and addition concentration for calcium aluminate cements. ANN models were used for representing non-linear temperature dependency of setting times and generalizing it to find exponential relationships between setting time and addition concentration.

ANN models with high accuracy were also developed to predict the degree of cement hydration [Basma et al 1999, Park et al 2005], the rate of cement heating evolution, the relative humidity and the total porosity of cement [Park et al 2005].

4.1.2 Strength of Hardened Concrete

Numerous studies have been explored to predict compressive strength of harden concrete using ANN models. The study of Wittmann and Martinola [1993] is one of the earliest ANN applications in connection with the prediction of concrete properties [Kasperkiewicz 2000]. They used BPNN to predict the compressive strength and the fracture parameter with water/cement (w/c) ratio and superplasticizer contents as input parameters. Mukherjee and Nag Biswas [1997] also examined ANN to predict stress and strain behavior of concrete at high temperature. There is a nonlinear and complex relationship which makes it difficult to include all the contributing factors in the traditional mathematical models.

Lai and Serra [1997] presented the BPNN models for the predictions of normal concrete compressive strengths using 8 input parameters (cement, fine sand, coarse sand, fine aggregate, coarse aggregate, cement weight, w/c, and plasticizer). They suggested that the neural network performance is independent of the number of neurons in the hidden layer in the range of 4 to 8 with 5% of precision. The efficiency of ANN for predictions of normal concrete strength with high accuracy is also demonstrated in following studies:

- The study by Ni and Wang [2000] who used 11 input parameters (type of cement, w/c, dosage of water, dosage of cement, the maximum size of coarse aggregate, the fine modulus of sand, the sand-aggregate ratio, the aggregate/cement ratio, the slump, the effect of admixtures, dosage of admixtures),
- The study by Kim et al [2004] who used concrete mix proportion parameters as inputs (slump, w/c, unit water content, unit cement content, unit aggregate content, fine aggregate and admixture content),
- The study by Gupta et al [2006] who considered concrete mix design, specimen geometry, curing and environmental conditions as inputs (shape and size of concrete specimen in terms of cross-sectional area, weight of concrete specimen, concrete grade, various curing techniques, curing period, average maximum temperature, average relative humidity, and average wind velocity), and
- The study by Kewalramani and Gupta [2006] who used weight and ultrasonic pulse velocity (UPV) as inputs.

Modified BPNN has been utilized for the prediction of compressive strengths. Yeh [1998a] employed a modified BPNN for modeling concrete strength with seven factors (w/c, water, cement, fine aggregate, coarse aggregate, maximum grain size, and age of testing). He added logarithm neurons and exponent neurons to input layer and the output layer in a standard BPNN. He also examined the efficiency and accuracy of a modified BPNN's in modeling concrete strength. The results showed that the logarithm neurons and exponent neurons in the network improved the performance of the networks for modeling the strength of concrete significantly. Dias and Pooliyadda [2001] evaluated various input data transforms for BPNN models to predict the strength and slump of ready mixed concrete and high strength concrete, in which chemical admixtures and/or mineral additives were used. They found that BPNN models trained with raw data on concrete mix make better predictions of strength and slump than those trained using non-dimensional ratios. Lee [2003] developed a modular ANN model with the multiple architectures composed of five BPNN to provide in-place strength information. The single architecture of ANN cannot appropriately predict the development of concrete strength as the curing condition changes each time. To solve this problem, each of the BPNN in the developed modular ANN model predicted the concrete strength under different curing conditions at a specific time which was within 24 hours after pouring or at 2nd to 28th day after pouring. This study demonstrates that the use of a modular ANN model with multiple BPNN architectures is very efficient for predicting the compressive strength development of concrete.

The use of alternative ANN for BPNN on prediction of compressive strength has been attempted. Kim et al [2005] applied probabilistic neural network (PNN) for predicting the compressive strength of concrete. The concrete mix proportions and the slump values of two types of ready mixed concrete were used as inputs. The study demonstrated that the PNN based models are very efficient and reasonable prediction of the compressive strength of concrete. Fazel Zarandi et al [2008] investigated the applicability of the fuzzy polynomial neural network (FPNN) in prediction of the compressive strength of the concrete. FPNN is a combination of fuzzy neural networks (FNN) and polynomial neural networks (PNN). As the If-Part of FPNN, FNN utilizes both BP algorithm and simplified fuzzy inference system. As Then-Part of FPNN, PNN is combined with FNN in two connection points. To enhance the performance of the network, back propagation (BP) and list square error (LSE) algorithms were utilized for training FNN and PNN. Two different architectures of FPNN were developed in this study. These models have six input variables including concrete mix components and one output variable, i.e., compressive strength of concrete. The results indicated FPNN as a potential tool for predicting the compressive strength of concrete mix-design.

ANN has been applied to characterize the properties of HPC, which meets the requirements of special combination of properties and constructability that cannot be achieved routinely with conventional constituents and normal mixing, placing, and curing procedures. The required properties of HPC include high strength, high modulus of elasticity, high durability, high resistance to mechanical and chemical attack, low permeability, volume stability, ease of placement and compaction without segregation. HPC can be made with selected high quality ingredients and optimized mixture design [Kosmatka et al 2002]. The selected materials often used in HPC to achieve special properties include fly ash, slag, silica fume, metakaolin and etc. HPC is a highly complex material, which makes characterization and prediction of its behavior difficult.

Kasperkiewicz et al [1995] presented the use of ART based networks for predicting strength properties of HPC mixes. They simplified the composition of HPC into six components (cement, silica, superplasticizer, and water, fine aggregate and coarse aggregate) and considered the 28-day compressive strength value as the only aim of the prediction. Obtained results indicate that the properties of concrete can be effectively prediction in a neural system, in spite of data complexity, incompleteness, and incoherence. Yeh [1998b], and Sergio and Mauro [1997] demonstrated that prediction of the strength of HPC is more accurately obtained through BPNN model as compared to regression analysis. Öztaş et al [2006] described applicability of BPNN to predict the compressive strength of HSC with suitable workability. The input parameters of ANN model were w/c, water content, fine aggregate ratio, fly ash content, air entraining agent, superplasticizer and silica fume replacement. The outputs of ANN model were compressive strength and slump. The results showed ANN to be a feasible tool for predicting compressive strength and slump values. Sebastiá et al [2003] focused on the application of ANN to predict compressive strength of coal fly ash–cement mixtures using several input parameters including cement, aggregate, water, additives, and fly ash. They confirmed the application of ANN to predict the strength of fly ash–cement mixtures. Pala et al [2007] applied BPNN models to evaluate the effects of fly ash and silica fume replacement content on the strength of concrete cured for a long-term period of time. The results of this study showed that ANN is a valuable tool for evaluating the effect of cementitious material on the compressive strength of concrete.

4.1.3 Durability of Hardened Concrete

The durability of concrete can be defined as the ability of concrete to resist weathering action, chemical attack, freezing/thawing, and abrasion while maintaining its desired engineering properties for the longer life of concrete [Kosmatka et al 2002]. The durability of concrete is seen to depend on many factors including the concrete ingredient, proportioning of these ingredients, interactions between ingredients and placing and curing practice. ANN models have been utilized to understand the effect of these factors on concrete durability.

Haj-Ali et al [2001] utilized ANN methodology to predict the long-term expansion response of concrete exposed to sulfate solution. The inputs parameters of ANN models developed were time and two mixture parameters (w/c and the tricalcium aluminate content of the cement). The expansion of the concrete was used as an output of models. It is shown that ANN can effectively learn and predict expansion of the concrete within a practical range of the two mixture parameters.

The study of Ukrainczyk and Ukrainczyk [2008] describes the use of an ANN method with fuzzy inferences to investigate the effects of the environmental conditions, structure and properties of concrete on the degree of damage caused by steel corrosion. The damage was classified into six categories based on the type of remedial work necessary and used as outputs of ANN models. They suggested that the ANN model developed could be used to predict the extent and severity of degradation in a structure during its service life, to plan the maintenance and to assist in the design and restoration of reinforced concrete structures.

Peng et al [2002] focused on ANN approach using the cascade correlation algorithm [Fahlman and Lebiere 1990] to predict the chloride diffusion causing deterioration of concrete structures. The cascade-correlation algorithm was selected because this can synthesize an appropriate architecture and train the network simultaneously while back-propagation type algorithm requires a selection of suitable architectures in advance by a costly trial-and-error approach for the network and convergence of the learning algorithm. They concluded that the cascade-correlation algorithm has the potential of becoming an effective tool in the prediction of durability problems due to its self-constructive capacity not available in typical back-propagation networks.

Parichatprecha and Nimityongskul [2009] analyzed the influence of mix proportion parameters on the durability of HPC by using ANN. The data used in the ANN model was arranged in a format of eight input parameters, which included the content of ordinary portland cement (OPC), fly ash (F), silica fume (SF), water (W), superplasticizer (SP), coarse aggregate (CA), fine aggregate (FA) and water–binder ratio (W/B). The one output parameter of ANN model was chloride ions permeability representing the durability of HPC. This study illustrated how ANN can be used to predict durability parameter of HPC, which is difficult with traditional regression analysis.

4.1.4 Concrete Mix Design

The concrete mix design or proportion is the process of determining the quantities of ingredient in concrete mixture to achieve the required and specified characteristics of fresh and hardened concrete. In this process, engineer meets the uncertainties of materials, temperature, site environmental situations, personal skillfulness, and errors in calculations and testing processes. Adjustments are also made for a proper mix proportion. Since this kind of concrete mix proportion and adjustments is somewhat complicated, time-consuming, and an uncertain tasks, the development of concrete mix design tool based on ANN methodology was tried out.

Oh et al [1999] and Uomoto et al [1998] applied ANN to minimize the uncertainties and errors of proportioning concrete mixes. Ji-Zong et al [1999] developed a knowledge-acquisition system based on ANN to design concrete mix. The main parts of the system were the mix-design model supported by the slump-prediction model and the strength-prediction model. This system not only makes full use of the mix designs but also provides the prediction of the slump and 28-day compressive strength of ready-made concrete.

Yeh [1999] presented a method of optimizing HPC mix proportion for a given workability and compressive strength by using ANN and nonlinear programming. The basic procedure of the methodology consists of three steps: (1) build accurate models for workability and strength using ANN and experimental data; (2) incorporate these models in software allowing an evaluation of the specified properties for a given mix; and (3) integrate the software in a nonlinear programming package allowing a search of the optimum proportion mix design. The proposed methodology can provide a guideline to select appropriate materials and mix proportions as a starting trial batch of HPC and reduce the number of trial mixes required.

4.1.5 Special Types of Concrete Behaviors

Special types of concrete are those with the extraordinary properties or those produced by unusual techniques with/without cement [Kosmatka et al 2002]. Models for properties of normal concrete are not always consistent special concrete because of having different ingredient and production process. ANN approach has been considered as alternative way to understand the properties and behavior of special concrete.

Nehdi et al [2001a] investigated the use of ANN to predict the performance of cellular concrete mixtures. Cellular concrete is a lightweight material consisting of portland cement paste or mortar with a homogeneous void or cell structure created by introducing air or gas in the form of small bubbles during the mixing process. A major concern with the production of cellular concrete is achieving product consistency and performance predictability. Research study shows that production yield, foamed density, unfoamed density, and compressive strength of cellular concrete mixtures can be predicted much more accurately using ANN as compared to existing methods.

Nehdi et al [2001b] presented that ANN could be used to predict the performance of self-compacting concrete (SCC) mixtures effectively. SCC is highly workable concrete that can flow through congested structural elements under its own weight and adequately fill voids without segregation and excessive bleeding. Because of its complex mixture proportions, limited models based on traditional regression analysis and statistical methods have been developed to describe its rheological and mechanical properties. Results show that the ANN method can accurately predict the slump flow, filling capacity, segregation, and compressive strength test results of SCC mixtures. Prasad et al [2009] demonstrated the use of ANN in training data on SCC with low volume of fly ash. Models not only predicted the strength and slump of SCC (with high volume of fly ash) but also the strength of HPC, which is very difficult to predict otherwise. Studies by Fletcher and Coveney [1995] and Topçu et al[2008] indicates that ANN is promising in characterizing nontraditional cement based materials such as oil field cements and blended cements.

ANN was also employed to predict the properties of concrete containing recycled materials. Topçu and Sarıdemir [2007] utilized ANN to predict properties of hardened concrete containing waste crushed autoclaved aerated concrete. For training the networks, seven mixture proportion parameters (cement, water, sand, crushed stone-I, crushed stone-II, fine autoclaved aerated concrete and coarse autoclaved aerated concrete) were entered as input while unit weight, compressive strength, ultrasound pulse velocity and dynamic modulus of elasticity values were the outputs. They also utilized ANN to predict the unit weight and flow property of the fresh concrete containing waste rubber [Topçu and Sarıdemir 2008]. For networks training, seven mixture proportion parameters (cement, water, sand, fine crushed stone, coarse crushed stone, fine rubber, and coarse rubber) were entered as input. Fresh concrete's unit weight and flow table values were the outputs. Both the studies indicate that the use of ANN models can determine the properties of concrete containing recycled materials without attempting any experiments.

4.2 Asphalt Concrete (AC)

Asphalt is a dark brown to black cementitious material either occurring in nature or produced through petroleum processing. AC is a mixture of asphalt and aggregate which can be spread and compacted. Hot mix Asphalt (HMA) is AC produced at batch or drum–mixing facility with elevated temperature. The primary use of AC is road construction. 94 % of U.S road are construct with AC [Asphalt Institute 1989]. The proper understanding of properties and behaviors of AC is essential to achieve long lasting AC pavement with desired performances.

AC can be described as a multiphase heterogeneous material composed of a viscoelastic asphalt binder, irregular rigid aggregate particles in high volume fraction, and small percentage of air voids. These component materials exhibiting various properties contribute complex mechanical behaviors of AC characterized as elastic, viscoelastic, and plastic properties under different conditions such as temperature, load applications and aging. Therefore, it is important to understand the behavior of both the individual properties of AC components as well as that of asphalt binder and aggregate acting together.

Since the property of AC is between the range of brittle and ductile behavior, it is very difficult to capture and model the complex behaviors of AC. Recently, some researchers have attempted to apply ANN approach in AC materials characterization.

4.2.1 Rheology Properties of Asphalt Binder

Rheology properties, the deformation and flow behaviors of asphalt binder, are important in determining asphalt pavement performance. In U.S, the measurement methods for rheology properties of asphalt binder have evolved from empirically derived tests to fundamental engineering parameters based tests to Strategic Highway Research Program (SHRP) researches in the 1980s and 1990s. Superpave (Superior Performing Asphalt Pavements) binder specification is one of final products of these SHRP researches and is a widespread methodology for characterizing asphalt binder based on engineering parameters. However, prediction of rheology properties of asphalt binder is particularly difficult because of the complexity of asphalt binder which consists of a wide variety of molecules such as paraffinics, naphthenics, and aromatics including heteroatoms [Michon et al 1997].

Michon et al [1997] investigated the effects of molecules on rheological properties of bitumens using ANN approaches. BPNN with three layers was used in this study. Two asphalt rheological properties, m (creep slope at low temperature) and $G^*/\sin\delta$ (stiffness at high temperature) were selected as output parameters, whereas the average molecular parameters which characterize the hydrocarbon skeleton of bitumens, were the inputs. This study showed that the skeleton information contained in the average molecular parameters was correlated with m value but not with $G^*/\sin\delta$. Specht et al [2007] presented modeling of asphalt-rubber rotational viscosity by statistical analysis and ANN. The architecture of ANN selected was the BPNN with two hidden layers. Input data used for training BPNN were all elements of the complete factorial matrix including rubber

content, rubber particle size and duration and temperature of mixture. The output was viscosity measured using rotational viscometer. The results of this study indicate that the ANN method provides a best prediction of viscosity than statistical methods.

4.2.2 Aggregate Behaviors in AC

Aggregate in AC should provide enough shear strength to resist load application [Asphalt Institute 2001]. The aggregate properties that enhance internal friction are desired for this purpose. While these properties are empirically characterized in practical use, image processing combined with ANN has been applied to characterize these aggregate properties with respect to the performance of AC mixture. Wilson et al [1995] quantified the shape and texture of individual aggregate particle using two dimensional image processing and classified these properties using ANN. Kutay et al [2008] utilized three dimensional image analysis and ANN to determine the approximate size and location of the coarse aggregates in AC mixture.

4.2.3 AC Properties and Behaviors

Dynamic modulus ($|E^*|$) is one of the AC stiffness measures that determines the strains and displacements in AC pavement structure when it is either loaded or unloaded. The AC stiffness can alternatively be characterized via the flexural stiffness, creep compliance, relaxation modulus and resilient modulus. $|E^*|$ is important AC property as it is the primary input in the new U.S. pavement design method, namely Mechanistic Empirical Pavement Design Guide (MEPDG) [NCHRP 2004]. Various $|E^*|$ predictive models have been developed over the last 50 years to estimate $|E^*|$ as an alternative to laboratory testing, which require days of specimen preparation, temperature equilibration, and loading. The most widely used models are the Witczak [Andrei et al. 1999, Bari and Witczak 2006] and the Hirsch $|E^*|$ [Christensen et al. 2003] predictive models based on conventional multivariate regression analysis of laboratory test data.

Researchers at Iowa State University (ISU) were the first to propose an advanced approach for predicting HMA $|E^*|$ using an ANN methodology [Ceylan et al 2007; Ceylan et al 2008; Ceylan et al 2009]. A typical four-layered (i.e., one input- two hidden–one output layer) BPNN architecture, as shown in Fig. 1 and Fig. 2, was used in development of the ANN $|E^*|$ predictive models. The eight input variables of the Witczak equations reported in 1999 and 2006 were used in the ANN 1999 and ANN 2006 models, respectively. The four input variables of the Hirsch equation were used in ANN Hirsch model. These input variables are aggregate gradation, mixture volumetrics, rheology properties of the asphalt binder (η for ANN 1999 model, $|G_b^*|$| for ANN 2006 and ANN Hirsch models, and δ for ANN 2006 model), loading frequency (f for ANN 1999 model), and one contact factor (P_c for ANN Hirsch model). The aggregate gradation variables for ANN 1999 and ANN 2006 models include percent passing #200 sieve ($\rho_{\#200}$), percent retained #4 sieve ($\rho_{\#4}$), percent retained 9.5 mm sieve ($\rho_{9.5mm}$), and percent retained 19 mm sieve (ρ_{19mm}). The mixture volumetrics for ANN 1999 and ANN 2006 models includes air void (V_a) and effective binder content (V_{beff}). The mixture volumetrics for ANN Hirsh model includes voids in mineral aggregate

(VMA) and voids filled with asphalt (VFA). The predicted dynamic modulus |E*| was the sole output variable in all of the ANN models. The 8-30-30-1 architecture (8 inputs, 30 and 30 hidden neurons, and 1 output neurons, respectively) was chosen as the best architecture for both the ANN 1999 and ANN 2006 models based on its lowest training and testing mean squared errors (MSEs). The error level of the ANN Hirsch model was minimized when the number of hidden nodes approached 40 (4-40-40-1 architecture).

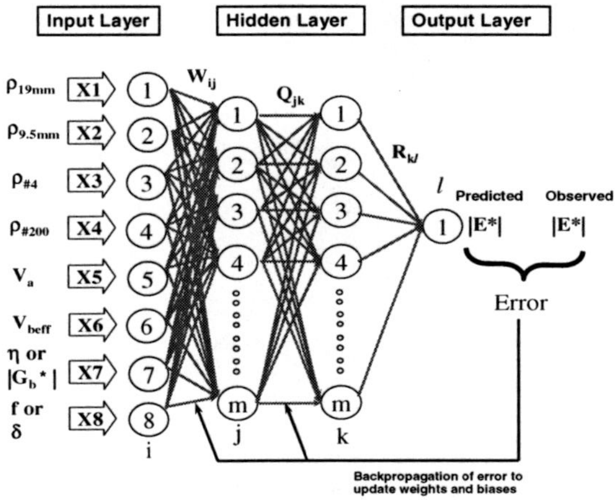

Fig. 1. ANN 1999 and ANN 2006 models.

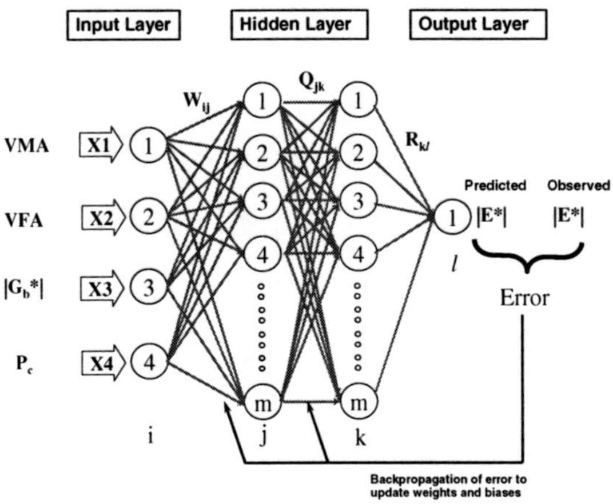

Fig. 2. ANN Hirsch model

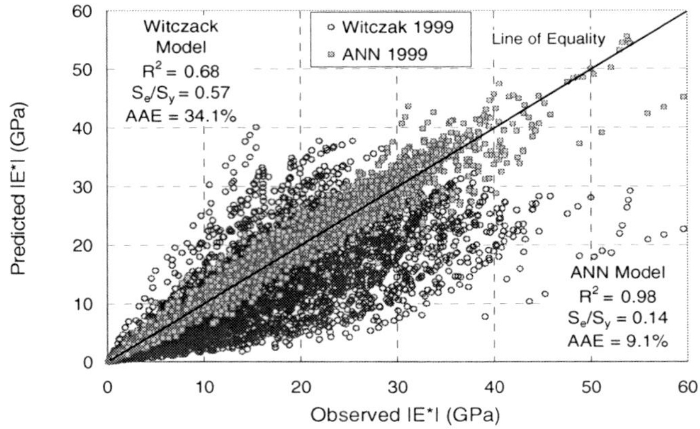

Fig. 3. Predicted vs. measured |E*| values by the Witczak 1999 and ANN 1999 models.

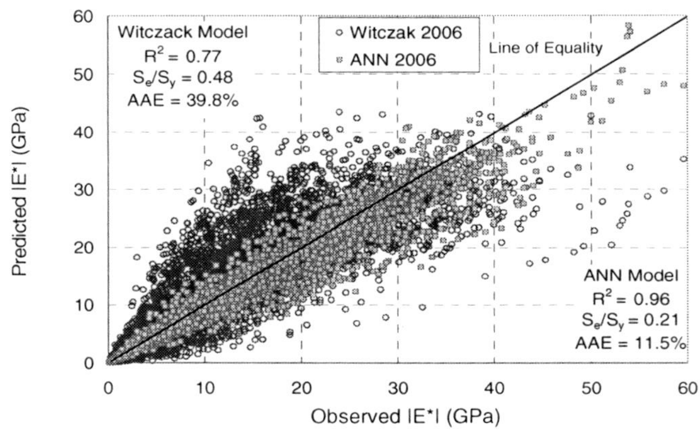

Fig. 4. Predicted vs. measured |E*| values by the Witczak 2006 and ANN 2006 models.

Fig. 3 summarizes the measured vs. predicted |E*| obtained from Witczak 1999 and ANN 1999 models. Fig. 4 and Fig. 5 provide similar comparisons for the Witczak 2006/ANN 2006 and Hirsch/ANN Hirsch model pairs.

It is clear that the ANN versions of the Witczak models (ANN 1999 and ANN 2006) have the highest overall accuracy, followed closely by the ANN version of the Hirsch model (ANN Hirsch) and somewhat more distantly by the Witczak models and the Hirsch model. The results also indicate that the ANN-based models have a much lower tendency at the lower and/or higher |E*| spectrum providing better predictions of distresses at these spectrum. These finding from ISU studies [Ceylan et al 2007; Ceylan et al 2008] were agreed by the study of Sakhaei Far et al [2009] using more expanded database. ANN approaches have been also applied to predict not only alternative stiffness measurement (creep compliance) of AC [Zeghal 2008] but also stiffness measurements of non traditional AC (emulsified AC [Ozsahin and Oruc 2008] and rubberized AC [Xiao and Amirkhanian 2009]) to a higher degree of accuracy.

ANN approaches has been explored to characterize volumetric nature of AC which are critical design and construction of AC materials. Tarefder et al [2005] develop BPNN with four-layer to predict permeability with five factors: (1) air voids (V_a) (2) the grain size through which 10% materials pass (d_{10}); (3) the grain size through which 30% materials pass (d_{30}); (4) saturation, or the CoreLok Infiltration Coefficient (CIC); and (5) effective asphalt to dust ratio (P_{be} /$P_{0.075}$). Commuri and Zaman [2008] focused on using ANN to determine the desired density during field construction of asphalt mixes. Based on the hypothesis that a vibratory compactor and the HMA mat form a coupled system with unique vibration properties, the measured vibrations of the compactor along with the process parameters such as lift thickness, mix type, mix temperature, and compaction pressure were used as inputs to predict the density of the asphalt mat. Hejazi et al [2008] selected BPNN to identify the effect of fiber parameters (as input neurons) on the fiber-reinforced asphalt concrete (FRAC) properties, the specific gravity and the stability and flow measured by Marshall test as output neurons). Five input neurons (fiber length, density, finesse, percentage, and melting point) with a hidden layer of 15 neurons and three output neurons or units (specific gravity, stability, and flow) formed the architecture of BPNN. All of these studies demonstrated that ANN is a useful approach to characterize volumetric nature of AC.

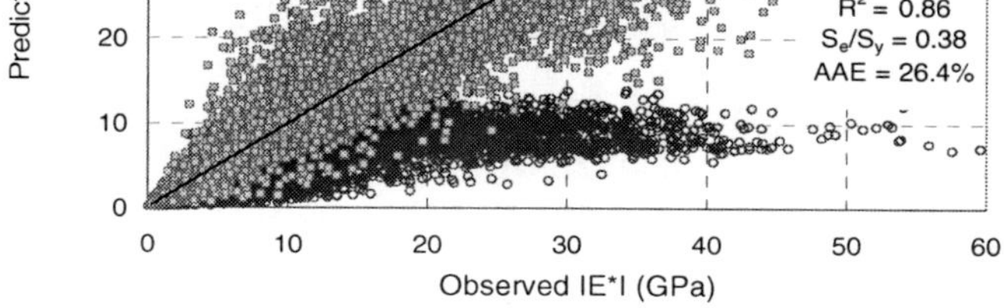

Fig. 5. Predicted vs. measured |E*| values by the Hirsch and ANN Hirsch models.

5 Concluding Remarks

Characterization of infrastructure materials is important in infrastructure design, construction and performance. The properties and behaviors of infrastructure materials have been characterized trough experimental tests requiring cost and time. By using classic mathematics and traditional techniques, various material models have been developed as alternative. However, most of these techniques simplify a multitude of tributary factors and the uncertainty involved in the experimental tests with strict assumptions such as normality, linearity, variable independence, etc. Consequently, many mathematical or physical models fail to predict the complex behavior of most infrastructure materials. In contrast, artificial neural networks (ANN) based model is developed from experimental dataset on which it has been trained without a priori assumptions about the material behavior.

It is evident from this review that ANN can provide analytical alternatives to classical mathematics and traditional techniques in infrastructure material problems. These include fresh portland cement concrete (PCC) properties, hardened PCC strength and durability, concrete mix design, special types of concrete, asphalt rheology properties, aggregate behavior in asphalt concrete (AC) and AC properties and behaviors. The application of ANN in infrastructure materials has been mainly focused on two tasks: (1) prediction of engineering properties, (2) analysis of variables effect on engineering properties. While ANN has been applied to many PCC materials problems since the 1990s, ANN applications to AC materials are very recent and still relative rare. Even in concrete materials, most applications are focused on PCC strength problems. These indicate that ANN could be applied to more infrastructure material problems. The feasible areas considering ANN applications to PCC include rheology properties, behavior under freezing and thawing, alkali and aggregate reaction, properties and behaviors of roller compacted concrete and other special concretes. The feasible areas considering ANN applications to AC include Superpave binder grade, AC mix design, durability properties and behaviors, and recycled asphalt behaviors.

A main barrier of application might be lack of reliable data. However, once reliable data is adequately available, ANN becomes a powerful and practical tool for various infrastructure material problems.

References

Adeli, H., Hung, S.L.: Machine learning: neural networks, genetic algorithms, and fuzzy systems. Wiley, New York (1995)

Adeli, H.: Neural networks in civil engineering: 1989-2000. Computer-Aided Civil and Infrastructure Engineering 16, 126–142 (2001)

Aleksander, I., Morton, H.: An introduction to neural computing. Van Nostrand Reinhold Co., New York (1990)

Andrei, D., Witczak, M.W., Mirza, M.W.: Development of a revised predictive model for the dynamic (complex) modulus of asphalt mixtures. NCHRP 1-37 A Inter Team Report, University of Maryland, College Park, Maryland (1999)

Asphalt Institute, The asphalt handbook. MS-4, Lexington, KY (1989)

Asphalt Institute, Superpave mix design. SP-2, Lexington, KY (2001)

Bari, J., Witczak, M.W.: Development of a new revised version of the Witczak E* predictive model for hot mix asphalt mixtures. Journal of the Association of Asphalt Paving Technologists 75, 381–424 (2006)

Bai, J., Wild, S., Ware, J.A., Sabir, B.B.: Using neural networks to predict workability of concrete incorporating metakaolin and fly ash. Advances in Engineering Software 34, 663–669 (2003)

Basma, A.A., Barakat, S., Al-Oraimi, S.: Prediction of cement degree of hydration using artificial neural networks. ACI Materials Journal 96(2), 167–172 (1999)

Bishop, C.M.: Neural networks for pattern recognition. Oxford University Press Inc., New York (1995)

Ceylan, H., Kim, S., Gopalakrishnan, K.: Hot mix asphalt dynamic modulus prediction models using neural network approach. In: Dagli, C.H. (ed.) Intelligent Engineering Systems through Artificial Neural Networks, Proceedings of the ANNIE 2007, vol. 17, pp. 117–124. American Society of Mechanical Engineers (2007)

Ceylan, H., Gopalakrishnan, K., Kim, S.: Advanced approaches to hot-mix asphalt dynamic modulus prediction. Canadian Journal of Civil Engineering 35(7), 699–707 (2008)

Ceylan, H., Schwartz, C.W., Kim, S., Gopalakrishnan, K.: Accuracy of predictive models for dynamic modulus of hot mix asphalt. ASCE Journal of Materials in Civil Engineering 21(6), 286–293 (2009)

Christensen, D.W., Pellinen, T., Bonaquist, R.F.: Hirsch model for estimating the modulus of asphalt concrete. Journal of the Association of Asphalt Paving Technologists 72, 97–121 (2003)

Commuri, S., Zaman, M.: A novel neural network-based asphalt compaction analyzer. International Journal of Pavement Engineering 9(3), 177–188 (2008)

Dias, W.P.S., Pooliyadda, S.P.: Neural networks for predicting properties of concretes with admixtures. Construction and Building Materials 15(7), 371–379 (2001)

El-Chabib, H., Nehdi, M., Sonebi, M.: Artificial intelligence model for flowable concrete mixtures used in underwater construction and repair. ACI Materials Journal 100(2), 165–173 (2003)

Fahlman, S.E., Lebiere, C.: The cascade-correlation learning architecture. Rep. CMU-CS-90-100. Carnegie Mellon Univ., Pittsburgh (1990)

Fausett, L.V.: Fundamentals of neural networks, 1st edn. Prentice-Hall, NJ (1994)

Fazel Zarandi, M.H., Türksen, I.B., Sobhani, J., Ramezanianpour, A.A.: Fuzzy polynomial neural networks for approximation of the compressive strength of concrete. Applied Soft Computing 8(1), 488–498 (2008)

Fletcher, P., Coveney, P.: Prediction of thickening times of oil field cements using artificial neural networks and fourier transform infrared spectroscopy. Advanced Cement Based Materials 2(1), 21–29 (1995)

Ghaboussi, J., Garrett Jr., J.H., Wu, X.: Knowledge-Based Modeling of Material Behavior with Neural Networks. ASCE J. Engrg. Mech. 117(1), 132–153 (1991)

Gupta, R., Kewalramani, M.A., Goel, A.: Prediction of concrete strength using neural-expert System. ASCE J. Mat. in Civ. Engrg. 18(3), 462–466 (2006)

Haj-Ali, R.M., Kurtis, K.E., Sthapit, A.R.: Neural network modeling of concrete expansion during long-term sulfate exposure. ACI Materials Journal 98(1), 36–43 (2001)

Haykin, S.: Neural networks: a comprehensive foundation, 2nd edn. Prentice Hall, Upper Saddle River (1998)

Hegazy, T., Fazio, P., Moselhi, O.: Developing practical neural network applications using backpropagation. Microcomputers in Civil Engineering 9(2), 145–159 (1994)

Hejazi, S.M., Abtahi, S.M., Sheikhzadeh, M., Semnani, D.: Introducing two simple models for predicting fiber-reinforced asphalt concrete behavior during longitudinal loads. Journal of Applied Polymer Science 109(5), 2872–2881 (2008)

Ji-Zong, W., Hong-Guang, H., Jin-Jun, H.: The application of automatic acquisition of knowledge to mix design of concrete. Cem. Concr. Res. 29(12), 1875–1880 (1999)

Kasperkiewicz, J., Racz, J., Dubrawski, A.: HPC strength prediction using artificial neural network. ASCE J. Comp. in Civ. Engrg. 9(4), 279–284 (1995)

Kasperkiewicz, J.: The applications of ANNs in certain materials-analysis problems. Journal of Materials Processing Technology 106, 74–79 (2000)

Kewalramani, M.A., Gupta, R.: Concrete compressive strength prediction using ultrasonic pulse velocity through artificial neural networks. Automation in Construction 15(3), 374–379 (2006)

Kim, D.K., Lee, J.J., Lee, J.H., Chang, S.K.: Application of probabilistic neural networks for prediction of concrete strength. ASCE J. Mat. in Civ. Engrg. 17(3), 353–362 (2005)

Kim, J.I., Kim, D.K., Feng, M.Q., Yazdani, F.: Application of neural networks for estimation of concrete strength. ASCE J. Mat. in Civ. Engrg. 16(3), 257–264 (2004)

Kosmatka, S.H., Kerkhoff, B., Panarese, W.C.: Design and control of concrete mixtures, 14th edn. Portland Cement Association, Skokie (2002)

Kutay, M.E., Arambula, E., Gibson, N., Youtcheff, J., Petros, K.: Use of artificial neural networks to detect aggregates in poor-quality X-ray CT images of asphalt concrete. In: Roesler, J.R., Bahia, H.U., Al-Qadi, I.L., Murrell, S.D. (eds.) Airfield and highway pavements: efficient pavements supporting transportation's future, Proceedings of the 2008 Airfield and Highway Pavements Conference, Bellevue, Washington, pp. 40–51 (2008)

Lai, S., Serra, M.: Concrete strength prediction by means of neural network. Construction and Building Materials 11(2), 93–98 (1997)

Lee, S.C.: Prediction of concrete strength using artificial neural networks. Engineering Structures 25(7), 849–857 (2003)

Livingstone, D.J., Manallack, D.T., Tetko, I.V.: Data modelling with neural networks: Advantages and limitations. Journal of Computer-Aided Molecular Design 11(2), 135–142 (1997)

Mehra, P., Wah, B.W.: Artificial neural networks: concepts and theory. IEEE Computer Society Press, Los Alamitos (1992)

Mehrotra, K., Mohan, C.K., Ranka, S.: Elements of artificial neural networks. MIT Press, Cambridge (1997)

Michon, L., Hanquet, B., Diawara, B., Martin, D., Planche, J.P.: Asphalt study by neuronal networks correlation between chemical and rheological properties. Energy Fuels 11(6), 1188–1193 (1997)

Mukherjee, A., Nag Biswas, S.: Artificial neural networks in prediction of mechanical behavior of concrete at high temperature. Nuclear engineering and design 178(1), 1–11 (1997)

NCHRP, Guide for mechanistic-empirical design of new and rehabilitated pavement structures. National Cooperative Highway Research Program 1-37 A Project Report, Transportation Research Board, National Research Council, Washington DC (2004)

Nehdi, M., Djebbar, Y., Khan, A.: Neural network model for preformed-foam cellular concrete. ACI Materials Journal 98(5), 402–409 (2001a)

Nehdi, M., Chabib, H.E., Naggar, M.H.E.: Predicting performance of self-compacting concrete mixtures using artificial neural networks. ACI Materials Journal 98(5), 394–401 (2001b)

Ni, H.-G., Wang, J.-Z.: Prediction of compressive strength of concrete by neural networks. Cement and Concrete Research 30(8), 1245–1250 (2000)

Oh, J., Lee, I., Kim, J., Lee, G.: Applications of neural networks for proportioning of concrete mixes. ACI Mater. J. 96(1), 51–59 (1999)

Ozsahin, T.S., Oruc, S.: Neural network model for resilient modulus of emulsified asphalt mixtures. Construction and Building Materials 22(7), 1436–1445 (2008)

Öztaş, A., Pala, M., Özbay, E., Kanca, E., Çagľar, N., Bhatti, M.A.: Predicting the compressive strength and slump of high strength concrete using neural network. Construction and Building Materials 20(9), 769–775 (2006)

Pala, M., Özbay, E., Öztaş, A., Yuce, M.I.: Appraisal of long-term effects of fly ash and silica fume on compressive strength of concrete by neural networks. Construction and Building Materials 21(2), 384–394 (2007)

Parichatprecha, R., Nimityongskul, P.: Analysis of durability of high performance concrete using artificial neural networks. Construction and Building Materials 23(2), 910–917 (2009)

Park, K.B., Noguchi, T., Plawsky, J.: Modeling of hydration reactions using neural networks to predict the average properties of cement paste. Cement and Concrete Research 35(9), 1676–1684 (2005)

Peng, J., Li, J., Ma, B.: Neural network analysis of chloride diffusion in concrete. ASCE J. Mat. in Civ. Engrg. 14(4), 327–333 (2002)

Prasad, B.K.R., Eskandari, H., Reddy, B.V.V.: Prediction of compressive strength of SCC and HPC with high volume fly ash using ANN. Construction and Building Materials 23(1), 117–128 (2009)

Rafiq, M.Y., Bugmann, G., Easterbrook, D.J.: Neural network design for engineering applications. Computers & Structures 79(17), 1541–1552 (2001)

Sakhaei Far, M.S., Underwood, B.S., Kim, R.: Application of artificial neural networks for estimating dynamic modulus of asphalt concrete. Paper #09-3799, Proceedings of Transportation Research Board Annual Meeting, Transportation Research Board, National Research Council, Washington DC (2009)

Sebastiá, M., Fernández Olmo, I., Irabien, A.: Neural network prediction of unconfined compressive strength of coal fly ash–cement mixtures. Cement and Concrete Research 33(8), 1137–1146 (2003)

Sergio, L., Mauro, S.: Concrete strength prediction by means of neural network. Constr. Build. Mater. 11(2), 93–98 (1997)

Shahin, M.A., Jaksa, M.B., Maier, H.R.: Artificial neural network applications in geotechnical engineering. Australian Geomechanics 36(1), 49–62 (2001)

Specht, L.P., Khatchatourian, O., Brito, L.A.T., Ceratti, J.A.P.: Modeling of asphalt-rubber rotational viscosity by statistical analysis and neural networks. Materials Research 10(1), 69–74 (2007)

Stegemann, J.A., Buenfeld, N.R.: Neural network modelling of the effects of inorganic impurities on calcium aluminate cement setting. Advances in Cement Research 13(3), 101–114 (2001)

Swingler, K.: Applying neural networks: a practical guide. Academic Press, London (1996)

Tarefder, R.A., White, L., Zaman, M.: Neural network model for asphalt concrete permeability. ASCE Journal of Materials in Civil Engineering 17, 19–27 (2005)

Topçu, İ.B., Sarıdemir, M.: Prediction of properties of waste AAC aggregate concrete using artificial neural network. Computational Materials Science 41(1), 117–125 (2007)

Topçu, İ.B., Sarıdemir, M.: Prediction of rubberized concrete properties using artificial neural network and fuzzy logic. Construction and Building Materials 22(4), 532–540 (2008)

Topçu, İ.B., Karakurt, C., Sarıdemira, M.: Predicting the strength development of cements produced with different pozzolans by neural network and fuzzy logic. Materials & Design 29(10), 1986–1991 (2008)

Topping, B.H.V., Bahreininejad, A.: Neural computing for structural mechanics. Saxe-Coburg Publications, Edinburgh (1997)

TRB Circular. Use of artificial neural networks in geomechanical and pavement systems. Number E-C012. Transportation Research Board, National Research Council, Washington DC (1999)
Tsoukalas, L.H., Uhrig, R.E.: Fuzzy and neural approaches in engineering. Wiley, New York (1997)
Tu, J.V.: Advantages and disadvantages of using artificial neural networks versus logistic regression for predicting medical outcomes. J. Clin. Epidemiol. 49(11), 1225–1231 (1996)
Uomoto, T., Ohya, T., Tsutsumi, T.: Development of new concrete mixing system using neural network. In: Proceedings of the CONSEC Conference, pp. 282–290 (1998)
Ukrainczyk, N., Ukrainczyk, V.: A neural network method for analysing concrete durability. Magazine of Concrete Research 60(7), 475–486 (2008)
Wilson, J.D., Koltz, L.D., Nagaraj, C.: Automated measurement of aggregate indices of shape. Report FHWA-RD-95-116. FHWA, U.S. Department of Transportation, Washington DC (1995)
Wittmann, F.H., Martinola, G.: Optimization of concrete properties by neural networks. In: Dhir, R.K., Jones, M.R. (eds.) Concrete 2000-economic and durable construction through excellence, Proceedings of the International Conference, Dundee, E &FN Spon, London, pp. 1889–1898 (1993)
Wu, X., Garrett Jr., J.H., Ghaboussi, J.: Representation of material behavior: neural network-based models. In: IJCNN, International Joint Conference on Neural Networks, vol. 1, pp. 229–234 (1990)
Xiao, F., Amirkhanian, S.N.: Effects of binders on resilient modulus of rubberized mixtures containing RAP using artificial neural network approach. ASTM Journal of Testing and Evaluation 37(2) (2009)
Yeh, I.-C.: Modeling concrete strength with augment-neuron networks. ASCE Journal of Materials in Civil Engineering 10(4), 263–268 (1998a)
Yeh, I.-C.: Modeling of strength of high-performance concrete using artificial neural networks. Cement and Concrete Research 28(12), 1797–1808 (1998b)
Yeh, I.-C.: Design of high-performance concrete mixture using neural networks and nonlinear programming. ASCE J. Comput. Civil Eng. 13(1), 36–42 (1999)
Zeghal, M.: Modeling the creep compliance of asphalt concrete using the artificial neural network technique. In: Annual Congress of the Geo-Institute of ASCE, GeoCongress 2008, New Orleans, pp. 1–7 (2008)

Backcalculation of Flexible Pavements Using Soft Computing

A. Hilmi Lav[1], A. Burak Goktepe[2], and M. Aysen Lav[3]

[1] Professor, Department of Civil Engineering, Istanbul Technical University, 34469, Ayazaga, Istanbul
lav@itu.edu.tr
[2] Project Coordinator, Egemen Construction Co., 06550, Ankara, Turkey
abgoktepe@gmail.com
[3] Associate Professor, Department of Civil Engineering, Istanbul Technical University, 34469, Ayazaga, Istanbul
lavay@itu.edu.tr

Abstract. Analysis of the mechanical properties of existing road pavements is crucial for pavement rehabilitation and management problems. Numerous studies have focused on developing an efficient method for determining the structural conditions of pavements. Nondestructive testing (NDT) methods can characterize stress-strain behavior of pavement layers at relatively low strain levels. However, the majority of NDT techniques are based on measuring the deflections caused by an applied load to determine the stress-strain behavior. Structural analysis techniques can also calculate deflections using material and loading properties where it is commonly necessary to make an inversion between measured deflections and mechanical properties using a back-calculation tool. Soft computing techniques, i.e. neural networks, fuzzy logic, genetic algorithms, and hybrid systems, have successfully been used to perform efficient and precise back-calculation analyses. This chapter explains the advances in pavement back-calculation methodologies based on soft computing approaches by presenting the concepts behind them and the fundamental advantages of each. An alternative utilization of soft computing techniques for pavement engineering is also presented.

1 Introduction

Structural analysis of pavement systems comprises valuable information for the estimation of the pavement condition and the selection of feasible rehabilitation and/or reconstruction strategy. Therefore, pavement engineers intend to utilize efficient and reliable methods for the determination of the physical condition of pavement section that is subject to evaluation. In order to decide on the physical condition of a pavement structure, mechanical properties of the system should be known. In this respect, nondestructive testing (NDT) methods emerged from the fact that they quickly and efficiently provide valuable information on the mechanical properties of a pavement structure in a nondestructive manner. NDT methods are typically categorized as: (i) deflection basin methods, and (ii) surface wave methods. Methods based on deflection rely on the measurement of surface

K. Gopalakrishnan et al. (Eds.): Intel. & Soft Comp. in Infra. Sys. Eng., SCI 259, pp. 67–106.
springerlink.com

deformations caused by applied loads which can lead to correlations between surface deflections and layer stiffness. Basically, the surface deflection is affected by loading conditions (type, magnitude, contact area, and duration), measurement location, and layer properties (thickness, mass, and stiffness). The most common approaches to deflection basin testing techniques are the Benkelman beam, the LaCroix deflectometer, the Dynaflect road rater, the falling weight deflectometer (FWD), and the rolling weight deflectometer (RWD) [1-5]. The second type of NDT methods, i.e. surface wave methods, is based on Rayleigh waves that emerge by applying load and propagate through the pavement surface. In this method, the travel times between successive receivers are computed for different excitation frequencies using collected wavelength data. The procedure is also referred to as spectral analysis of surface waves (SASW) and depends on the phase velocities and the excitation frequencies [4-7].

Among all NDT methods, FWD is probably the most widely used technique because its can successfully simulate traffic loads and rapidly produce a large amount of data. Basically, FWD measures time-domain deflections emerging from the impulse load utilizing different deflection sensors which are mounted radially from the center of the load plate [3, 8-10]. Deflection data recorded by FWD is usually utilized for the backcalculation of mechanical pavement properties using numerical techniques. Typically, there are two calculation directions in the analysis: forward and backward. In the forward process, deflections are calculated for traffic loads being considered, pavement structure, and initial mechanical parameters using structural analyses techniques. Structural analyses, such as layered elastic theory, finite element method, and finite difference method utilize stiffness properties of pavement layers to calculate deformations. For backward direction, obtained deflections are compared with measured deflections, and new mechanical properties are estimated by a parameter identification routine. Consequently, the optimization is performed until computed and measured deflections are matched within a predefined level of accuracy. Obviously, the iterative process is time-consuming and requires excessive computational effort [4-5].

Backcalculation methods can be divided into three basic categories; i.e. static, dynamic, and adaptive [4]. Static and dynamic methods coincide with the type of loading, and require models for the structural pavement response. In this respect, these methods involve forward and backward calculations separately. It should be mentioned that, the inverse process can be performed by several techniques, such as the least-squares, the gradient descent method, database search, and genetic algorithms [4-5, 11-14]. Adaptive processes utilize neural network (NN) and adaptive neuro-fuzzy inference (ANFIS) methodologies to establish an inverse map of the model for the structural pavement response or experimental data directly. Therefore, a pavement response model is not necessary in adaptive backcalculation methods, since the approach relies on the simulation of inverse mapping through learned target behavior via known input-output data patterns [4-5, 15-20].

NN methodology is a field of artificial intelligence, which is focused on building intelligent codes that mimic the learning mechanism of a human brain by constituting a parallel-connected network model [21-23]. In a NN model, once the system is trained, network can calculate output as a functional mapper using last updated network parameters. This is also the reason why NNs are called by *universal functional approximaters* [23]. It is possible to simulate the considered inverse mapping in real time when NN models are used for pavement backcalculation problems [15-17]. Hence, significant savings in time and money can be achieved when evaluating the performance of a pavement.

It is possible to simulate nonlinear mapping defined by a known input-output data, using a fuzzy inference system. This is an unconstrained optimization problem which depends on seeking the optimal model parameters. Jang [24] presented an adaptive network approach to solve such an unconstrained optimization problem, namely the adaptive neuro-fuzzy inference system (ANFIS). From the modeling point of view, ANFIS can also be employed for backcalculation of the mechanical properties of a pavement in a supervised manner [20].

On the other hand, soft computing (SC) emerged from the fact that conventional computing techniques could not solve complex problems having imprecision and uncertainty. Unlike conventional computing techniques, SC methods focus on partial exactness through an approximation with a tolerance of imprecision. SC models exploit biological processes, predicate logic, the partial belongingness concept, parallel processing, and techniques which mimicking human mind as well as the nature. There are several methods considered in the context of SI, i.e. fuzzy systems, NN, evolutionary computation, swarm intelligence, machine learning, chaos theory and probabilistic reasoning [22].

NN and ANFIS are first SC methodologies used in backcalculation problems. Both techniques have potential to learn load – deformation behaviors, which are characterized by either synthetic (obtained via numerical analysis) or testing data. Therefore, developed model can be used to evaluate the pavement condition utilizing known deflection measurements. On the other hand, there is also potential for other SC methods for use with the backcalculation problem. In this context, Optimization methods (evolutionary or swarm intelligence based) can also be employed for the parameter identification process of pavement backcalculation problem. In simple words, if the fitness function calculates output values using a structural analyzing program, SC based optimization method can determine the model parameters. In this context, GA employed for the solution of pavement backcalculation problem by several researchers [13-14].

In the second section of this chapter, the NDT methods and pavement backcalculation problem are explained in detail. A comprehensive literature review is also included to support the main idea behind pavement analysis using NDT data. In the third and forth sections, NN, NN-based learning philosophy, and backpropagation algorithm are considered to make good understanding of NN-based learning process. In the fifth section, fuzzy logic and ANFIS methodologies are briefly explained. In the sixth and seventh sections, NN and ANFIS methodologies, as branches of SC, are considered as powerful pavement backcalculation tools. Furthermore, several NN and ANFIS based backcalculation (also referred to as adaptive backcalculation) examples and detailed comparisons are presented in these sections. In the eighth section, susceptibility of SC based optimization methods are evaluated, and previously performed examples are explained. In the ninth section, pitfalls for SC based backcalculation analyses are considered in detail. In this manner, several practical and guiding keypoints are underlined. Finally, in the tenth section, the advances in pavement backcalculation methodologies based on soft computing approaches and alternative utilization of soft computing techniques for this problem are explained. Furthermore, necessary conclusions are drawn and presented for future soft computing based pavement backcalculation analyses.

2 Nondestructive Testing Methods and Backcalculation

Highway engineers consider two major problems during the structural evaluation of a pavement structure:

- How long can a pavement structure serve under estimated traffic loads?
- What is the optimum maintenance or overlay decision for a pavement structure?

Answers of these questions rely on solving an optimization problem which depends on a combination of information and approaches such as material behavior, mechanical modeling, statistical evaluation, and economical equilibriums. In this context, the determination of mechanical material properties is essential in describing the problem statement. As mentioned before, NDT, by which mechanical properties of flexible pavement layers can be measured at low strain levels, is the most common approach in obtaining mechanical material properties. The philosophy behinds NDT methodology is that the structural performance of the pavement system is inversely proportional with the amount of surface deflections observed by an applied load. The exception to this method is the SASW method, in which pavement stiffness is obtained by Rayleigh wave velocities [4-5, 6, 11, 17].

Fundamental variation among deflection based NDT methods comes from the fact that loading type and deflection measurement locations are different for each method. In general, applied loads are divided into three categories: static, steady-state vibratory and time domain impulse. Static loading is the simplest case, which cannot simulate the nature of actual traffic loads. Benkelman Beam and La Croix Deflectometer are examples of such static type of loading. Obviously, dynamic loading is more precise and realistic in simulating the traffic affects. Steady-state dynamic case is the simplest way of simulating the actual effect of vehicle passing over a pavement. Dynaflect and Road Rater are two popular steady-state dynamic loading devices. In time domain impulse loading, an impulse load is applied on pavement surface and deflection data is recorded in time domain. Generally, there are several sensors to measure the deflection values on different points of pavement surface. Falling Weight Deflectometer (FWD) is an impulse loading device. In FWD test, a falling mass applies an impact on the pavement surface, and transient deflection data is recorded by each sensor. The impulse load of FWD is applied with a circular plate and a rubber seal is placed between plate and pavement surface in order to reduce the instant impact effect. Additionally, transient surface deflections are measured at different locations by sensors. Consequently, peak values for each sensor are used to plot deflection basin curve. A typical FWD result is illustrated in Figure 1 [3-5, 7, 25-28].

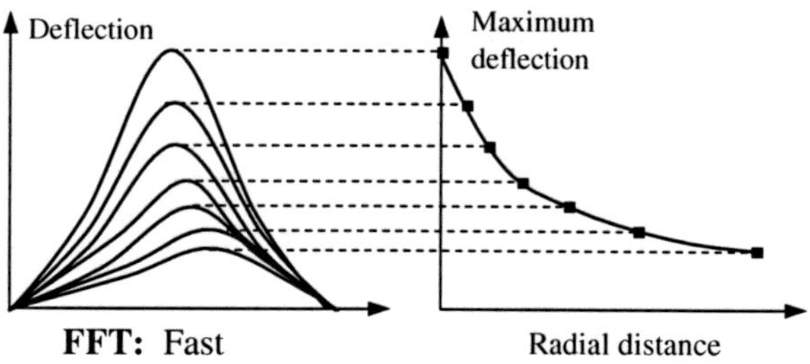

Fig. 1. Typical FWD deflection graph

On the other hand, there are a number of latest types of NDT devices for analyzing pavement structures since FWD operation is both time consuming and costly. Main feature of the recent devices is to measure deflection bowls at a certain operation speed. These devices comply with the features above are Rolling Dynamic Deflectometer (RDD), Rolling Weight Deflectometer (RWD), Rolling Deflection Testing (RDT), and High Speed Deflectograph (HSD). All of these devices can accomplish continuous deflection testing in non-destructive manner. Based on a device evaluation [29], the RDD is a good device yet not appropriate for measurement at high speeds. Although the RWD has been successfully utilized, the device measures one deflection at a time; obviously, not suitable to assess the condition of pavements. The RDT also has the same feature. On the other hand, the RDT is the unique device available giving more than one deflection measurement at a time. In conclusion, latest types of NDT devices introduced above are not commercially available except the RDT. Moreover, their measurement abilities seem not as accurate as FWD [29].

Backcalculation problems, which are also referred to as "parameter identification problems", are used in many scientific disciplines. Basically, it is an optimization problem performed to characterize the inverse relationship of a known mapping established by discrete or continuous data points. On the other hand, backcalculation process in pavement system is the numerical analysis of measured surface deflections, which is carried out for the estimation of layer stiffness parameters. In order to accomplish this, measured deflections are matched with calculated deflections, which are obtained by structural analyzing technique. The matching is continued iteratively until a close match between measured and calculated deflection values is reached [5, 11-12, 30]. Numerous backcalculation techniques are developed for the pavement backcalculation problem so far. The main principle of all these developed methods is based on the establishment of a correlation between surface deflections and layer moduli [4-5, 11-12; 31-37]. The fundamental discrepancies among backcalculation models are the type of forward response model and optimization procedure carried out for the determination of appropriate layer moduli. In this context, following classification is proposed for the backcalculation of pavement systems [5]:

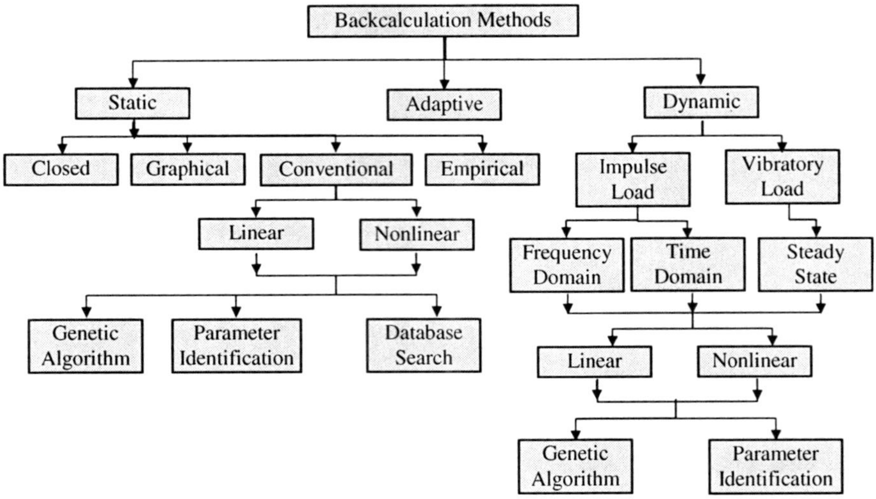

Fig. 2. Overview of backcalculation methods [5]

Referring to Fig.2, structural pavement analysis can be considered as either static or dynamic in its approach. Preliminary studies on backcalculation problem are focused on static analyses, and the inverse mapping is characterized by functional, statistical, and empirical techniques. Due to the complex and nonlinear nature of the problem, these initial attempts were not successful. As can be seen from Fig.2, optimization processes can be performed by a parameter identification algorithm (PIA), such as nonlinear least squares, database search (DSA), and genetic algorithm (GA). GA is an artificial intelligence (AI) based model-free optimization technique, which mimics the theory of evolution [3, 5, 12-14, 25-26, 36-37]. Schematic illustration of static linear conventional backcalculation with is given in Fig.3.

The difference between linear and nonlinear static backcalculation techniques are shown in Fig. 4. Similar to static linear backcalculation, stiffness of each layer (i.e. pavement moduli, E) is changed by a parameter identification algorithm (PIA) to match calculated (Δ) and measured deflections (δ). It should be noted that, Poisson's ratio (ν) and thickness (h) of each layer can be considered to be constant when utilizing this algorithm [4-5, 12, 36].

Dynamic pavement response models are used in backcalculation procedures in order to increase the quality of the results [4-5, 32, 35-37]. In dynamic response analysis, loading can be either impulsive or vibratory. Deflection data can be obtained in frequency domain or in time domain for impulse loads, and in steady state for vibratory loads. Fourier analyses are usually carried out for the transformation of the domain in order to use the dynamic loading data. Successively, elasto-dynamic numerical integration methods (such as Green function solution)

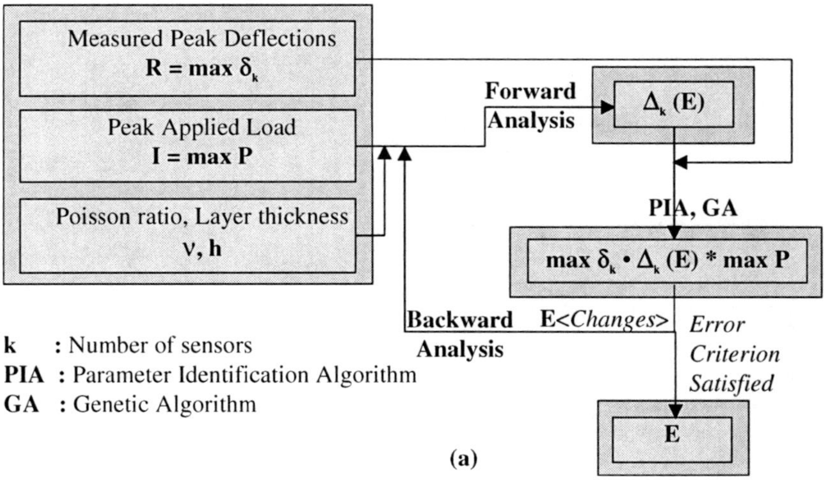

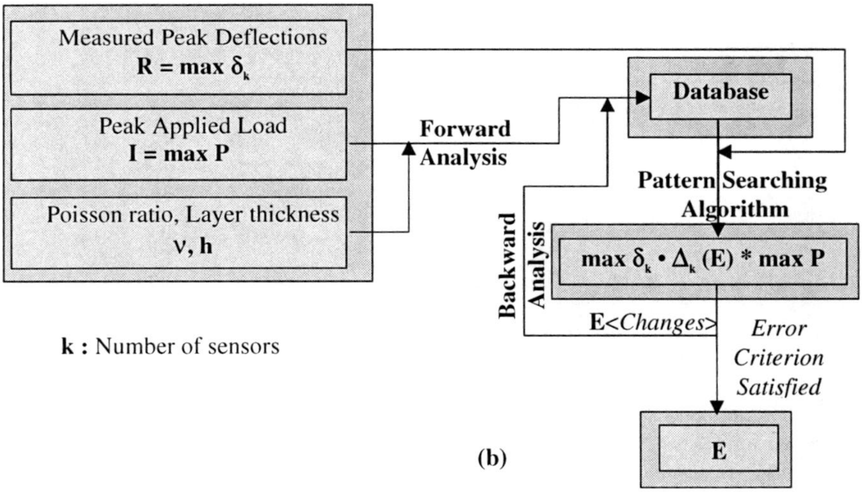

Fig. 3. Static linear backcalculation with (a) PIA-GA and (b) DSA [5]

or dynamic FEMs are performed to calculate surface deflections. In order to reduce the computational complexities, material behaviors can be considered as nonlinear similar to the static case. Furthermore, optimization process of dynamic backcalculation can be performed by suitable parameter identification routine using artificial layer moduli [4-5, 37].

Basically, dynamic response of a pavement depends on the elastic moduli, thickness, Poisson's ratios, mass densities (ρ), and damping ratios (β) of each layer. The variations in Poisson's ratios, mass densities, and damping ratios have small effects on the dynamic response of the pavement. The unknown parameters in a dynamic backcalculation analysis are the complex moduli (G*) and the thickness of the pavements layers.

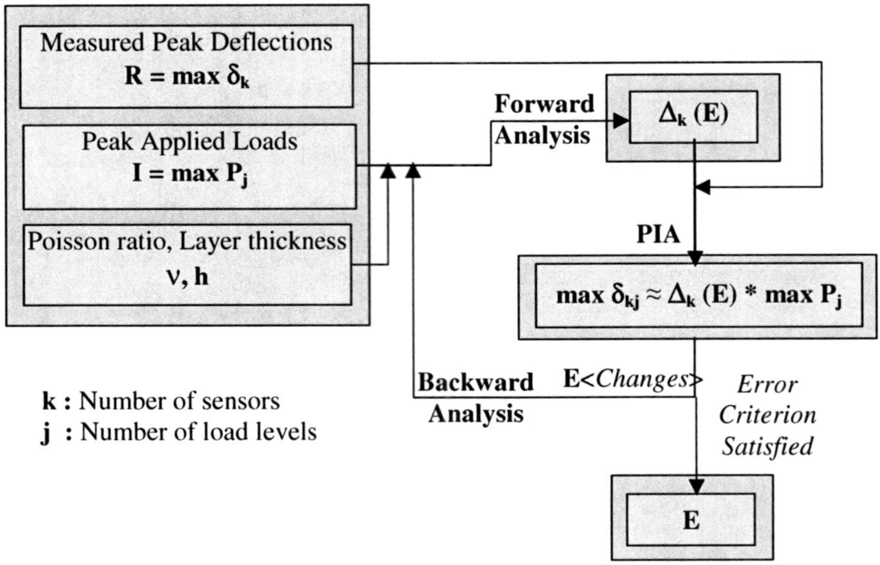

Fig. 4. Static nonlinear backcalculation with PIA [5]

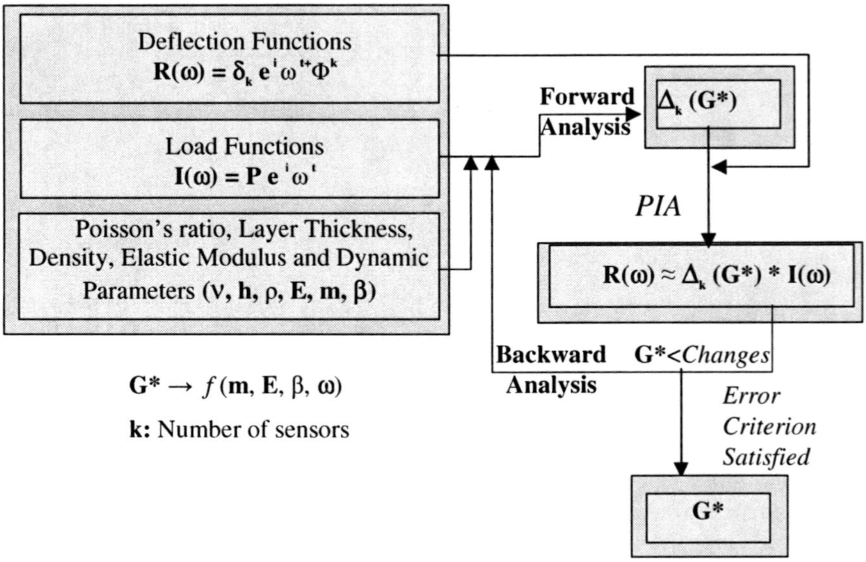

Fig. 5. Schematic Representation of dynamic steady-state vibratory load backcalculation [5]

The complex modulus is the function of angular frequency (ω) and the three material properties, i.e. slope of creep compliance curve (m_s), internal damping ratio, and Young's modulus (E). It should be noted that, creep compliance is a viscoelastic property that is related with asphalt concrete (AC) layer as well as internal damping is the function of inertia, which is considered for base, subbase, and subgrade layers in elastodynamic analyses [4-5, 12, 38-40]. The viscoelastic properties of AC layer can be characterized by creep compliance defined in the time domain, and the dynamic complex modulus can be considered in the frequency domain [4-5, 28, 36-42, 43-47]. For base and subgrade layers, complex modulus is generally assumed to be independent from the frequency, and several material models, such as Kelvin and Maxwell, are utilized to characterize these layers [37-38,48-50].

Loading is performed either impulsive or vibratory in dynamic structural analyses. In this context, deflection data can be recorded in time domain for time domain impulse loadings, and in the frequency domain for steady state vibratory loadings. In order to use the dynamic loading data, Fourier analyses are performed for transformation of the domain. In Fig.5, schematical representation of dynamic

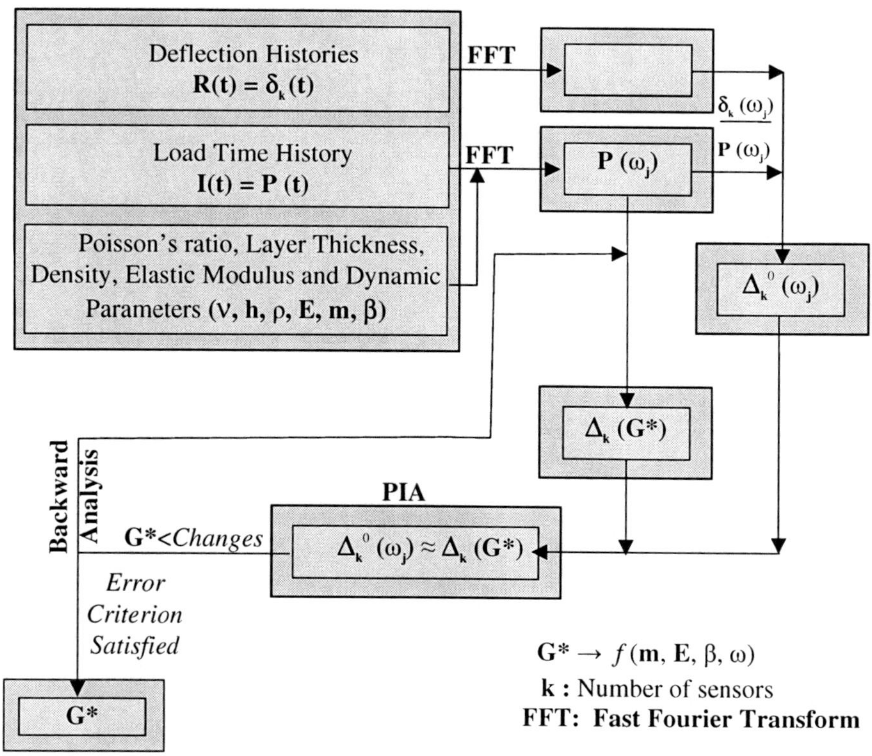

Fig. 6. Schematic representation of dynamic the frequency domain fitting for impulse load backcalculation [5]

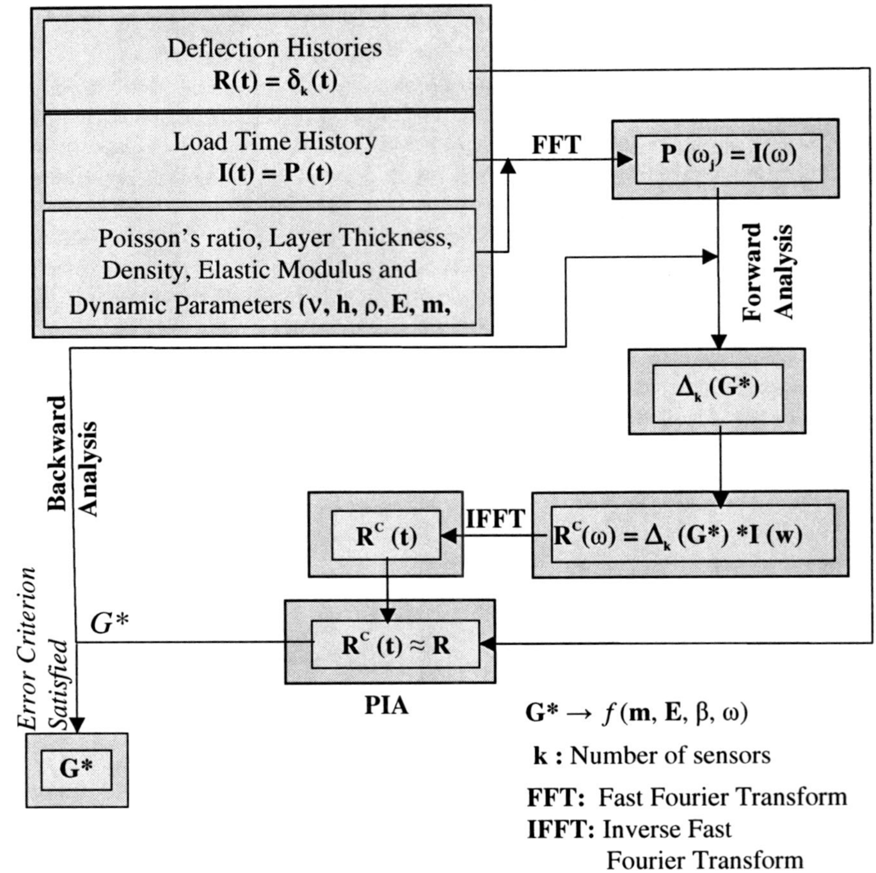

Fig. 7. Schematic representation of dynamic the time domain fitting for impulse load backcalculation [5]

backcalculation for steady-state vibratory loads is presented [4-5, 12, 37]. For impulse loads as in FWD test, it is required to transform the time (t) domain data to the frequency (ω) domain data. In this context, there can be two possibilities, i.e. (a) frequency domain fitting and (b) time domain fitting. In time domain fitting, since computed deflections are in the frequency domain, inverse Fourier transformation should be carried out to compare calculated and measured deflections [4-5, 12, 37]. Details of impulse load dynamic backcalculation for both cases are explained in Fig.6 and Fig.7.

In order to perform an efficient and precise backcalculation analysis, axisymmetric layer and half-space spectral elements can be used to characterize the dynamic behavior of flexible pavements [51]. Spectral element technique is the combination of FEM with wave propagation basics. In spectral element method, each layer is characterized by one element instead of subdivisions, which is its fundamental difference from FEM. In terms of computational requirements, this approach overcomes the drawback of FEM for use with inverse problems [51-52]. Consequently, spectral element method is powerful to simulate dynamic pavement response efficiently considering viscoelastic and poroelastic aspects [52-54].

Besides existing advantages of dynamic approach, it has several obstacles coming from the complexity and time consumption problem of dynamic analyses. Furthermore, in many problems, it is hard to get all necessary data for dynamic analysis. For these reasons, in the majority of pavement backcalculation problems, static approaches are preferred because of their simplicity and acceptable error range within the scope of the considered problem.

On the other hand, the last alternative to pavement backcalculation techniques is adaptive backcalculation which is performed by SC techniques, namely NN and ANFIS. In this type of analysis, problem is considered as one-step analysis and the database characterizing the target behavior is taught to the adaptive system. Supervising database can be obtained either by test procedure or structural analyzing program. In this respect, aim of the adaptive system is only to map the inverse relationship directly [4, 20].

3 Neural Networks

Neural networks (NN) are parallel connectionist structures, which simulate the working network of neurons in human brain. NN are widely used in many scientific areas such as, learning, classification, simulation, forecasting, and pattern recognition. Basically, as in human brain, NN consist of neurons (or processing units), which are parallel connected to each other by synapses. Neurons consist of two basic parts, namely, *dendrite* and *axon*. Dendrites establish connection between two neurons, and signal processing is performed via *synapses* (areas where

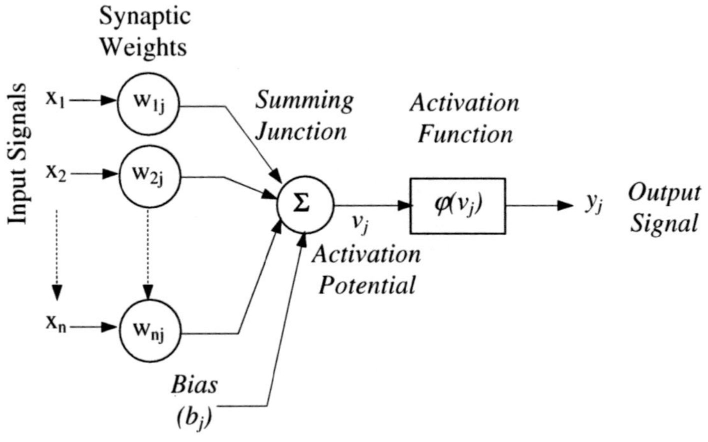

Fig. 8. An Artificial Neuron Model

messages pass). Synapses exhibit either excitatory or inhibitory attributes when processing the signals. Axons are long branches constituting the body of a neuron. Simply, a neuron is the element processing input signals and producing output signals. In this context, neurons increase or decrease the sum of incoming signals with the consideration of a bias value. In the Fig.8, schematic representation of a neuron is illustrated [22-23]:

Referring again to Fig.8, input signals are accumulated (summed) after being incorporated with synaptic weights. Later, total impulse is compared with bias term (or threshold) and *activation potential* (v_j) is calculated. In the last step, output signal is produced by the normalization of activation potential to a certain range ([0,1] or [-1,1]). In other words, activation potential is scaled by an *activation function* (φ) to reduce mapping range; otherwise, network may never converge to a solution. Mathematical representation of an artificial neuron is given below [23]:

$$y_k = \varphi(v_j) = \varphi\left(\sum_{i=1}^{n} x_i\, w_{ij} - b_j\right) \tag{1}$$

where, x_i is input signal, w_{ij} is synaptic weight, b_j is bias value, v_j is activation potential, $\varphi()$ is activation potential (or induced local field), y_k output signal, n is the number of neurons for previous layer, and k is the indice of processing neuron.

On the other hand, the term *perceptron* is equated with a processing unit including a single neuron, synaptic weights, and bias term. In addition, perceptrons can classify linearly separable patterns. Multilayer perceptrons (MLPs), also referred as multi layer feedforward neural networks, involve an input layer, one or more hidden layer, and an output layer. Each layer has a number of perceptrons that are parallelly connected to previous and successive layers [23]. In Fig.9, schematic representation of a MLP is illustrated. Actually, there are various types of NN (such as, self-organizing maps and Boltzman machines), which are successfully used for learning complex and nonlinear mappings. Therefore, MLPs are also referred to as universal function approximators [23].

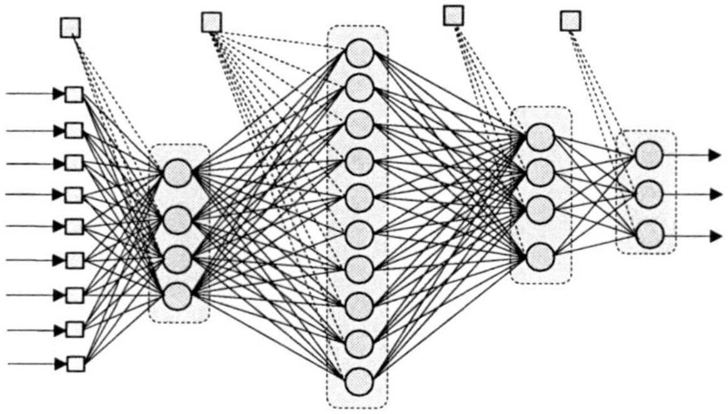

Fig. 9. Multilayer NN Structure

From modeling viewpoint, MLPs are successful in the solution of inverse problems. In detail, if input and output patterns are known, it is possible for NN to learn the inverse relation by shifting input and output spaces. In the light of this, the preferred NN type is MLP within the scope of this study; henceforth, MLP and NN are used synonymously.

4 Learning Algorithms and Backpropagation

The popularity of NN has begun to increase after the development of *backpropagation* algorithm [55]. This algorithm, also known as gradient-descent backpropagation, utilizes *Widrow-Hoff (Delta)* learning rule and free parameters (synaptic weights) are updated by the *gradient (steepest) descent* optimization technique [23]. In addition, *error energy* (generalized value of all errors in output layer) is calculated by a *least squares* based formulation. Commonly, error energy (E) is defined by mean of least squares as follows:

$$E = \frac{1}{m\,N}\sum_{k=1}^{N}\sum_{j=1}^{m}\left(y_j^k - t_j^k\right)^2 \tag{2}$$

where, m is the number of neurons in output layer, N is number of training patterns, t_j^k is the target value of processing neuron. Virtually, this algorithm changes (updates) synaptic weight along the negative gradient of error energy functions. Namely, weight changes are proportional with the magnitude of error energy. In this manner, the local error gradient is defined by:

$$\delta_n = \frac{\partial E_n}{\partial w_n} \tag{3}$$

The formulation of weight update is given as:

$$\Delta w_n = \eta\,\delta_n\, y_n \tag{4}$$

Where, Δw is weight update, η is learning rate parameter, δ_n is local error gradient, y is output signal and n represents processing neuron. It should be noted that, learning rate varies within [0,1], and *shapes* the convergence trajectory in weight space.

On the other hand, standard backpropagation algorithm may converge to local minima for some examples; thus, system could never be stable and oscillates forever. In order to avoid this, momentum term (α) is added to the standard formulation of the algorithm [23, 55]. With the addition of momentum term, Equation 3 is modified as follows:

$$\Delta w_n = \alpha\,\Delta w_{n-1} + \eta\,\delta_n\, y_n \tag{5}$$

Where, α indicates momentum term, and varies in the range of [0, 1]. This coefficient has a *stabilizing effect* for optimization direction in weight space, and enables the convergence to global minima instead of a local minimum. Momentum modification is a *heuristic* approach, and became a standard for gradient-based backpropagation [23, 56].

After the presentation of gradient-descent backpropagation algorithm, numerous researchers focused on the development of new learning algorithms exhibiting better performance in terms of precision and speed. Within this context, the most popular learning algorithms are summarized below:

In standard gradient descent method (with or without momentum term), the learning rate parameter (η) is constant throughout learning process. However, the optimal value of this parameter, which prevents oscillation at a local minimum, changes with the gradient's location and the trajectory (path) on error surface. In order to overcome this obstacle, an adaptive learning rate parameter is utilized in gradient-descent backpropagation algorithm. Thereby, the *variable learning rate backpropagation algorithm* was developed in this way [57].

The variable learning rate algorithm is a heuristic approach which begins with an initial value of the learning rate. For successive iterations, the value of learning rate is either increased or decreased until the network can learn without large changes in error. Namely, this algorithm is an adaptive technique using *delta-bar-delta* methodology for learning rate modification to reduce the amount of error energy. In this methodology, a *near-optimal* η value is adopted in Equation 5 as follows:

$$\Delta w_n = \alpha \, \Delta w_{n-1} + \eta_n \delta_n \, y_n \tag{6}$$

where, η_n indicates variable learning rate parameter for each node. In standard backpropagation method, weight update is considered by the gradient descent. Average local error gradient ($\delta^{avg}{}_n$) and convex weight factor (θ_c) are utilized to calculate learning rate update ($\Delta\eta_n$) as follows:

$$\Delta\eta_n = \begin{cases} \eta_{inc}, & \delta^{avg}{}_{n-1}\,\delta_n > 0 \\ -\eta_{dec}\,\eta_n, & \delta^{avg}{}_{n-1}\,\delta_n < 0 \\ 0, & \delta^{avg}{}_{n-1}\,\delta_n = 0 \end{cases} \tag{7}$$

where, η_{inc} is learning rate increase factor, and η_{dec} is learning rate decrease factor. As can be seen from Equation 6, learning rate increment is linear; but, decrement is nonlinearly applied [57].

Reidmiller and Braun [58] made another *heuristic* contribution to gradient-based backpropagation algorithm. Resilient backpropagation is based on the elimination of drawbacks of partial derivatives existing in gradient-descent approach.

Generally, sigmoidal (logistic or hyperbolic tangent) activation functions are utilized in NN to compress the infinite input range into a finite output range. In addition, these activation functions converge to zero as the input gets large quantities; thus, a significant drawback occurs when using descent approach. Because, gradients may have very small magnitudes, and this may cause small changes throughout the adaptation of free parameters even though the parameters are far from their optimal values [58].

In this method, the size of weight update is determined by an initial update parameter (Δ). If the signs of the derivatives with respect to synaptic weights are same for two successive iterations, then free parameters are increased by a factor (η_{inc}); otherwise, they are decreased by another factor (η_{dec}). In short, if the weights are oscillating, then weight changes are reduced, and visa versa. Δ_n is determined by:

$$\Delta_n = \begin{cases} \eta_{inc}\, \Delta_{n-1}, & \delta_{n-1}\,\delta_n > 0 \\ \eta_{dec}\, \Delta_{n-1}, & \delta_{n-1}\,\delta_n < 0 \\ \Delta_{n-1}, & \delta_{n-1}\,\delta_n = 0 \end{cases} \tag{8}$$

where, Δ_n is update parameter. Then, the update value of weights (Δw_n) is calculated using following equation [58]:

$$\Delta w_n = \begin{cases} -\Delta_n, & \delta_n > 0 \\ +\Delta_n, & \delta_n < 0 \\ 0, & \delta_n = 0 \end{cases} \tag{9}$$

Apart from heuristic modifications on gradient descent based backpropagation algorithm, several researchers focused on the integration of better numerical optimization techniques into backpropagation methodology. In general, following numerical optimization categories can be listed to be replaced by gradient (steepest) descent method [56]:

- Quasi-Newton methods,
    - BFGS algorithm,
    - DFP algorithm,
    - One-step secant algorithm,
- Conjugate gradient methods,
    - Fletcher-Reeves update,
    - Polak-Ribiere update,
    - Powell-Beale update,
    - Scaled conjugate gradient algorithm,
- Lavenberg-Marquardt algorithm.

Newton's method is a good choice for a precise optimization. Basically, Newton's method utilizes the Hessian matrix of the error energy. The Hessian matrix (H_n) is used to adjust (update) of synaptic weights with the following formulation:

$$\Delta w_n = -H_n^{-1}\delta_n \tag{10}$$

However, the calculation of Hessian matrix is an extremely time-consuming and computationally inefficient process. Therefore, quasi-Newton (or secant) methods, which do not require the calculation of second derivatives, were developed by updating the Hessian matrix at the each iteration of the algorithm.

Quasi-Newton methods are based on the gradient concept and a quadratic error energy minimization. In quasi-Newton methods, apart from gradient descent method utilizing a linear approximation technique for weight modification, following higher order approximation equation is used for adjusting the synaptic weights [23]:

$$\Delta w_n = w_{n+1} - w_n = \eta_n s_n \tag{11}$$

In which, s_n is search direction vector, and η_n is variable learning rate parameter. It should be noted that, learning-rate parameter is not a constant in this equation and enables the convergence increment along searching direction. In detail, quasi-Newton methods, involve *second-order* information about the error surface without the consideration of Hessian matrix using following approximation [23]:

$$H_n \approx q_n \Delta w_n \tag{12}$$

Where, q_n is the curvature parameter. Discrepancies among all quasi-Newton methods come from the iterative definition of S_n vector and η_n parameter. There are several quasi-Newton methods such as *Davison-Fletcher-Powell (DFP) algorithm*, *Broyden-Fletcher-Golfarb-Shanno (BFGS) algorithm*, and one *step secant algorithm* [23, 59-61].

Another attempt for a better optimization performance is related with conjugate gradient methods. They are second-order optimization methods as a member of the *conjugate-direction methods* family. In short, the conjugate-direction minimizes the quadratic performance function over progressively expanding the linear vector space that eventually including the global minimum of performance function [23, 60].

The philosophy of conjugate gradient methods is that, although it provides the decreases along gradient descent direction most rapidly, this does not produce the fastest convergence trajectory every time. In conjugate gradient algorithms, the search is performed along conjugate directions that are linearly independent. In the conjugate-gradient methods, successive direction vectors are generated as conjugate versions of the successive gradient vectors of performance function. Furthermore, NN learning process generally requires large number of free parameter adjustments. From this perspective, conjugate gradient methods are successful in optimization of large-scale problems [23, 56, 62].

In detail, conjugate gradient algorithms use an adaptive learning-rate parameter that determines the step size of weight update to reach the global minimum of the performance function. Furthermore, the step size is adjusted separately in each step in accordance with the conjugate direction. This adaptation process is performed by the following expression [62]:

$$\Delta w_n = w_{n+1} - w_n = \alpha_n s_n \tag{13}$$

Where, α_n is step size and s_n is search direction vector. Searching (conjugate) direction is determined by a recursive process, which using following expression:

$$p_{n+1} = r_{n+1} + \beta_n p_n \tag{14}$$

Where, β_n is scaling factor, which is the fundamental difference among conjugate gradient methods [23, 62].

It is obvious that, the conjugate gradient method is an optimization algorithm approximating the step size utilizing a line search routine. The objective is to obtain the step size by minimizing error energy along the line $w_n + \alpha p_n$. In other words, line search is performed for the determination of optimal distance along conjugate direction. For this reason, it is necessary to utilize a line search routine in each step of the optimization process. Nevertheless, this is a drawback making the algorithm tedious to use. Under the light of this fact, *Scaled Conjugate Algorithm*, which doesn't require any line searching routine, was developed [62]. Scaled conjugate gradient algorithm combines the *model trust region approach* with the conjugate gradient approach [62]. In scaled conjugate algorithm, Hessian matrix should be positive and definite; nevertheless, this is proposition not valid for every situation. In order to avoid this danger, scalar parameters (Lavenberg and Marquardt parameters) are set into the algorithm.

As mentioned before, this algorithm is only reliable within the territories of a small region around the searching point (*model-reliable region*). Actually, the extent of reliable region is controlled by Marquardt parameter, and the Marquardt parameter is changed gradually to regulate the indefiniteness of the Hessian matrix [62].

The last alternative numerical optimization technique that will be explained here is based on the Lavenberg-Marquardt second-order numerical optimization technique. It is a *model trust region approach* designed to increase the training speed without having to compute the Hessian matrix. The basis of this algorithm is on the *maximum neighborhood principle*, and the least-squares optimization technique [23, 62]. In general, this algorithm combines the advantages of Gauss-Newton and Steepest-Descent algorithms [63, 71].

In this method, Jacobian matrix, J_n, is computed through a standard backpropagation technique that is less complex than computing the Hessian matrix. Mathematically speaking, if the error function is in the form of sum of squares, then the Hessian matrix H_n can be approximated as follows:

$$H_n = 2\, J_n^T\, J_n + 3^{rd} \text{ and higher order terms} \tag{15}$$

Additionally, the Jacobian matrix is defined by:

$$J_n = \frac{\partial e_n}{\partial w_n} \tag{16}$$

where, J_n is the Jacobian matrix that contains first derivatives of network errors with respect to the free parameters, and e_n is error signal vector. The Jacobian matrix can be computed through a standard backpropagation technique, which is simpler than computing the Hessian matrix. The Lavenberg-Marquardt algorithm uses this approximation technique in the following Newton-like update:

$$\Delta w_{n+1} = -[J_n^T J_n + \lambda \mathbf{I}]^{-1} J_n^T e_n \qquad (17)$$

in which, **I** is the identity matrix, and λ is Marquardt parameter. In this method, Jacobian matrix is assumed to be definite, and the goal is to calculate the synaptic weights, when error function is minimized [23, 63, 71].

5 Adaptive Neuro-fuzzy Inference

Fuzzy inference system (FIS), which establishes a functional mapping between input/output spaces with the help of fuzzy logic and linguistic rule-base, is a powerful tool for simulating the complex behavior. In the literature, there are different inference techniques developed for fuzzy rule-based systems, such as Mamdani [64] and Sugeno [65]. Mamdani FIS is the first inference methodology, in which inputs and outputs are represented by fuzzy relational equations in a canonical rule-based form. In Sugeno FIS, output of the fuzzy rule is characterized by a crisp function. Typical representation of a fuzzy rule in a Sugeno FIS is given by:

$$\text{IF } x \text{ is } \underline{A}_l \text{ AND } y \text{ is } \underline{B}_l \text{ THEN } z = f(x, y) \qquad (18)$$

where $\underline{A}$ and $\underline{B}$ are fuzzy sets and z is a crisp function. In Sugeno FIS, the outcome of each rule is a crisp value, and the result of all rules is calculated by weighted average. This is the major advantage of Sugeno FIS which enables a derivative computation possible that are important for optimization techniques. Mathematical definition of the inference of a Sugeno FIS (f_{FS}) can be written as follows [66]:

$$f_{FS} = \frac{\sum_{i=1}^{m} w_i \prod_{j=1}^{n} \mu_{A_j^i}(x_j)}{\sum_{i=1}^{m} \prod_{j=1}^{n} \mu_{A_j^i}(x_j)} \qquad (19)$$

Where, m is the number of rules, n defines the number of data points, and μ_A is the membership function of fuzzy set $\underline{A}$.

The simulation of the nonlinear mapping defined by known input-output data is an unconstrained parameter identification problem based on the searching for optimal model parameters that can simulate target behavior. Jang [66] presented an adaptive network approach to solve this unconstrained optimization problem which is named ANFIS. Learning process in ANFIS methodology, namely adaptation of membership functions, is commonly performed by two techniques, i.e. backpropagation and hybrid learning algorithms. In hybrid learning algorithm, consequent parameters are identified in forward computation using LSE algorithm, and premise parameters are adjusted in backward computation with the help of backpropagation algorithm. In LSE methodology, the output of a linear model is expressed by [20, 67]:

$$y = \theta_1 f_1(u) + \theta_2 f_2(u) + ... + \theta_n f_n(u) + \varepsilon \tag{20}$$

Where, u $(u_1....u_n)$ is input vector, f $(f_1....f_n)$ are known functions, y $(y_1....y_m)$ is output vector, and θ $(\theta_1.... \theta_n)$ is unknown parameter vector, and error is denoted by ε. The objective is to find LSE (θ) that minimizes the sum of squared error [20, 67].

Adaptive backcalculation, which involves using of SC techniques such as NN and ANFIS, is fundamentally different from traditional techniques. In adaptive backcalculation, two steps (forward and backward) of traditional backcalculation are reduced to a single step with the help of a supervised learning algorithm. In this respect, the system is taught by known data pattern that is derived from structural analyzing technique or experimental study to simulate the nonlinear mapping between input and output spaces. A basic illustration of an adaptive system is shown in Fig.10 [20]. The idea of adaptive backcalculation was first introduced by Meier and Rix [15], who used NN for the SASW test data inversion for flexible pavement layers. Later, Meier and Rix [16] considered the susceptibility of NN for pavement moduli backcalculation utilizing FWD data. They finally published a complementary article comprising the dynamic aspects and the rigid bottom depth concepts [17]. Apart from these, several other studies were carried out focusing on NN-based pavement backcalculation models [18-20, 35, 68-70]. Schematic illustration of NN-based backcalculation model for nonlinear elastic material behavior and static loading is given in Fig.11a. It should be noted that, NN can solely learn the mapping characterized by input-output patterns; therefore, underlying material model and structural analysis basics do not exist in NN-based backcalculation. In other words, the performance of NN-based backcalculation is based on quality and quantity of training data [20, 68].

Adaptive neuro-fuzzy inference can also be used for the backcalculation of pavement moduli with previously determined input-output data patterns [4, 20]. Schematical representation of ANFIS-based backcalculation procedure is presented

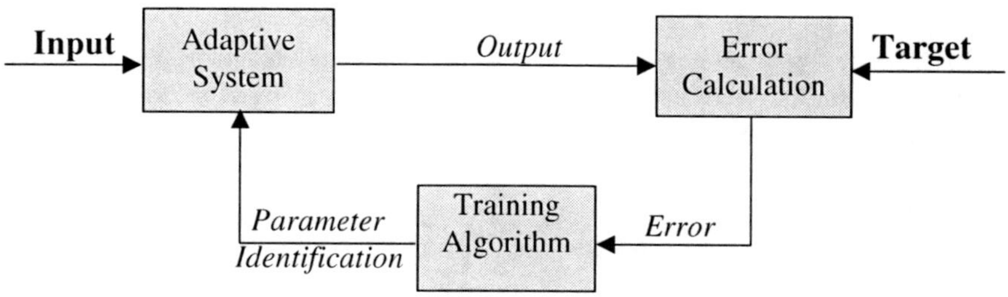

Fig. 10. Basics of Adaptive System

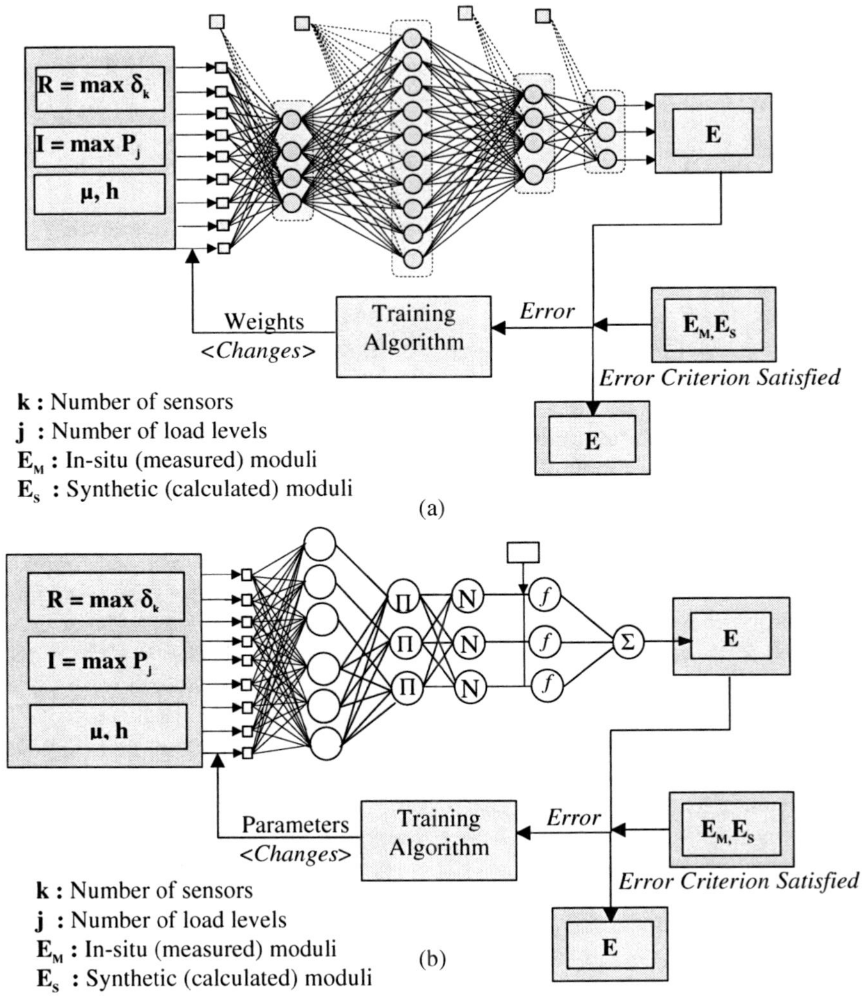

Fig. 11. Illustration of (a) NN and (b) ANFIS-Based Static Nonlinear Backcalculation [4]

in Fig.11b. It should be noted that, due to the computational expense, fuzzy inference methodology is not appropriate for large number of input-output patterns. As an alternative to NN-based backcalculation, it can be a good choice for small amount of training data involving considerable amount of uncertainty [4, 20].

As mentioned before, the training process can be performed by either experimental data to characterize specific test section or synthetically collected (with the help of a structural analyzing method) data to inversely simulate the pavement response model. The fundamental advantages of adaptive backcalculation methodologies are that they present real-time backcalculation ability and the precise results. Nevertheless, the performance of the model is solely based on the quality and the quantity of the training data in terms of the correct characterization of the relationship between input/output patterns. In this context, outliers in the training database as well as irrational test results in terms of material behavior basics should also be considered carefully for not to train the system with incorrect information [4, 20].

6 NN Based Backcalculation

NN-based pavement backcalculation basics and keypoints are considered in the following example, which is provided by Goktepe et al. [72]. Several NN models are trained and tested with synthetic database generated by structural analyses. In this respect, inverse mapping of the structural analysis using layered elastic theory is established via NN methodology. Loading is applied with 80kN of Equivalent Single Axle Load (ESAL) for dual wheels. The synthetic database consists of 1440 patterns and ranges of the database are summarized in Table 1 [71]. After the generation of synthetic database, 1100 patterns were randomly selected for the training data set, and the remaining patterns are kept for the testing process. It should be stressed that, random selection methodology is the correct technique when data quantity is considered. Nevertheless, distribution and scatting of the data points

Table 1. Details of the synthetic database

Layer	Thick. (cm)	Elastic Modulus (MPA)	Poisson's Ratio
Surface	5 - 20	1000 - 15000	0.350
Base	15 - 50	35 – 300	0.350
Subgrade	∞	30 – 200	0.350
Mean (SG)	12.459	7291	-
Mean (BS)	32.705	162	-
Mean (SB)	-	114	-

Table 2. Classification of NN models [70]

Model	Category	Learning algorithm
68	Steepest descent	Gradient descent with mom.
43	Steepest descent	Variable learning rate
37	Steepest descent	Resilient backpropagation
61	Conjugate gradient	Scaled conjugate gradient
33	Quasi-Newton	BFGS
42	Nonlin. least-square	Lavenberg-Marquardt

must be checked and evaluated in order to be sure about data characterization ability of the database. This practical issue is crucial for correct backcalculation analyses, especially when they are conducted with a limited amount of data.

In this example, six different learning algorithms are considered for their effect on backcalculation analyses in terms of precision and speed. Furthermore, different network architectures and network parameters, such as learning rate, momentum term, increment factors, and Lavenberg parameter are considered to observe their impact as well. Summary of 284 different NN models is given in Table 2 [71]. Apart from these, the scaling was performed in accordance with the hyperbolic tangential activation function and the error energy is measured by MSE (Mean Squared Error) based formulation as given in Eq.2.

It is worth mentioning that it is crucial for NN based learning procedures to pre-process the input data and synaptic weights is crucial to obtain successful results as well as to speed up the training sessions. In this context, prior to training process, input data points should be normalized to a certain range, i.e. either [0, 1] or [-1, 1], using an activation function, which provides similar or close values for out of certain range inputs so that they are considered to be same or similar by the NN. For the Sigmoid activation function, the following normalization expression can be applied to input data before training to facilitate this feature [23, 71]:

$$X_{new} = \frac{X - x_{\min}}{x_{\max} - x_{\min}} \tag{21}$$

where, X_{new} is normalized value, X is original value, and x_{max}, x_{min} are the maximum and the minimum values in the dataset, respectively. It should be noted that, normalization range is [0, 1] since Sigmoid activation function produces outputs in same range. On the other hand, following normalization equation can be used if tangential activation function is selected:

$$X_{new} = 2 \times \frac{X - x_{\min}}{x_{\max} - x_{\min}} - 1 \tag{22}$$

Another pre-processing should be performed for the selection of initial synaptic weights and bias values in order to achieve better NN performances. The most common approach for the selection of initial weight and bias values is to select them randomly from normal distribution curve [23]. Alternatively, Nguyen-Widrow method can be used for the selection of initial values of synaptic weight and bias values. Small random values are selected as the initial weights in utilizing the Nguyen-Widrow algorithm. Then, the weights are modified by dividing the regions into small intervals. Therefore, the training process can be faster by setting the initial weights of the first layer by assigning each node to its own interval at the beginning of training. Throughout the training of the network, each hidden node can be adjusted to its interval size and location [72].

Table 3. Summary of the results of selected training and testing sessions [71]

Algorithm	Architecture	Epoch	Tr (R^2)	Ts (R^2)	Er^1	Er^2	Er^3
GDM	70	10000	0.697	0.609	268	21	18
GDM	500	10000	0.746	0.633	297	19	16
GDM	40x40	10000	0.689	0.614	282	17	15
GDM	60x60	10000	0.698	0.619	204	18	17
VLR	40	10000	0.815	0.682	241	16	13
VLR	500	10000	0.791	0.597	262	17	12
VLR	20x30	10000	0.804	0.605	279	15	12
VLR	20x60	10000	0.845	0.644	271	14	11
RB	10x30	10000	0.923	0.824	82	10	7
RB	10x70	10000	0.919	0.832	74	9	8
RB	250x250	10000	0.920	0.814	75	10	7
RB	70x70x70	10000	0.925	0.813	79	11	10
SCG	10x60	10000	0.939	0.858	35	8	8
SCG	20x70	10000	0.941	0.860	41	9	8
SCG	70x70x70	10000	0.943	0.852	48	10	9
SCG	100x100x100	10000	0.942	0.855	52	9	7
BFGS	10x40	5443	0.937	0.867	26	6	6
BFGS	20x40	3847	0.939	0.871	28	5	5
BFGS	30x50	10000	0.940	0.872	29	7	4
BFGS	30x30x30	10000	0.934	0.865	22	6	5
LM	70	1000	0.948	0.897	17	4	2
LM	100	1000	0.945	0.823	19	6	3
LM	10x40	1000	0.947	0.893	12	5	2
LM	20x50	1000	0.949	0.875	14	4	2

[1, 2, 3] Max. abs. errors for AC, BS, and SG layers, respectively

Results of the study showed that different learning algorithms and network structures may produce outputs varying within a wide range. The results of selected NN models are summarized in Table 3 [71]. As can be derived from Table 3, LM, BFGS, and SCG algorithms exhibited better performances than other techniques. In Fig.12, learning graphs of selected networks (GDM-40x40, VLR-20x60, RB-10x70, SCG-20x70, BFGS-20x40, and LM-70) are shown for training and testing patterns separately. It can finally be concluded that the training performance of LM algorithm is better than other algorithms; however, testing performance of the NN learned with LM algorithm commences to decrease after 300 epochs.

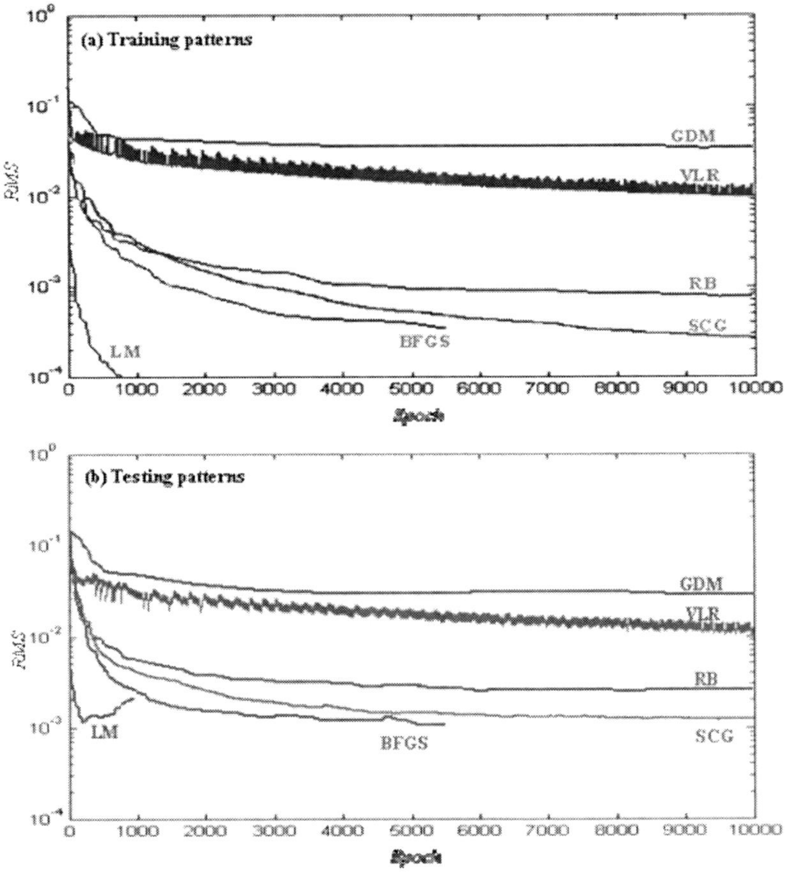

Fig. 12. Visualizing the performances of selected NNs for (a) training and (b) testing patterns [71]

7 Comparison of NN and ANFIS Methodologies for Pavement Backcalculation

NN and ANFIS methodologies are considered for the solution of the pavement backcalculation problem, in the following example, which was given by Goktepe et.al [20]. In the first step, synthetic training and testing databases were generated by Finite Element Method (FEM). Then, NN and ANFIS models were trained and tested using the databases. In the second phase of the study, small sized and poorly distributed synthetic database was generated using the same methodology in order to investigate the role of the size and the number of training patterns in both methodologies. Moreover, results of both adaptive techniques were compared with those of FEM and conventional backcalculation software based on least squares [71].

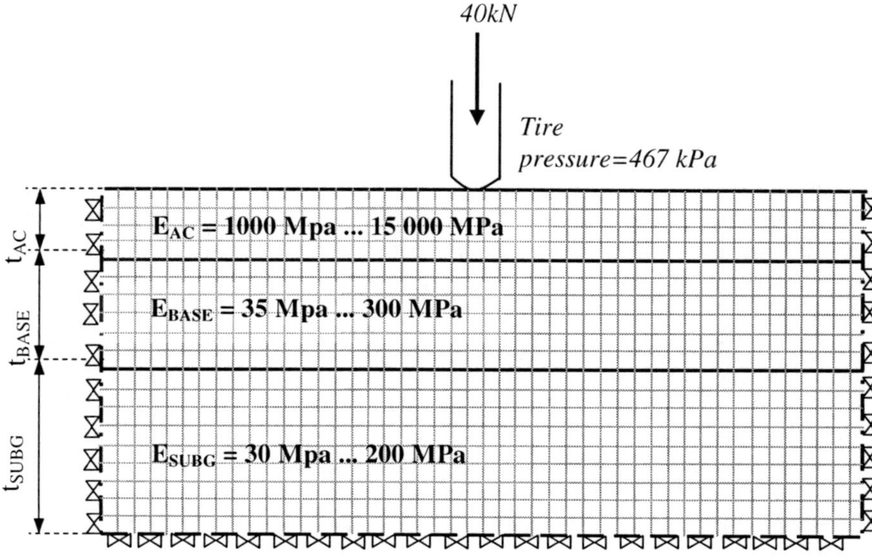

Fig. 13. Illustration of FEM model for structural analysis [20]

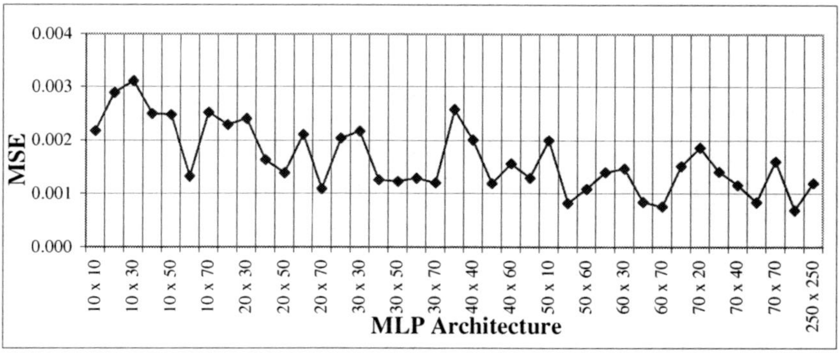

Fig. 14. Effect of network architecture on NN's performance [20]

7.1 Backcalculation with Sufficient Flexible Pavement Data

In the first part of the analysis, synthetic training and testing data sets generated by FEM, involving 1250 and 250 data patterns are utilized, respectively. A three-layered flexible pavement system is depicted in Fig.13. For loading condition, 80kN of Equivalent Single Axle Load (ESAL) for dual wheels is applied in the model. Pavement layers were assumed to be linearly elastic [20].

In NN models, scaled conjugate gradient learning algorithm was used for 10000 epochs. In order to choose the NN architecture, trial-and-error approach is used, and results of different architectures are given in Fig.14. The two learning parameters (λ and σ) are selected by 0.00005 and 0.00007, respectively [20].

In the second part, the same database is used for ANFIS-based backcalculation. Input parameters were partitioned employing the grid partitioning technique for input variables fuzzified to 3 partitions. Gaussian membership function is chosen. Furthermore, the first order Sugeno FIS with linear output function was selected as the inference system. Consequently, hybrid learning algorithm is preferred for the adaption process [20]. In the rule-base, fuzzy variables were connected with T-norm (fuzzy AND) operators and rules were associated using max-min decomposition technique. Training continued for over 1000 epochs and process terminated by the observation of the stability in error decrement. Considering the computational effort and the duration of training process, ANFIS is found to be not appropriate for such a database.

In the last step, MICHBACK computer program was used for same data to observe the performance of traditional backcalculation analysis. MICHBACK is traditional backcalculation software involving nonlinear least-square optimization technique [73]. Results of MLP, ANFIS, and MICHBACK based backcalculation analyses are summarized in Table 4. As can be seen from this table, NN exhibited better performance over other methods. Although MICHBACK is less precise than NN, it also produced satisfactory results. Apart from that, ANFIS was unsuccessful for the considered backcalculation problem in terms of modeling ability and extremely high computational expense. Consequently, three methods were

Table 4. Results of training and testing sessions [20]

Method	Session	Duration (hour)	Asphalt	Base	Subgrade
			Coef. of det. (R^2)	Coef. of det. (R^2)	Coef. of det. (R^2)
MLP (9x50x40x3)	Training	0.92	0.97	0.92	0.96
	Testing	real-time	0.91	0.88	0.92
ANFIS	Training	73	0.71	0.62	0.69
	Testing	9	0.64	0.58	0.60
MICHBACK	Training	0.27	0.94	0.90	0.91
	Testing	0.14	0.92	0.87	0.90

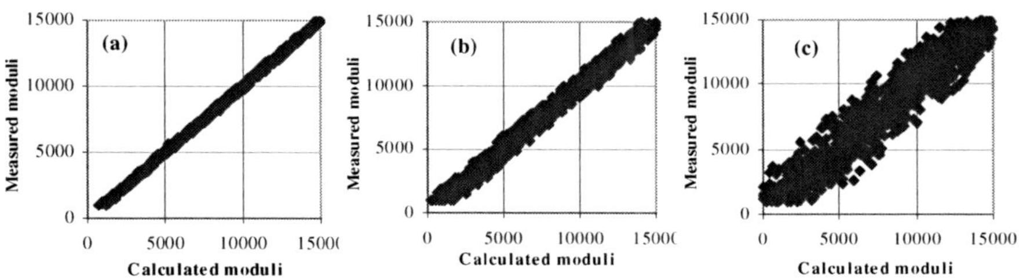

Fig. 15. Scatter plots of AC layer for training data (a) NN, (b)MICHBACK, (c)ANFIS [20]

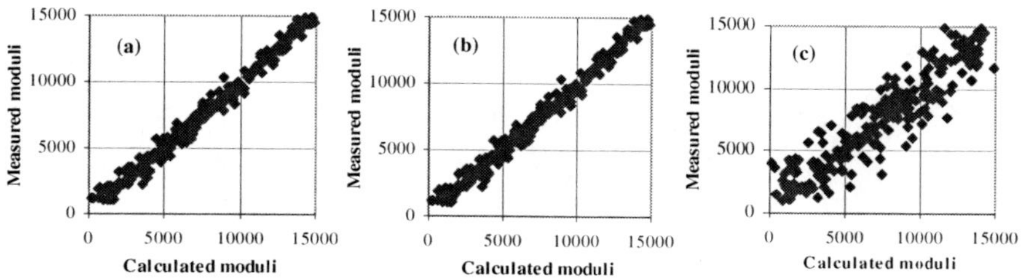

Fig. 16. Scatter plots of AC layer for testing data (a) NN, (b) MICHBACK, (c) ANFIS [20]

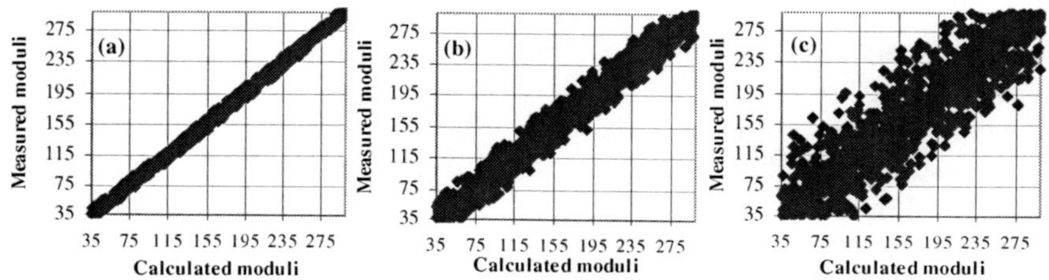

Fig. 17. Scatter plots of base layer for training data (a)NN,(b)MICHBACK,(c)ANFIS [20]

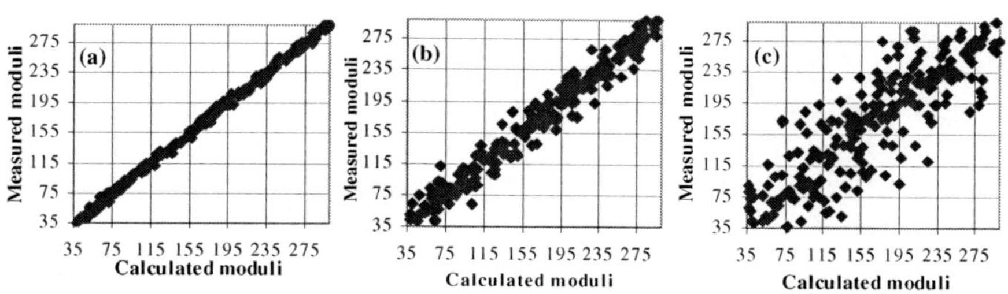

Fig. 18. Scatter plots of base layer for testing data (a) NN, (b) MICHBACK, (c) ANFIS [20]

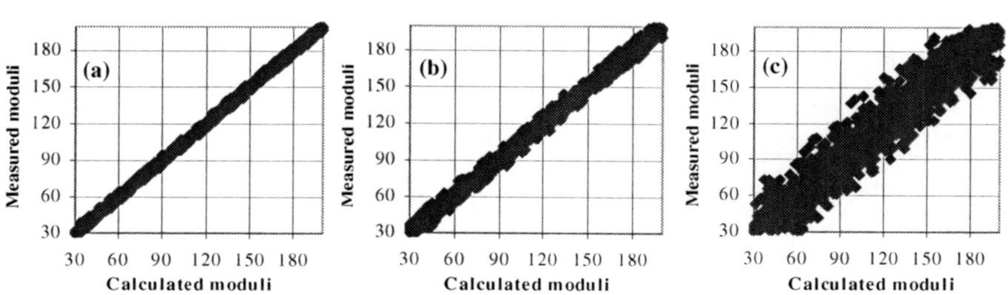

Fig. 19. Scatter plots of subgrade for training data (a) NN, (b) MICHBACK, (c) ANFIS [20]

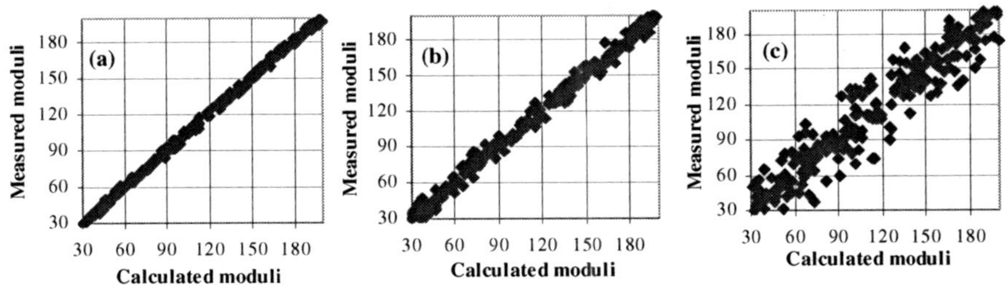

Fig. 20. Scatter plots of subgrade for testing data (a) NN, (b) MICHBACK, (c) ANFIS [20]

tested with 250 unseen data patterns. Results of this session are given in Table 4. It can be derived from the table that NN exhibited poorer performance in the testing session when comparing to the training session. NN is the most successful when backcalculation from a large amount of data. It also has a real-time backcalculation ability. In order to illustrate the modeling performance for pavement layers, scatter graphs between calculated and target deflections are shown from Fig.15 to Fig.20, respectively [20].

7.2 Backcalculation of Incomplete Pavement Data

Goktepe et al. [20] also carried out another comparative analysis to evaluate the performances of considered methodologies, specifically on incomplete flexible pavement data. In this context, there may not be a large amount of data or the data distribution may not be uniform for the proper characterization of the behavior. Therefore, it may be misleading to prefer NN over other methods for backcalculation under such conditions. In other words, NN method is quite sensitive to the quality and the quantity of data, and network cannot produce meaningful outcomes for unrecognized inputs [20, 68]. In order to evaluate this, another hypothetical flexible pavement system was designed, and input-output data points were constrained intentionally. The summary of model parameters for this analysis is given in Table 5. The numbers of training and testing patterns are 76 and 24, respectively [20].

Table 5. Ranges of training and testing variables for the second analysis [20]

Layer	Thickness (m)	Young's Modulus (MPa)	Poisson's Ratio
Surface	0.10 (fixed)	1000 - 15000	0.35 (fixed)
Base	0.30 (fixed)	35 – 300	0.40 (fixed)
Subgrade	15.00 (fixed)	30 – 200	0.45 (fixed)
FWD deflections (δ_0, δ_1, δ_2, δ_3, δ_4)			

Table 6. Results of second training and testing sessions [20]

Method	Session	Duration (hour)	Asphalt	Base	Subgrade
			Coef. of det. (R^2)	Coef. of det. (R^2)	Coef. of det. (R^2)
MLP (9x50x40x3)	Training	0.12	0.89	0.85	0.88
	Testing	real-time	0.68	0.61	0.71
ANFIS	Training	0.68	0.90	0.88	0.90
	Testing	0.02	0.81	0.80	0.82
MICHBACK	Training	0.09	0.92	0.87	0.89
	Testing	0.04	0.91	0.85	0.87

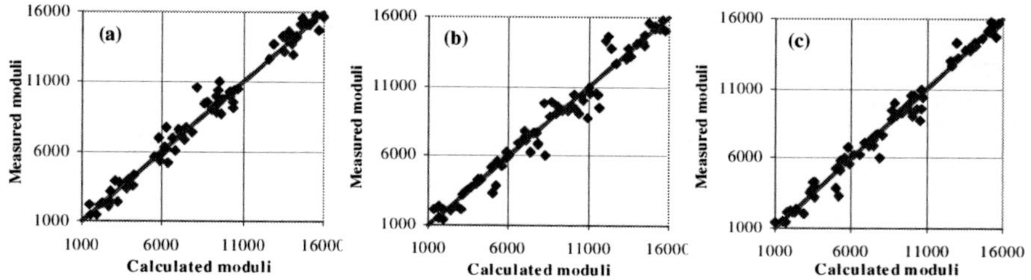

Fig. 21. Scatter plots of AC layer for 2nd train data (a)MLP,(b)MICHBACK,(c)ANFIS [20]

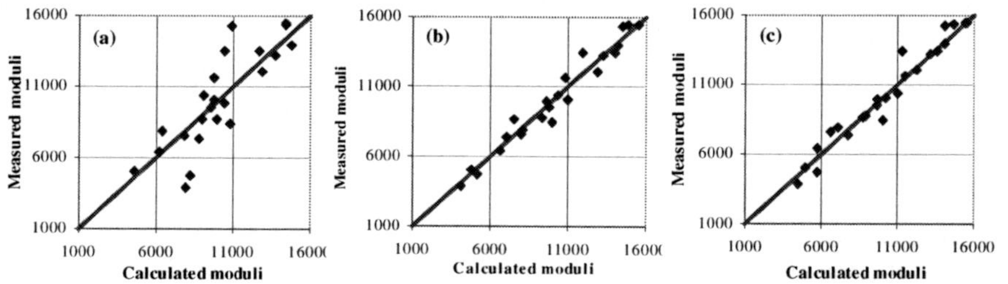

Fig. 22. Scatter plots of AC layer for 2nd test data (a) MLP, (b) MICHBACK, (c) ANFIS [20]

In Table 6, results of training and testing sessions are given. Contrary to the first analysis, ANFIS model exhibited better performance than NN. Especially for testing session, NN-based backcalculation exhibited poorer performance on unrecognized data patterns than ANFIS-based backcalculation. Scatter graphs between calculated and measured deflections for AC layer are given in Fig.21 and Fig.22 [20].

In summary, NN-based backcalculation results in the poorest performance of all methods considered. Although training results are not quite different, outcomes of testing session indicate that NN is unsuccessful for testing session, and fails for unrecognized patterns. ANFIS model exhibited better performance than NN, but slightly poorer performance when compared with MICHBACK [20].

8 Susceptibility of SC Based Optimization Methods in Pavement Backcalculation Problems

As mentioned before, SC based optimization methods (evolutionary or swarm intelligence based) can also be employed for parameter identification process of a pavement backcalculation problem. In other words, if the fitness function calculates output values using a structural analyzing program, SC based optimization method can determine the model parameters. In this context, GA is employed for the solution of pavement backcalculation problem by several researchers [13, 74-77].

GA, a field of AI and SC, is a powerful heuristic searching technique using the evolution theory's "survival of the fittest" rule. From a mathematical point of view, GA is a direct search method to find exact or approximate solutions to global optimization problems. GA is also a branch of evolutionary computation which has a unique searching ability for problems involving many local minima, complex constraints, and/or nonlinear objective functions. Therefore, they can be employed for the solution of complex optimization problems [13]. Due to the theory of evolution, the evolution usually commences from a random population and happens in next generations. Due to the theory of evolution, the fitness of every individual in the population is evaluated, multiple individuals are elected from the current population, and the population is modified (mutated) to generate new offsprings [76-77].

Methodologically, finite number of solution alternatives is created and performances of different solutions are compared with each other in the first step. Free parameters are first represented by genes that are in the form of binary strings. It should be noted that solutions are represented in binary as strings (0 and 1); nevertheless, other encodings can also be used. Parent solution, which is characterized by a set of genes, is referred to as chromosome. In the successive steps, a new parent solution, also referred to as offspring, is iteratively generated utilizing the previous parent solution. Offspring is generated with three-phased process, i.e. (a) reproduction, (b) crossover, and (c) mutation. In the reproduction, different possible solution alternatives are created and tested with fitness (objective) function's output. Basically, the fitness function measures the success of the solution numerically. In crossover, a part of the good solution is selected from solution alternatives, and unselected choices are eliminated. In mutation process, offsprings are obtained. Commonly, GA is terminated when either a maximum number of generations or a predefined fitness level is reached [13, 74-77]. Flow chart of a typical GA is presented in Fig. 23.

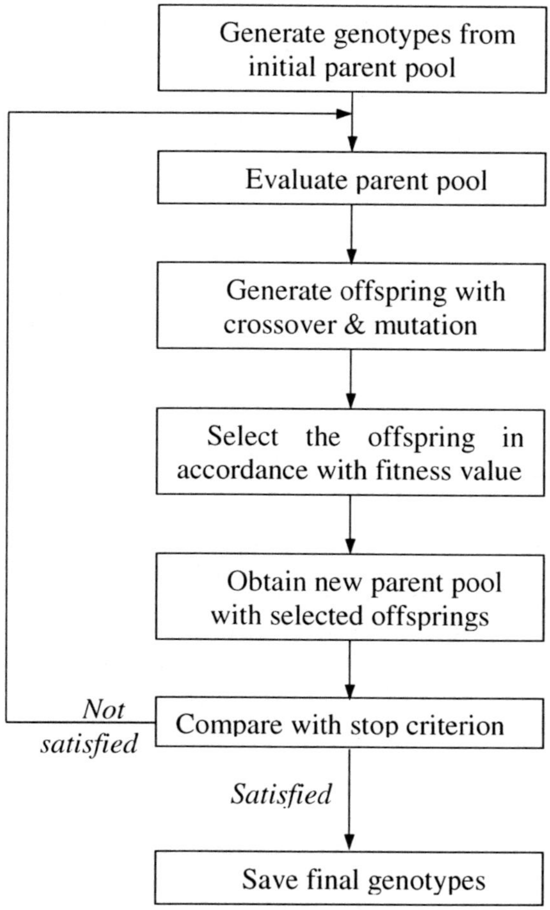

Fig. 23. Flow chart of the Genetic Algorithm

The first application of GA in pavement backcalculation problem is performed by [13]. In essence, pavement mechanical parameters are identified by GA approach to result in measured deflection values. Although GA gives precise results, computational expense is higher than classical optimization techniques [13, 14, 36, 74-77]. In the studies utilizing GA approach for pavement backcalculation problem, following conclusions were drawn:

- Accuracy of GA based backcalculation is higher than conventional techniques. Especially, conventional backcalculation techniques may lead to misleading results because of the premature convergence and local optima problems.
- Large populations should be used for the model parameters.

- Major advantage of GA-based pavement backcalculation is when the layer number increases. For flexible pavements and subgrade system having higher number of layers (for example, more than 4 layers) GA-based backcalculation gives accurate and efficient results. For lower layered systems, computational expense of GA-based backcalculation is open to question.

9 Pitfalls for SC Methods in the Backcalculation

After the development of backpropagation algorithm, NN has emerged in several engineering disciplines. In this context, numerous studies were performed to employ NN methodology in material science due to its universal approximation ability and powerful learning attributes. In many studies, as an alternative to traditional statistical approaches such as regression analysis, the relationships between input/output spaces were established via NN learning process and their results on unseen data were evaluated. Rather than summarizing the behavior characterized by learning data, NN can produce meaningful outcomes for unseen data as a modeling tool. In essence, utilization of NN in pavement backcalculation problem is based on learning the relationship between defections (or other measured parameters) and pavement system; therefore, it can produce outcomes as a functional mapper in real time. Thereby, the performance of NN-based pavement backcalculation (or any material modeling process) is fundamentally based on the quantity and the quality of the data [4, 68]. Nevertheless, modeling data may have such deficiencies as:

- Model data may be insufficient to characterize the target behavior in terms of scatting and/or quantity.
- The data may be misleading in terms of ranges should be covered. Therefore, it would be uncertain how NN produce outcomes for unseen input data which is out of the range of supervising input data.
- There may be noise in the modeling data. Consequently, outliers in the data must be considered and/or filtered carefully.
- Because of several reasons, the best fit to existing modeling data may be irrational in terms of material behavior. Therefore, the best fit may not be the best backcalculation result considering the behavior of the data.
- Model data includes certain amount of measurement uncertainty. High amount of uncertainty may result in undesirable mistakes is modeling process.
- Sequence of modeling data may be not appropriate for efficient training. In other words, similar data points could be successively as data groups; thus, the uniformity of data sequence and training efficiency would be problematic.

As a result of the factors explained above, the model data may be misleading to establish the target relationship correctly. These drawbacks are especially major for the backcalculation analyses performed by test data. In case of using a synthetic training data, the aim of adaptive methods is only to characterize the inverse mapping of a structural analyzing software. It should be noted that, this drawback is valid for all adaptive backcalculation methods using SC techniques. Therefore, following precautions should be taken before and after the training process to develop a proper adaptive backcalculation model.

- Scatter of the modeling data must be observed and evaluated before training. Therefore, necessary data plots and statistics should be drawn and performed before training sessions. The validity, the modeling ranges, and the precision of the backcalculation analysis must be considered under the light of these evaluations. If possible, necessary corrections, such as gathering new information and changing modeling ranges, must be carried out at this stage.
- After necessary scatter plots are drawn, outliers should be determined. Such noisy data should be removed or rechecked for a proper training process. This is especially important for modeling the data obtained by in-situ or laboratory tests.
- Considering the basics and the theories of structural analysis, irrational or meaningless relationships, which were identified by adaptive technique, must be realized. It may be necessary to prefer classical methods or forced adaptive models for such incoherent mappings.
- Uncertainty analysis could be carried out for the determination of measurement uncertainty; however, this is useful in understanding the precision of the model as well as to comprehend the boundaries of NN-based (or adaptive) modeling in terms of its data-driven philosophy. In other words, NN methodology cannot be considered as a structural analyzing tool as FEM or layered elastic theory, and its success depends on the quality and the quantity of the input database.
- The data sequence should be evaluated before training. In order to remove this NN-based problem, training mode, i.e. batch, sequential, etc., should be considered carefully and/or the data points can be chosen randomly again.

On the other hand researchers implemented several solutions for the drawbacks of GA-based pavement backcalculation problem [13, 14, 74]. The following problems and solutions can be listed:

- If small amount of populations are chosen for the model parameters, precision of the model decreases. This should be evaluated during the model development, and large populations should be used.
- Accuracy is very important for any backcalculation model. More accurate model is mostly preferred by analyzers. However, it must be considered that excessive level of precision may be meaningless when dealing with the amount of uncertainty in pavement systems and soil substructure.
- For lower layered systems, even though there is no considerable accuracy advantage, computational expense of GA-based backcalculation is higher than traditional methods.
- There is an automation problem with GA-based pavement backcalculation. This can be overcome by Dynamic Parameterless Genetic Algorithm (DPGA).

10 Conclusions

This chapter addresses the advances in pavement back-calculation methodologies based on SC approaches by presenting the concepts behind them and the fundamental advantages of each. In this context, following conclusions can be drawn from this study:

- Adaptive backcalculation methods, developed by SC methodologies, can perform real-time backcalculation analyses as a functional mapper. However, there is no underlying mechanical background in these techniques; therefore, they must be applied carefully. It should be stressed that adaptive backcalculation is not an alternative to theoretical structural analyzing techniques, such as FEM, and elasto-dynamic theory.
- ANFIS is appropriate for limited amount of data including considerable uncertainty. For large training databases, NN looks much more promising than ANFIS.
- Utilization of NN and ANFIS enables real-time backcalculation analyses. Furthermore, two phases of the backcalculation problem, namely structural analysis and the parameter identification (or optimization), reduces one phase, and the relationship can be established directly.
- Precision of adaptive backcalculation models using soft computing techniques are outstanding. However, the level of precision should be considered with the level of uncertainty in the problem.
- Especially for backcalculation analyses performed to model test results, soft computing based backcalculation methods looks promising. Nevertheless, there are several pitfalls for adaptive backcalculation methods due to the nature of the SC methodologies. These advantages and drawbacks must be taken into account carefully before and after the modeling process.
- From practical viewpoint, adaptive methods using SC techniques cannot give a simple formulation to the analyzer. The NN and ANFIS methods, can give several matrices as a result of training sessions, and calculations can be conducted using the matrices and algebraic operations. However, this drawback can be overcome by developing graphical user interfaces including necessary routines and training values.
- Optimization technique is also a complementary issue for pavement backcalculation problems. Preference of an optimization algorithm is important in terms of speed and precision of the analysis. In this context, GA gives good opportunity for pavement systems having more layer numbers without sacrificing from the precision.

References

[1] Callop, A.C., Cebon, D.: Stiffness reductions of flexible pavements due to cumulative fatigue damage. J. Transp. Eng.- ASCE 122, 131–139 (1996)

[2] Briggs, R.C., Lukanen, E.O.: Variations in backcalculated pavement layer moduli. In: Tayabji, S.D., Lukanen, E.O. (eds.) LTTP seasonal monitoring sites, NDT of Pavements and Backcalculation of moduli, vol. 3, pp. 113–128. Special Technical Publication, STP 1375, ASTM Publication, Pennsylvania (2000)

[3] Tawfiq, K., Armaghani, J., Sobanjo, J.: Seismic pavement analyzer vs. FWD for pavement evaluation: comparative study. In: Tayabji, S.D., Lukanen, E.O. (eds.) NDT of Pavements and Backcalculation of Moduli, vol. 3, pp. 327–345. Special Technical Publication, STP 1375, ASTM Publication, Pennsylvania (2000)
[4] Ullidtz, P.: Will nonlinear backcalculation help. In: Tayabji, S.D., Lukanen, E.O. (eds.) NDT of Pavements and Backcalculation of Moduli, vol. 3, pp. 14–22. Special Technical Publication, STP 1375, ASTM Publication, Pennsylvania (2000)
[5] Goktepe, A.B., Agar, E., Lav, A.H.: Advances in Backcalculating the Mechanical Properties of Flexible Pavements. Adv. Eng. Softw. 37, 421–431 (2006)
[6] Nazarian, S., Stokoe, K.H.: Nondestructive evaluation of pavements by surface wave method. In: Bush, A.J., Baladi, G.Y. (eds.) NDT of Pavements and Backcalculation of Moduli, vol. 1, pp. 119–137. Special Technical Publication, STP 1026, ASTM Publication, Pennsylvania (1989)
[7] Seng, C.R., Stokoe, K.H., Roesset, J.M.: Effect of depth to bedrock on the accuracy of backcalculated moduli obtained with Dynaflect and FWD Tests, Special Report, SP 1175-5,Center for Transportation Research. University of Texas at Austin. Texas (1993)
[8] Zhou, H.: Comparison of backcalculated and laboratory measured moduli on AC and granular base layer materials. In: Tayabji, S.D., Lukanen, E.O. (eds.) NDT of Pavements and Backcalculation of Moduli, vol. 3(3-37), pp. 161–172. Special Technical Publication, STP 1375, ASTM Publication, Pennsylvania (2000)
[9] Hoffman, M.S., Thompson, M.R.: Backcalculating nonlinear resilient moduli from deflection data, Transportation Research Record, 852, TRB, pp. 42–51. National Research Council, Washington DC (1982)
[10] Tholen, O., Sharma, J., Terrel, R.L.: Comparison of FWD with other deflection testing devices, Transportation Research Record 1007, TRB, pp. 20–26. National Research Council, Washington, DC (1985)
[11] Lytton, R.L.: Backcalculation of layer moduli, state of the art. In: Bush, A.J., Baladi, G.Y. (eds.) NDT of Pavements and Backcalculation of Moduli, vol. 1, pp. 7–38. Special Technical Publication, STP 1026, ASTM Publication, Pennsylvania (1989)
[12] Uzan, J.: Advanced backcalculation techniques. In: Von Quintus, H.L., Bush, A.J., Baladi, G.Y. (eds.) NDT of Pavements and Backcalculation of Moduli, vol. 2, pp. 3–37. Special Technical Publication, STP 1198, ASTM Publication, Pennsylvania (1994)
[13] Fwa, T.F., Tan, C.Y., Chan, W.T.: Backcalculation analysis of pavement-layer moduli using genetic algorithms, Transportation Research Record 1570, TRB, pp. 134–142. National Research Council, Washington DC (1997)
[14] Reddy, M.A., Reddy, K.S., Pandey, B.B.: Backcalculation of pavement moduli using genetic algorithms. J. High Res. Board 66, 1–10 (2002)
[15] Meier, R.W., Rix, G.J.: An initial study of surface wave inversion using artificial neural networks. Geotech. Test J. 16, 425–431 (1993)
[16] Meier, R.W., Rix, G.J.: Backcalculation of flexible pavement moduli using artificial neural networks, Transportation Research Record, 1448, TRB, pp. 75–82. National Research Council, Washington, DC (1994)
[17] Meier, R.W., Rix, G.J.: Backcalculation of flexible pavement moduli from dynamic deflection basins using artificial neural networks, Transportation Research Record, 1473, TRB, pp. 72–81. National Research Council, Washington, DC (1995)

[18] Tutumluer, E., Seyhan, U.: Neural network modeling of anisotropic aggregate behavior from repeated load triaxial tests, Transportation Research Record, 1615, pp. 86–93. National Research Council, Washington, DC (1998)
[19] Kim, Y.R., Xu, B., Kim, Y.: A new backcalculation procedure based on dispersion analysis of FWD time history deflections and surface wave measurements using ANNs. In: Tayabji, S.D., Lukanen, E.O. (eds.) NDT of Pavements and Backcalculation of Moduli, vol. 3, pp. 297–312. ASTM Publication, STP 1375, Pennsylvania (2000)
[20] Goktepe, A.B., Agar, E., Lav, A.H.: Comparison of multilayer perceptron and adaptive neuro-fuzzy system on backcalculating the mechanical properties of flexible pavements. ARI Bull. of Ist Tech. Uni. 54, 65–77 (2005)
[21] Principle, J.C., Euliano, N.R., Lefebvre, W.C.: Neural and Adaptive Systems. John Wiley and Sons Co., New York (2000)
[22] Kecman, V.: Learning and Soft Computing. MIT Press, Massachusetts (2001)
[23] Haykin, S.: Neural Networks: A Comprehensive Foundation. Prentice-Hall Inc., New Jersey (1999)
[24] Jang, J.S.R.: ANFIS: Adaptive-network-based fuzzy inference systems. IEEE T Syst. Man Cy. 23, 665–685 (1993)
[25] Mamlouk, M.S.: Use of dynamic analysis in predicting field multilayer pavement moduli, Transportation Research Record, 1043, TRB, pp. 113–119. National Research Council, Washington DC (1985)
[26] Sebaaly, B., Davies, T.G., Mamlouk, M.S.: Dynamics of falling weight deflectometer. J. Transp. Eng-ASCE 111, 618–632 (1985)
[27] Stolle, D.F.E.: Modeling of dynamic response of pavements to impact loading. Comput. Geotech. 11, 83–94 (1991)
[28] Huang, Y.H.: Pavement Analysis and Design. Prentice Hall Inc., New Jersey (1993)
[29] Arora, J., Tandon, V., Nazarian, S.: Continuous deflection testing of Texas Pavements, Project Summary Report No: 0-4380-S, Center for Transportation Infrastructure Systems. The University of Texas El Paso (2006)
[30] Sivaneswaran, N., Kramer, S.L., Mahoney, J.P.: Advanced backcalculation using a nonlinear least squares optimization technique, Transportation Research Record, 1293, TRB, pp. 93–102. National Research Council, Washington, DC (1991)
[31] Harichandran, R.S., Yeh, M.S., Baladi, G.Y.: MICH-PAVE: A nonlinear finite element program for analysis of flexible pavements, Transportation Research Record 1286, TRB, pp. 123–131. National Research Council, Washington DC (1990)
[32] Magnuson, A.H., Lytton, R.L., Briggs, R.C.: Comparison of Computer Predictions and Field Data for Dynamic Analysis of Falling Weight Deflectometer Data, Transportation Research Record, 1293, TRB, pp. 124–135. National Res. Council, Washington, DC (1991)
[33] Magnuson, A.H.: Computer analysis of falling weight deflectometer data, part I: Vertical displacement computations on the surface of a uniform surface pressure distribution, Research Report No: 1215-1F, Texas Transportation Institute, Collage Station (1988)
[34] Newcomb, D.E.: Development and evaluation of regression method to interpret dynamic pavement deflections, Ph.D. Dissertation, Department of Civil Engineering. University of Washington, Seattle (1986)
[35] Reddy, M.A., Reddy, K.S., Pandey, B.B.: Selection of genetic algorithm parameters for backcalculation of pavement moduli. The Int. J. Pave Eng. 5, 81–90 (2004)

[36] Mamlouk, M.S., Davies, G.D.: Elasto-dynamic analysis of pavement deflections. J. Transp. Eng- ASCE 110, 536–567 (1984)
[37] Uzan, J.: Dynamic linear backcalculation of pavement material parameters. J. Transp. Eng-ASCE 120, 109–125 (1994)
[38] Stubbs, N., Torpunuri, V.S., Lytton, R.L., Magnuson, A.H.: A methodology to identify material properties in pavements modeled as layered viscoelastic half-spaces. In: Von Quintus, H.L., Bush, A.J., Baladi, G.Y. (eds.) NDT of Pavements and Backcalculation of Moduli, vol. 2, pp. 159–169. Special Technical Publication, STP 1198, ASTM Publication, Pennsylvania (1994)
[39] Sousa, J.B., Weissman, S.L., Sackman, J.L., Monismith, C.L.: Nonlinear elastic viscous with damage model to predict permanent deformation of asphalt concrete mixes, Transportation Research Record, 1136, TRB, pp. 57–68. National Research Council, Washington, DC (1993)
[40] Chang, D.W.: Nonlinear effects on dynamic response of pavements using the NDT technique, PhD Dissertation, Department of Civil Engineering. University of Texas at Austin, Austin (1991)
[41] Sousa, J.B., Monismith, C.L.: Dynamic response of paving materials, Transportation Research Record, 1136, TRB, pp. 57–68. National Res. Council, Washington, DC (1987)
[42] Shao, K.Y.: Dynamic interpretation of Dynaflect, Falling Weight Deflectometer and spectral analysis of surface waves tests on pavement system, PhD Dissertation. University of Texas at Austin, Texas (1985)
[43] Roesset, J.M.: Stiffness and damping coefficients of foundations. In: O'Neal, M.W., Dobry, K. (eds.) Proceedings of ASCE National Convention on Dynamic Response of Pile Foundations, Analytical Aspects, New York, pp. 1–30 (1980)
[44] Kausel, E., Roesset, J.M.: Stiffness matrices for layered soils. Bul. Seism. Soc. Am. 71, 1743–1761 (1981)
[45] Haskell, N.A.: The dispersion of surface waves on multilayered media. Bul. Seism. Soc. Am. 71, 17–34 (1953)
[46] Lysmer, J.: Lumped mass method for Rayleigh waves. Bul. Seism. Soc. Am. 60, 89–104 (1970)
[47] Founquinos, R., Roesset, J.M., Stokoe II, K.H.: Response of pavement systems to dynamic loads imposed by nondestructive tests, Transp. Res. Rec. 1504, TRB, Washington, DC, pp. 57–67 (1995)
[48] Nazarian, S., Stokoe, K.H.: Use of surface waves in pavement evaluation, Transp. Res. Rec. 1070, TRB, pp. 132–144. National Res. Council, Washington, DC (1986)
[49] Nazarian, S., Stokoe, K.H., Hudson, W.R.: Use of spectral analysis of surface waves method for determination of moduli and thicknesses of pavement systems, Transp. Res. Rec. 921, TRB, pp. 38–45. National Res. Council, Washington, DC (1983)
[50] Liang, R.Y., Zhu, J.X.: Efficient computational algorithms for forward and backward analysis of a dynamic pavement system. Comput. Struct. 69, 255–263 (1998)
[51] Al-Khoury, R., Scarpas, A., Kasbergen, C., Blaauwendraad, J.: Spectral element technique for efficient parameter identification of layered media. I. Forward calculation. Int. J. Solids Struct. 38, 1605–1623 (2001)
[52] Al-Khoury, R., Scarpas, A., Kasbergen, C., Blaauwendraad, J.: Spectral element technique for efficient parameter identification of layered media. II. Inverse calculation. Int. J. Solids Struct. 38, 8753–8772 (2001)

[53] Al-Khoury, R., Scarpas, A., Kasbergen, C., Blaauwendraad, J.: Spectral element technique for efficient parameter identification of layered media. III. Viscoelastic aspects. Int. J. Solids Struct. 39, 2189–2201 (2002)
[54] Al-Khoury, R., Kasbergen, C., Scarpas, A., Blaauwendraad, J.: Poroelastic spectral element for wave propagation and parameter identification in multi-layer systems 39, 4073–4091 (2002)
[55] Rumelhart, D.E., McClelland, J.L.: Parallel distributed processing: explorations in the microstructure of cognition. MIT Press, Massachusetts (1986)
[56] Hagan, M.T., Demuth, H., Beale, M.: Neural Network Design. PWS Publishing Company, New York (1996)
[57] Jacobs, R.A.: Increased rates of convergence through learning rate adaptation. Neural Networks 1, 295–307 (1988)
[58] Reidmiller, M., Braun, H.: A direct adaptive method for faster backpropagation learning: the RPROP algorithm. In: Proceedings of the IEEE International Conference on Neural Networks, New York, May 10-15, 1993, pp. 586–591 (1993)
[59] Battiti, R.: First and second order methods for learning: between steepest descent and Newton's method. Neural Comput. 4, 141–166 (1992)
[60] Fletcher, R.: Practical Methods of Optimization. John Wiley and Sons Co., New York (1987)
[61] Bertsekas, D.P.: Nonlinear Programming. Athena Scientific Publishing, Massachusetts (1995)
[62] Moller, M.F.: A scaled conjugate gradient algorithm for fast supervised learning. Neural Networks 6, 525–533 (1993)
[63] Hagan, M.T., Menhaj, M.: Training feedforward networks with the Marquardt algorithm. IEEE T. Neural Network 5, 989–993 (1994)
[64] Mamdani, E.H., Assilian, S.: An experiment in linguistic synthesis with a fuzzy logic controller. Int. J. Man-Mach Studies 7, 1–13 (1975)
[65] Takagi, T., Sugeno, M.: Fuzzy identification of systems and its applications to modeling and control. IEEE T. Syst. Man Cy. 15, 116–132 (1985)
[66] Jang, J.S.R.: ANFIS: Adaptive-network-based fuzzy inference systems. IEEE T. Syst Man Cy. 23, 665–685 (1993)
[67] Jang, J.S.R., Sun, C.T., Mizutani, E.: Neuro-Fuzzy and Soft Computing: A Computational Approach to Learning and Machine Intelligence. Prentice Hall, New Jersey (1997)
[68] Tutumluer, E., Meier, R.W.: Attempt at resilient modulus modeling using artificial neural networks. Transport Res. Rec. 1560, TRB, Washington, DC, pp. 1-6 (1996)
[69] Abdallah, I., Ferregut, C., Nazarian, S., Lucero, M.O.: Prediction of remaining life of flexible pavements with artificial neural network models. In: Tayabji, S.D., Lukanen, E.O. (eds.) NDT of Pavements and Backcalculation of Moduli, vol. 3, pp. 484–498. Special Technical Publication, STP 1375, ASTM Publication, Pennsylvania (2000)
[70] Saltan, M., Tigdemir, M., Karasahin, M.: Artificial neural network application for flexible pavement thickness modeling. Turkish J. Eng. and Env. Sci. 26, 243–248 (2002)
[71] Goktepe, A.B., Agar, E., Lav, A.H.: Role of Learning Algorithm in Neural Network Based Backcalculation of Flexible Pavements. ASCE J. Comput. Civil Eng. 20, 370–373 (2006)
[72] Nguyen, D., Widrow, B.: Improving the learning speed of 2-layer neural Networks by choosing initial values of the adaptive weights. In: Proceedings of the International Joint Conference on Neural Networks, vol. 3, pp. 21–26 (1990)

[73] Harichandran, R.S., Ramon, C.M., Mahmood, T., Baladi, G.Y.: MICHBACK user's manual. Michigan Department of Transportation, Michigan (1994)
[74] Alkasawneh, W.: Backcalculation pavement moduli using genetic algorithms, PhD. Distertation, The University of Akron (2007)
[75] Pan, E., Chen, E., Alkasawneh, W.: Layered Flexible Pavement Studies: Challenges in Forward and Inverse Programs. Int. J. Pave. Res. Techn. 1, 12–16 (2008)
[76] Goldberg, D.E.: Genetic algorithms in search, optimization and machine learning. Addison-Wesley Co., Reading (1989)
[77] Chambers, L.: The practical handbook of genetic algorithms. Chapman & Hall / CRC Press, Washington, D.C (2001)

Abbreviations

AC	asphalt concrete
AI	Artificial intelligence
ANFIS	adaptive neuro-fuzzy inference system
$\underline{A}, \underline{B}$	**Fuzzy sets**
BFGS	Broyden-Fletcher-Golfarb-Shanno algorithm
b_j	bias value
DSA	database search
DFP	Davison-Fletcher-Powell algorithm
E	pavement modulus
e_n	error signal vector
ESAL	Equivalent Single Axle Load
FEM	finite element method
FIS	Fuzzy inference system
FFT	fast Fourier Transform
FWD	falling weight deflectometer
f	unknown function
f_{FS}	inference of a Sugeno FIS
GA	genetic algorithm
G*	complex moduli
h	layer thickness
HSD	High Speed Deflectograph
H_n	Hessian matrix
I	Identity matrix
IFFT	inverse fast Fourier Transform
J_n	**Jacobian matrix**
k	indice of processing neuron
m	number of neurons in output layer
m_s	slope of creep compliance curve
LSE	least sum of exponentials
MLP	multilayer perceptron
MSE	mean squared error
n	processing neuron
N	number of training patterns

NDT	nondestructive testing
NN	neural network
PIA	parameter identification algorithm
q_n	curvature parameter
RDD	Rolling Dynamic Deflectometer
RDT	Rolling Deflection Testing
RWD	rolling weight deflectometer
SASW	spectral analysis of surface waves
SC	soft computing
SCG	sensivity conjugate gradients
s_n	search direction vector
t	time
t_j^k	target value of processing neuron
X_{new}	normalized value
X	original value to be normalized
x_i	input signal
x_{max}	**maximum value in the dataset**
x_{min}	minimum value in the dataset
y	output signal
y_k	activation potential
z	crisp function
α	momentum term
β	damping ratio
β_n	**scaling factor**
Δ	calculated deflection
Δ_n	update parameter
$\Delta\eta_n$	learning rate update
Δw	weight update
ε	error energy
δ	measured deflection
δ_n	local error gradient
$\delta^{avg}{}_n$	average local error gradient
φ	activation function
η	learning rate parameter
η_{dec}	learning rate decrease factor
η_{inc}	learning rate increase factor
λ	Marquardt parameter
μ_A	membership function of fuzzy set $\underline{A}$
ν	Poisson's ratio
v_j	activation potential
ω	angular frequency
w_{ij}	synaptic weight
ρ	mass density
θ	unknown parameter vector
θ_c	convex weight factor

Knowledge Discovery and Data Mining Using Artificial Intelligence to Unravel Porous Asphalt Concrete in the Netherlands

Maryam Miradi[1], Andre A.A. Molenaar[2], and Martin F.C. van de Ven[3]

[1] Business Intelligence, Logica, George Hintzenweg 89,
3068 AX Rotterdam, the Netherlands
maryam.miradi@logica.com

[2] Dep. of Road and Railway Engineering, Delft University of Technology,
Stevinweg 1, 2628 CN, Delft, the Netherlands
a.a.a.molenaar@tudelft.nl

[3] Dep. of Road and Railway Engineering, Delft University of Technology,
Stevinweg 1, 2628 CN, Delft, the Netherlands
m.f.c.vdven@tudelft.nl

Abstract. The main goal of this study was to discover knowledge from data about Porous Asphalt Concrete (PAC) roads to achieve a better understanding of the behavior of them and via this understanding improve pavement quality and enhance its lifespan. The knowledge discovery process includes five steps, being understanding the problem, understanding the data, data preparation, data mining (modeling), and the interpretation/evaluation of the results of the models. At the moment, almost 75% of the Dutch motorways network has a PAC top layer. The main damage of PAC is raveling, which is when the top layer of the road loses stones. The SHRP-NL databases provided ten years of material property data from PAC roads. The data for climate and traffic were obtained from databases of the Royal Dutch Meteorological Institute (KNMI) and the Ministry of Transport and Water Management, respectively. Due to the low number of data points (74 data points), an extensive variable selection was performed using eight different methods to determine the four or five most influential input variables and consequently reduce the input dimension. These methods were decision trees, genetic polynomial, artificial neural network, rough set theory, correlation based variable selection with bidirectional and genetic search, wrappers of neural network with genetic search, and relief ranking filter. The modeling step resulted in 8 intelligent models which were developed using two prediction techniques, being artificial neural networks and support vector machines and two rule-based techniques, being decision trees and rough set theory. Taking the low number of data points into account, the prediction models showed a good performance (R2 = 0.95). The rule based models were transparent and easy to interpret but performed less.

1 Introduction

In many fields, data are being collected at a dramatic speed. By themselves data mean nothing. To extract useful information (knowledge) from the rapidly growing volumes of data, usage of computational theories and tools is necessary. Employing

K. Gopalakrishnan et al. (Eds.): Intel. & Soft Comp. in Infra. Sys. Eng., SCI 259, pp. 107–176.
springerlink.com

these tools to extract knowledge from data is both scientific and economic. For instance, data we capture about our environment are the basic evidence we use to build scientific theories and models of the universe we live in. Business use data as well, for example, to gain competitive advantage, increase efficiency, and provide more valuable services to customers. This scientific/economical process of extracting knowledge from data is called knowledge discovery. Different tools can be used for mining data in order to discover knowledge, but the newest generation of tools belongs to the field of artificial intelligence (AI). AI based tools attempt to mimic the human intelligence. Because of their ability to solve complex problems, they rapidly replace the classical statistical tools during the last decades.

Data are almost always gathered for a specific problem that we attempt to understand and solve. The problem considered in this dissertation are related to porous asphalt concrete road pavements. This road pavements are extensively used in the Netherlands. The goal of this study was to discover knowledge about PAC road pavements using AI-based techniques to achieve a better understanding of the behavior of these road pavements and via this understanding improve their quality and enhance their lifespan.

Artificial intelligence and pavement engineering are two completely different fields. The experts from one field have little knowledge about the other one. Therefore, after thorough consideration, it was decided to explain the basics of both fields to make the dissertation readable for the readers from both fields.

The remainder of the chapter is organized as follows: Section 2 gives an overview of the knowledge discovery process, including its five steps. This section also explains AI and machine learning. After a brief literature review from 70 studies in Section 3, Section 4 deals with the approach of this study, including description of machine learning techniques being used in this study. Section 5 discusses the problem of PAC and its main damage, being raveling. Section 6 explains which data sources are used for this study. The choices made regarding mixture properties, climatic and traffic data are discussed in this section. From Section 7 to 12, different steps of knowledge discovery for PAC data are given. Section 7 includes data cleaning and data scaling and variable selection. For variable selection (determination of the most influential variables) in this study, eight different methods are employed : decision trees, genetic polynomial, artificial neural network, rough set theory, correlation based variable selection with bidirectional and genetic search, wrappers of neural network with genetic search, and relief ranking filter. According to literature study (Section 3), none of the existing studies have employed so many variable selection methods. Section 8 formulates the data mining question with the four/five selected variables. Sections 9 to 12 presents the result of data mining using four techniques, respectively: artificial neural networks and support vector machines, decision trees, and rough set theory. It should mentioned that this study is the first which uses support vector machine for road pavement problems. The summary and conclusions of the results are given in Section 13.

2 Artificial Intelligence Based Knowledge Discovery

2.1 Knowledge Discovery from Data, Data Mining

Knowledge discovery is the nontrivial process of identifying valid, novel, potentially useful, and ultimately understandable patterns in data (Fayyad et al 1996). The term *process* implies that knowledge discovery comprises many steps. *Nontrivial* means that some search is involved and that it is not a straightforward computation of predefined quantities like computing the average value of a set of numbers. Here, *data* are a set of observations (measurements, cases, etc.), and *pattern* is an expression describing a subset of data or a model applicable to the subset of data (pattern ≈ model). Hence, extracting a pattern designates fitting a model to data, finding structure from data, or in general, making any high-level description of a set of data. The discovered pattern should be valid for new data with some degree of certainty. In many cases, it is possible to define measures of certainty (for example, estimated prediction accuracy for new data). A pattern is considered to be knowledge if its measure of certainty exceeds some threshold (pass the evaluation phase).

Knowledge discovery is an interactive and iterative process, involving numerous steps with many decisions made by the user. Figure 1 (Fayyad et. al., 1996) shows the steps involved in knowledge discovery.

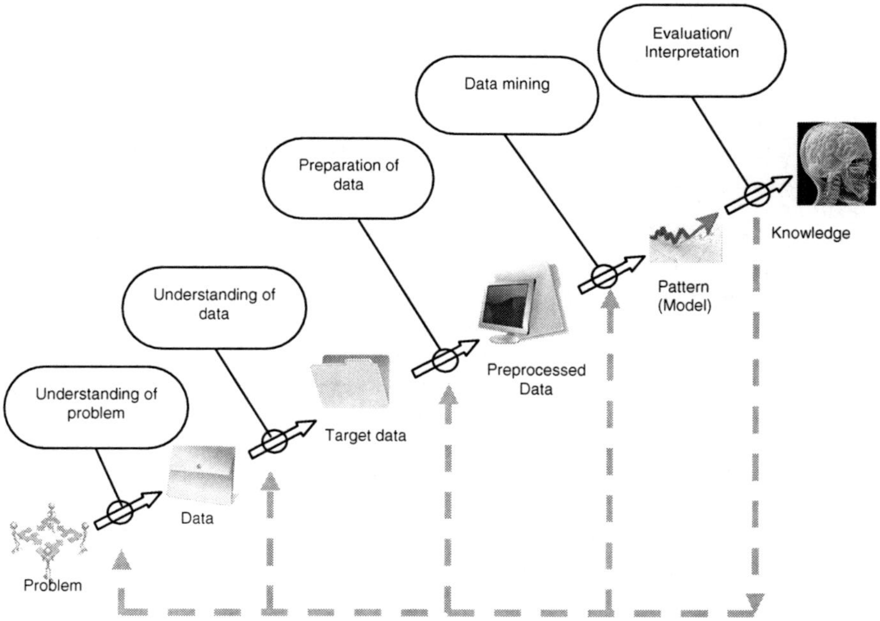

Fig. 1. The steps of knowledge discovery.

A detailed explanation of these steps is given by many researchers (Brachman and Anand 1994; Fayyad et al. 1996; Aboney et al. 2005; Cios et al 2007). To make these steps clear for the reader of this dissertation, a brief review of each step is given here:

Understanding the problem. First, an understanding of the application domain and the relevant prior knowledge should be developed.

Understanding the data. In the second step, the target database(s) is created by selecting the proper dataset, or focusing on subsets of variables per data samples, on which discovery is to be performed.

Data preparation. The third step concerns deciding which data will be used as input for the subsequent step (data mining). It involves sampling, running correlation and significance tests, and data cleaning, which includes checking the completeness of data records, removing or correcting for noise and missing values, etc. The cleaned data may be further processed by variable selection and extraction algorithms to reduce variable dimensionality. The main idea of variable selection is to choose a subset of input variables by eliminating variables with little or no predictive information. Variable selection can significantly improve the comprehensibility of the resulting models and often build a model that generalizes better to unseen data points. Further, it is often the case that finding the correct subset of predictive variables is an important problem in its own right (Dy and Brodley 2004). Finally, data preparation may include data transformation such as scaling of data.

Data mining (modeling). This is an important and time consuming step, which can be divided into three sub-steps:

4.1) Determination of data mining task. In this step, we should determine what kind of task we want to carry out with data mining. The most common data mining tasks are classification and regression.

- Classification: It is learning a function that maps (classifies) a data item into one of several predefined classes (Weiss and Kulikowski 1991). Examples of classification methods used as part of knowledge discovery applications include the classifying of trends in financial markets (Apte and Hong 1996) and the automated identification of objects of interest in large image databases (Fayyad et al. 1996).
- Regression: It is learning a function that maps a data item to a real-value prediction variable. There are many regression applications. Some examples are predicting the amount of biomass present in a forest given remotely sensed microwave measurements, estimating the probability that a patient will survive given the results of some diagnostics tests, or predicting consumer demand of a new product as a function of advertising expenditure.

Other possible data mining tasks are as follows:

- Clustering: Identification of a finite set of categories or clusters to describe the data. Closely related to clustering is the method of probability density estimation. Clustering quantizes[1] the available input-output data to get a set of prototypes and use the obtained prototypes (signatures, templates, etc.) as model parameters.

---

[1] Quantization is the procedure of constraining something from a continuous set of values to a discrete set.

- Summation: finding a compact description for a subset of data, e.g. the derivation of summary for association of rules and the use of multivariate visualization techniques.
- Dependency modeling: finding a model which describes significant dependencies between variables (e.g. learning of belief networks).
- Change and Deviation Detection: Discovering the most significant changes in the data from previously measured or normative values.

4.2) Choosing the data mining algorithm(s). The next sub-step is to select algorithms for searching patterns in the data (fit a model to data). This includes deciding which parameters may be appropriate and matching a particular algorithm with the overall criteria of the knowledge discovery (e.g. the end-user may be more interested in understanding the model than in its predictive capabilities.) One can identify three primary components in any data mining algorithm: model representation, model evaluation, and search.

- Model representation is the language used to describe the discoverable patterns. If the representation is too limited, then no amount of training time or examples will produce an accurate model for the data. Note that a more powerful representation of models increases the danger of overfitting the training data resulting in reduced prediction accuracy on unseen data. Overfitting simply means that the model fits to each single data point in the dataset (Figure 2(b)) instead of finding a general pattern from data (Figure 2(a)). It is important that the data analysis fully comprehend the representational assumptions which may be inherent in a particular technique.
- Model evaluation criteria are qualitative statements or fit functions of how well a particular pattern (a model and its parameters) meets the goals of the knowledge discovery. For example, predictive models can often be evaluated by testing their prediction accuracy using a part of the dataset, which is called test set. Descriptive models can be evaluated along the dimensions of predictive accuracy, novelty, utility, and understandability of the fitted model.
- Search method consists of two components, being parameter search and model search. Once the model representation and the model evaluation criteria are fixed, then the data mining problem has been reduced to purely an optimization task. This task is to find the parameters/models for the selected category which optimize the evaluation criteria given the observed data and the fixed model representation. Model search occurs as a loop over the parameter search method (Aboney et al. 2005).

4.3) Data mining. In this sub-step the algorithm chosen in the step 4.2 with the selected model parameters will be applied to the data.

Evaluation/Interpretation of mined pattern (model). This includes understanding the results, checking whether the discovered knowledge is novel and interesting, interpretation of the results by domain experts, and checking the impact of the discovered knowledge. Only approved models are retained, and the entire process is revisited to identify which alternative actions could have been taken to improve the results. A list of errors made in the process is prepared. Interpretation involves visualization of the extracted patterns and models or visualization of the data given the extracted model.

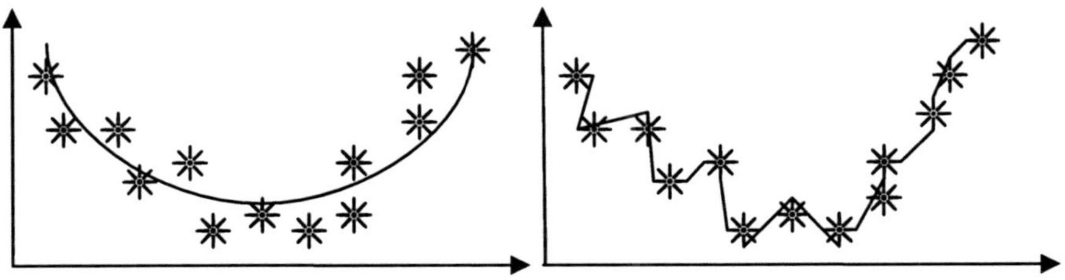

(a) (b)

Fig. 2. Fitting a model to data (a) and model overfitting (b).

2.2 Artificial Intelligence, Machine Learning

As mentioned before, data mining is an important step in knowledge discovery. The major distinguishing characteristic of data mining is that it is *data driven*, as opposed to other approaches that are often *model driven*. The heart of data mining is to find a good model from the data, which at the same time is easy to understand. We need to keep in mind, however, that almost always we will look for a compromise between model completeness and model complexity.

The earliest data mining tools dealing with data analysis were statistical tools. With the advent of the computer, the level of application of statistics increased. In parallel, other disciplines began to develop tools for data analysis, with different aims and objectives from statistics. In statistics, problems have been dealt from the perspective of inference, which was always at the base of statistics. However, new tools appeared on the scene originally not with the aim of analyzing data per se, but rather with the aim of simulating the way natural intelligent systems work, and then with the simple aim of building systems which could learn. In other words, it was attempted to create intelligent systems with learning ability for data mining.

These attempts resulted in the field *artificial intelligence (AI)*, which is now a collection of several intelligent analytical tools. Dictionaries define intelligence as the ability to comprehend, to understand and profit from experience, or having the capacity for thought and reason (especially to a high degree). In a technical level, often, the techniques and algorithms that can *learn* from data are characterized as intelligent. *Learning* means acquiring knowledge about a previously unknown or hardly-known system or concept. The human capability of learning, generalizing, memorizing, and predicting is the foundation of any AI system. AI has many sub-fields but one of the broadest sub-field of AI is machine learning.

Machine learning (ML) concerns a collection of techniques that develop models, which learn from data. Learning from data can result in rules, functions, relations, equation systems, probability distributions, and other knowledge representations. The results explain data and can be used for supporting decisions concerning the underlying process (e.g., forecasting, diagnostic, control, validation, and simulations).

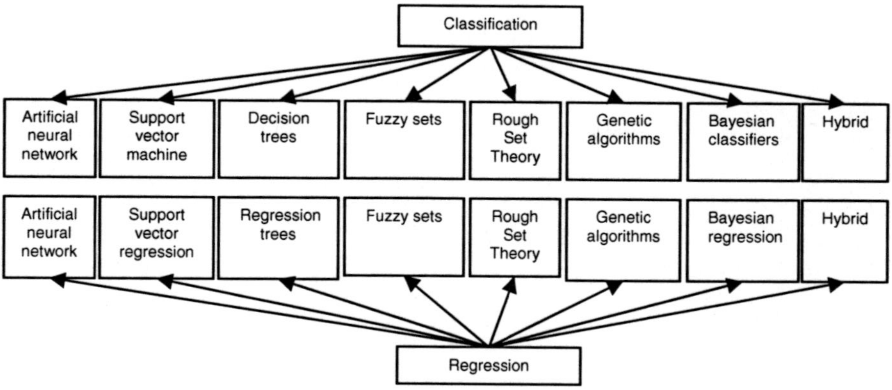

Fig. 3. A taxonomy of machine learning techniques.

As mentioned before, the most common data mining tasks are classification and regression. In machine learning, several techniques are used for classification and regression. Figure 3 shows the ML techniques that are most frequently mentioned in literature (Kononenko and Kutar 2007).

3 Literature Review

In this study, 10 traditional (Table 1) and 60 intelligent based studies (Tables 2 and 3) on knowledge discovery from pavement data were reviewed from the perspective of knowledge discovery for each of its five steps. The problems investigated by all reviewed studies were focused on cracking, rutting, roughness, and stiffness of pavement layers. The problem which received little attention was raveling, with only one publication on raveling of dense asphalt concrete and no

Table 1. Studies on traditional knowledge discovery.

Index	Name of author(s)	Year	Problem
1	Carey and Irick	1960	Serviceability
2	Way and Eisenberg	1980	Cracking, Roughness
3	Parsley and Robinson	1982	Cracking, Roughness
4	Geipot	1982	Cracking, Roughness
5	Lytton et al.	1982	Cracking, Roughness
6	Karan and Haas	1976	Pavement maintenance
7	Butt et al.	1994	Pavement maintenance
8	Li et al.	1996	Pavement deterioration rate
9	Huang	1997	Pavement deterioration rate
10	Hong and Wang	2003	Pavement deterioration rate

Table 2. Studies on application of intelligent techniques for the knowledge discovery from cracking, rutting, roughness, and stiffness/elastic modulus of pavements.

Index	Name of author(s)	Year	Problem
1	Hoffman and Chou	1994	Cracking, Rutting, Roughness
2	Eldin and Senouci	1996	Cracking, Rutting, Roughness
3	Hsu and Tsai	1997	Cracking, Rutting, Roughness
4	Roberts and Attoh-Okine	1998	Cracking, Rutting, Roughness
5	Loia et al.	2000	Cracking, Rutting, Roughness
6	Attoh-Okine	2002	Cracking, Rutting, Roughness
7	Chang et al.	2003	Cracking, Rutting, Roughness
8	Mu-yu and Shao-yi	2003	Cracking, Rutting, Roughness
9	Yang et al.	2003	Cracking, Rutting, Roughness
10	Chang et al.	2004	Cracking, Rutting, Roughness
11	Nakatsuji et al.	2005	Cracking, Rutting, Roughness
12	Karlaftis & Loizos	2006	Cracking, Rutting, Roughness
13	Terzi	2006	Cracking, Rutting, Roughness
14	Bosurgi et al.	2007	Cracking, Rutting, Roughness
15	Terzi	2007	Cracking, Rutting, Roughness
16	Kaur and Pulugurta	2007	Cracking, Rutting, Roughness
17	Meier and Rix	1995	Stiffness/Elastic modulus
18	Ferregut et al.	1999	Stiffness/Elastic modulus
19	Kaur and Chou	1999	Stiffness/Elastic modulus
20	Kim et al.	2000	Stiffness/Elastic modulus
21	Abdallah et al.	2001	Stiffness/Elastic modulus
22	Saltan et al.	2002	Stiffness/Elastic modulus
23	Terzi et al.	2003	Stiffness/Elastic modulus
24	Bredenhann and van de Ven	2004	Stiffness/Elastic modulus
25	Goktepe et al.	2004	Stiffness/Elastic modulus
26	Reddy et al.	2004	Stiffness/Elastic modulus
27	Ceylan et al.	2005a	Stiffness/Elastic modulus
28	Ceylan et al.	2005b	Stiffness/Elastic modulus
29	Chang et al.	2006	Stiffness/Elastic modulus
30	Goktepe and Altun	2006	Stiffness/Elastic modulus
31	Gopalakrishnan et al.	2006	Stiffness/Elastic modulus
32	Rakesh et al.	2006	Stiffness/Elastic modulus
33	Saltan and Sezgin	2006	Stiffness/Elastic modulus
34	Burak and Altun	2007	Stiffness/Elastic modulus

Table 2. (*continued*)

Index	Name of author(s)	Year	Problem
35	Ceylan et al.	2007	Stiffness/Elastic modulus
36	Demir	2007	Stiffness/Elastic modulus
37	Guclu and Ceylan	2007	Stiffness/Elastic modulus
38	Lee et al.	2007	Stiffness/Elastic modulus
39	Loizos et al.	2007	Stiffness/Elastic modulus
40	Ozashin and Oruc	2007	Stiffness/Elastic modulus
41	Saltan and Terzi	2007	Stiffness/Elastic modulus
42	Pekcan et al.	2007	Stiffness/Elastic modulus

Table 3. Studies on application of intelligent techniques for the knowledge discovery from raveling, cracking, rutting, and roughness.

Index	Name of author(s)	Year	Problem
43	Thube and Thube	2007	Raveling
44	Kaur and Tekkedil	2000	Rutting
45	Tarefder et al.	2005	Rutting
46	Chou et al.	1994	Cracking
47	Meignen et al.	1997	Cracking
48	Lou et al.	1999	Cracking
49	Lee and Lee	2003	Cracking
50	Avila et al.	2004	Cracking
51	Lea and Harvey	2004	Cracking
52	Mei et al.	2004	Cracking
53	Rababaah et al.	2005	Cracking
54	Bray et al.	2006	Cracking
55	Xiao et al.	2006	Cracking
56	Huang et al.	2007	Cracking
57	Ozbay and Laub	2001	Roughness
58	Aultman-hall et al.	2004	Roughness
59	Bayrak et al.	2004	Roughness
60	Choi et al.	2004	Roughness

studies on raveling of porous asphalt concrete. Not enough attention was given to the discussion of missing data and outliers. Moreover, for variable selection/reduction only 4 studies used variable selection despite the importance of this step. Two used PCA for variable reduction, one ANN for variable selection, and another one rough set theory. Concerning the data mining technique, it was noticed that about 60% of the reviewed studies have applied artificial neural networks, none of the studies applied support vector machines, and only a few employed techniques such as decision tree/regression trees or rough set theory. About 18% of the studies used a hybrid technique which was mainly some combination of neural network, genetic algorithm, or fuzzy sets. Next to that, the number of techniques which extract/generate rules from pavement data was considerably low. Also, most of the studies apply only one technique (except for the hybrid studies), while running a number of techniques on the data and comparing their results can lead us to very relevant information about the problems being investigated. Concerning the parameter model/selection, the simplest version of cross validation, hold-out, was used. Despite the high reliability of K-fold and leave-one-out cross validation, only a few researcher tried these methods. This can be blamed on the fact that they are computationally expensive methods. Moreover, despite the fact that many studies employed an artificial neural network, an optimal parameter selection for this powerful prediction/analysis technique was missing in many of these studies. Considerable performance improvement can be gained by a correct parameter selection. Finally, for implementation of data mining, the software MATLAB was used most of the time.

The majority of studies in the literature review have used a separate testing dataset to test the performance of the model. Almost none of the studies have analyzed the influence of input variables on the data mining step. This analysis could reveal relevant information for instance about the reasons for the pavement problem.

4 Approach

Based on the industrial needs/academic importance of the PAC problem the approach for this study was determined. The pictorial summary of this approach is presented in Figure 4.

Selecting the most important input variables is done with different techniques. Comparing their results allows determining whether there was consistency in the variable selection of different methods. If that is the case, the most often selected variables are then used for modeling purposes. Figure 5 shows the variable selection methods which are used in this study.

In this study, the parameter and model selection is done using cross validation. The cross validation method uses a part of the training set, which is called the validation set, to find the best model or the best parameters. Because for a lower number of data points (around 100 or less), leave-one-out cross validation delivers the most reliable results, this method is employed.

4.1 Data Mining Techniques

As mentioned before, this study applies four machine learning methods, being artificial neural network, support vector machines, decision trees, and rough set theory. To give an impression of these techniques, brief descriptions of them are given hereafter.

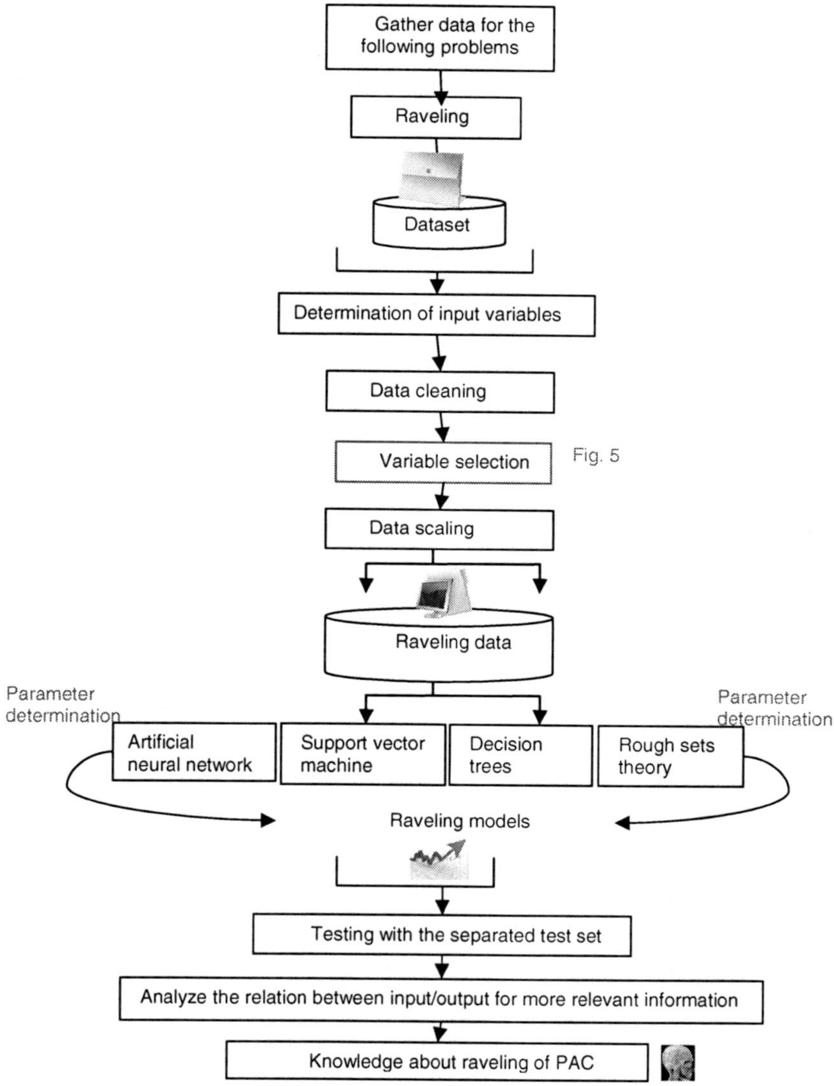

Fig. 4. The research approach.

4.1.1 Artificial Neural Network

An artificial neural network (ANN) (Engelbrecht 2007) is a layered network of artificial neurons (ANs). Each AN receives signals from input variables or from other ANs, gathers these signals and, when needed, transmits a signal to all connected ANs. Figure 6(a) is a representation of an artificial neuron. Input signals are

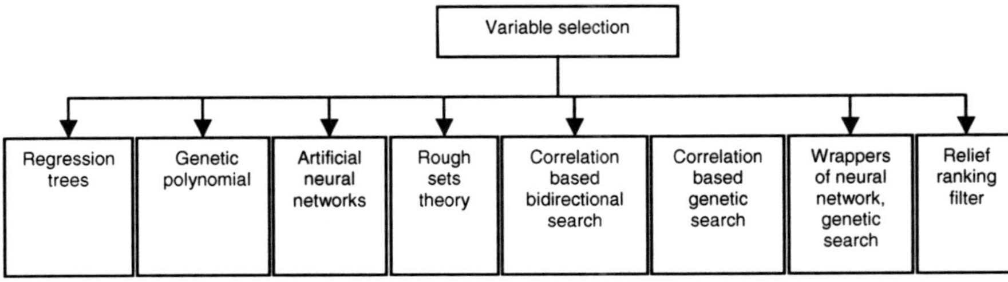

Fig. 5. The variable selection methods.

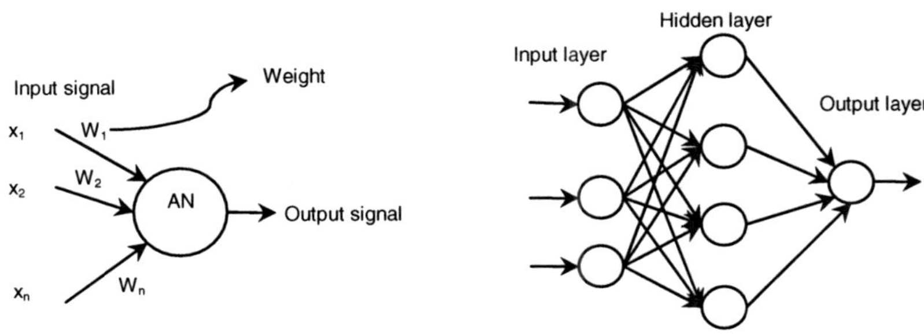

(a) (b)

Fig. 6. Illustration of an artificial neuron (a) and a three layer artificial neural network (b).

inhibited or excited through negative or positive numerical weights associated with each connection to the AN. The strength of an existing signal is controlled via a function, referred to as the activation function, which calculates the output signal of the AN. The role of this function is to bring nonlinearity to ANN. An ANN may consist of an input layer, hidden layer(s), and an output layer. ANs in one layer are connected, fully or partially, to the ANs in the next layer. A typical ANN structure is depicted in Figure 6(b). ANN can be employed for different data mining tasks such as regression and classification as well as for variable selection in the data preparation step of knowledge discovery.

4.1.2 Support Vector Machines

Support vector machines (SVM) *(Bishop and Tipping 2003)* ultimately make predictions based on the following function

$$f(x,w) = sign(\sum_{i=1}^{N} w_i K(x,x_i) - b) \tag{1}$$

The key feature of the SVM is that, in the binary classification case (only two classes available), its target function attempts to minimize a measure of error on the training set while simultaneously maximizing the distance (*margin*) between the two classes by a separating plane ($f(x,w)$). To calculate the margin, two parallel planes, one on each side of the separating plane, which are "pushed up against" the data points of two classes. These data points are called *support vectors*. Intuitively, a good separation is achieved by the plane that has the largest margin to the neighboring data points of both classes, since in general the larger the margin the better the performance of the SVM. This is an effective mechanism leading to good generalization because the training depends only on a subset of data points, namely the *support vectors* that lie on the margin. Next to this, SVM uses the kernel trick (kernel = $K(x,x_i)$) which makes the SVM construction independent on the dimensionality of the input space. Kernels are generally highly nonlinear functions such as a radial basis function, a two-layer neural network or a high degree polynomial, which enables SVM to solve complex nonlinear problems. Vector w_i is the orientation of the separating plane and b is the offset of the plane from the origin. Both w_i and b are automatically calculated during the construction of the separating plane. Figure 7 shows the structure of an SVM with three inputs for a classification task. SVMs have also been extended for regression application.

4.1.3 Decision Trees

Decision trees (DT) (Jang et al. 1997) partition the input space of a dataset into mutually exclusive regions, each of which is assigned a label, a value, or an action to characterize its data points. The decision tree mechanism is transparent and we can follow a tree structure easily to explain how a decision is made. Therefore, the decision tree has been used extensively in machine learning. It is perhaps the most highly developed technique for partitioning data into a collection of decision rules.

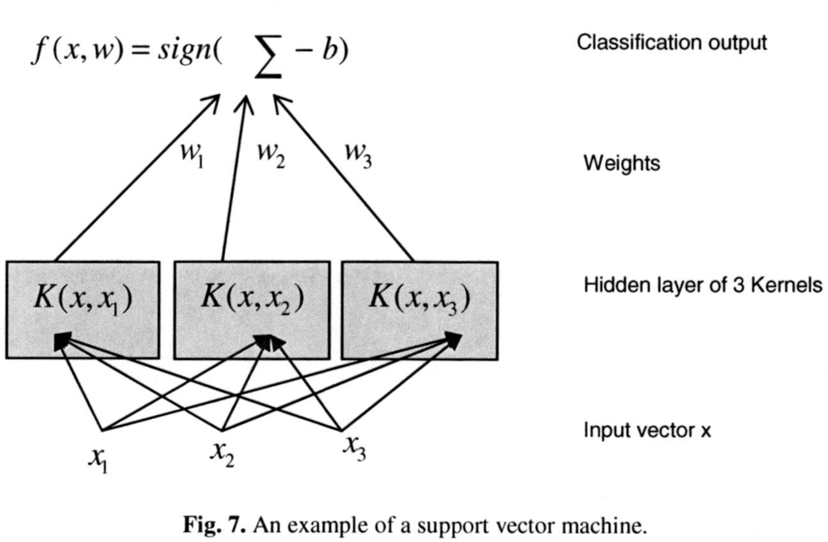

Fig. 7. An example of a support vector machine.

A decision tree is a tree structure consisting of internal and external nodes connected by branches. An internal node is a decision making unit that evaluates a decision function to determine which child node to visit next. In contrast, an external node, also known as a leaf or terminal node, has no child nodes and is associated with a label or value that characterizes the given data that lead to it being visited. Decision trees used for classification problems are often called classification trees, and each terminal node contains a label that indicates the predicted class. In the same way, decision trees used for regression problems are often called regression trees, and the terminal node labels may be constants or equations that specify the predicted output value of a given input.

4.1.4 Rough Set Theory

FS is the first theoretical treatment of the problem of vagueness and uncertainty, and has many successful implementations. FS is, however, not the only theoretical logic that addresses these concepts. Pawlak (1991) developed a new theoretical framework to work with vague concepts and uncertainty, which is called rough set theory (RST) (Engelbrecht 2007). While RST is somewhat related to fuzzy set theory, there are major differences. RST is based on the assumption that some information or knowledge about the data is initially available. This is contrary to fuzzy set theory where no such prior information is assumed. The basic idea of rough sets rests in the discernibility between data points. If two data points are indiscernible over a set of variables, it means that if their output variables have the same value the input variables should be the same as well. RST is a desirable technique for real-world applications because of its robustness to situations where data is incomplete. RST clarifies the set-theoretic characteristics of classes over combinational patterns of the variables. In doing so, RST also performs automatic variable selection by finding the smallest set of input variables necessary to discern between classes. Therefore RST can also be used for variable selection in the data preparation step of knowledge discovery.

4.2 Variable Selection Methods

Figure 5 showed the variable selection methods. A brief explanation about these methods will be given hereafter.

4.2.1 Regression Trees

After creating a regression tree, the variable present at the root node of the tree can be seen as the most important variable. The variables close to top nodes are the more important ones.

4.2.2 Genetic Polynomial Regression

One of the methods used in this study was the hybrid method combining genetic algorithms and polynomial models. This method is introduced by Maertens et al. (2006). The method selects a subset of relevant input variables such that the data are well approximated by a polynomial model structure represented by Equation (2). In this equation, a_0 and a_i are constant coefficients.

$$y = P(X; d, p, A) = \sum_{i=1}^{p} a_i \prod_{\forall j \in s_i^d} x_j + a_0 \quad (2)$$

The genetic algorithm evolves 'in parallel' a large number of (e.g. 100) of model structures with p different d^{th} order polynomial terms that are selected from n potential input variable $X_{sel}(n)$. The parameters are determined by the least-square method[2] and the fitness value of every model structure is calculated from the corresponding root mean square error[3] (RMSE).

If one of the n regressor variables from the selection $X_{sel}(n)$ is not present in any of the polynomial terms s_i^d, it is removed from the regressor set and the procedure is repeated with a smaller number of regressor $n-1$ until the desired minimal number $n_{\min}$ of input is reached. In case all n candidate variables are present in the p regressor combinations s_i^d, the number of polynomial terms is reduced to $p-1$ and the genetic polynomial regression process is repeated. At the end, a subset of $n_{\min}$ variables is retrained from the initial set of n_{total} potential input variables.

The user has to choose the initial number of polynomial terms (typically $p \approx 1.5n_{total}$ to $2n_{total}$) and the final number of regressors $n_{\min}$ (selected variables). A large initial set of polynomial terms and a low final number of regressor variables will imply large calculation times, but will reduce the need for repetitions of the selection algorithm. A large polynomial degree d will increase the calculation time and should be avoided to prevent overfitting (d = 2 is therefore typically used). Figure 8 gives an overview of the backward selection procedure as implemented in MATLAB.

4.2.3 Artificial Neural Network

A commonly used method, which is called weighted weight factor (WWF), seeks for relative importance of the input variables for the output variable in an artificial neural network. In this way, the most relevant input variables can be chosen and the rest can be removed or a subset of n most important variables can be selected. The WWF approach works as follows: once the training of the neural network is completed, the relative importance of the input x on the output y is computed by using the information encapsulated in the weight matrix. Assume for example a three layer neural network as depicted in Figure 9, which is composed of an input

[2] Least square is a method of fitting data. The best fit in this method is when the model has the lowest sum of squared error.

[3] For calculation of RMSE, the sum of the square of the deviations of data points from their true position should be calculated (the difference between actual output and the predicted output) and then be divided by the total number of data points.

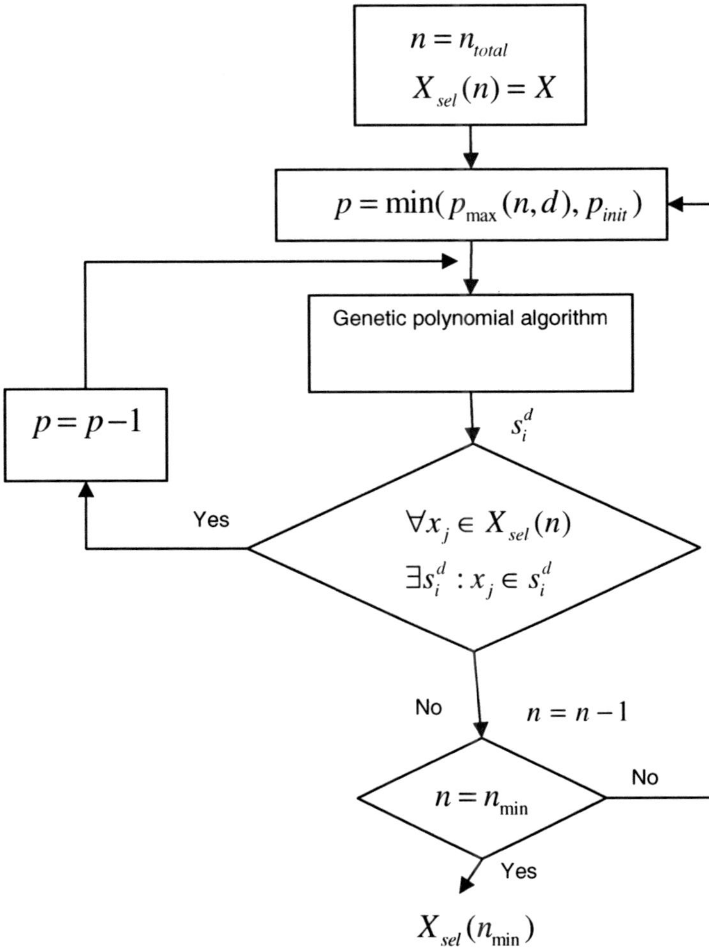

Fig. 8. Layout of the genetic polynomial regression process (after Maerten et al. 2006).

layer, a hidden layer and an output layer. Without loss of generality only two input variables, three neurons in the hidden layer, and one output variable are considered here. The computation is done in a forward direction as indicated below. The importance of each input variable for the output variable is evaluated for each neuron in the hidden layer.

The relative importance of input x_1 on y_1 is evaluated as follows:

Step1: Compute the influence of x_1 on y_1, using the first neuron path, h_1

$$W_{1_{x_1 y_1}} = \left(\frac{W_{x_1 h_1}}{W_{x_1 h_1} + W_{x_2 h_1}} \right) \left(\frac{W_{h_1 y_1}}{W_{h_1 y_1} + W_{h_2 y_1} + W_{h_3 y_1}} \right) \quad (3)$$

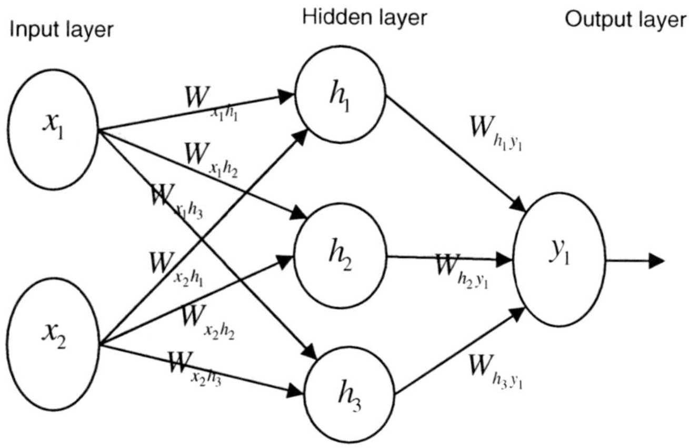

Fig. 9. Three layer artificial neural network, an example for calculating weighted weight factor.

Step2: Compute the influence of x_1 on y_1, using the second neuron path, h_2

$$W_{2_{x_1y_1}} = \left(\frac{W_{x_1h_2}}{W_{x_1h_2}+W_{x_2h_2}}\right)\left(\frac{W_{h_2y_1}}{W_{h_1y_1}+W_{h_2y_1}+W_{h_3y_1}}\right) \tag{4}$$

Step3: Compute the influence of x_1 on y_1, using the third neuron path, h_3

$$W_{3_{x_1y_1}} = \left(\frac{W_{x_1h_3}}{W_{x_1h_3}+W_{x_2h_3}}\right)\left(\frac{W_{h_3y_1}}{W_{h_1y_1}+W_{h_2y_1}+W_{h_3y_1}}\right) \tag{5}$$

Step4: Compute the overall influence of x_1 on y_1, using all the neurons path.

$$W_{1_{total}} = \left(\frac{1}{W_{h_1y_1}+W_{h_2y_1}+W_{h_3y_1}}\right)\left(\frac{W_{x_1h_1}W_{h_1y_1}}{W_{x_1h_1}+W_{x_2h_1}}+\frac{W_{x_1h_2}W_{h_2y_1}}{W_{x_1h_2}+W_{x_2h_2}}+\frac{W_{x_1h_3}W_{h_3y_1}}{W_{x_1h_3}+W_{x_2h_3}}\right) \tag{6}$$

The same procedure can be applied for the input variables in the neural network. This is a nonlinear variable selection. The search strategy of this method is heuristic; it has a multivariate variable generation and has a random direction in its search.

4.2.4 Rough Set Theory

The basic idea of rough sets rests in the discernibility[4] between data points. RST clarifies the set-theoretic characteristics of classes over combinational patterns of the variables. In doing so, RST also performs automatic variable selection by finding the smallest set of input variables necessary to discern between classes.

4.2.5 Correlation Based Variable Selection Using Bidirectional Search and Genetic Search

The central hypothesis of this method is that good variable sets contain variables that are highly correlated with the output, yet uncorrelated with each other. This is a filter variable selection with correlation as measure and a heuristic search strategy. In this dissertation, both bidirectional and genetic (random) search direction are used for this method.

4.2.6 Wrapper of Artificial Neural Network Using Genetic Search

This is another hybrid algorithm, combining neural network and genetic algorithm. In a wrapper model after variable selection, a learning algorithm is used. The learning algorithm used in this wrapper method is an artificial neural network. As mentioned before, variable selection for neural networks is multivariate. The optimal subset of variables is searched here using a genetic algorithm. This method combines the strength of artificial neural networks and genetic algorithms for finding the optimal variable subset.

4.2.7 Relief Ranking Filter

This is a filter model which ranks variables according to their separating power" in the context of other variables". The Relief algorithm (Kira and Rendell 1992) uses an approach based on the nearest-neighbor algorithm. Using a nearest-neighbor algorithm, for each data point, the closest data point with the same output (nearest hit) and the closest data points with a different output (nearest miss) are selected. The score of the ith variable/variable is computed as the average over all examples of the magnitude of the difference between the distance to the nearest hit and the distance to the nearest miss, in projection on the ith variable.

5 Problem of Porous Asphalt Concrete

Since the late 1980s, single layer porous asphalt concrete (PAC) is widely used on Dutch motorways. Later, two-layer PAC was developed in the Netherlands as well (DWW 2005). PAC is used as top layer on pavements. It is a mixture consisting of crushed stone, crushed sand, filler with 25% calcium hydroxide, and bitumen with penetration grade 70/100. The composition of PAC should satisfy the specifications given in Table 4 (CROW 2005). As can be seen in Table 4, standard PAC has a maximum grain size of 16 mm.

---

[4] If two data points are indiscernible over a set of variables, it means that if their output variable has the same value the input variable should be the same as well.

Table 4. Gradation of porous asphalt concrete 0/16.

Sieve size [mm]	Desired mass % on sieve	Minimum mass % on sieve	Maximum mass % on sieve
C16	-	0	7
C11.2	-	15	30
C8	-	50	65
C5.6	-	70	85
2	85	-	-
0.063	95.5	-	-

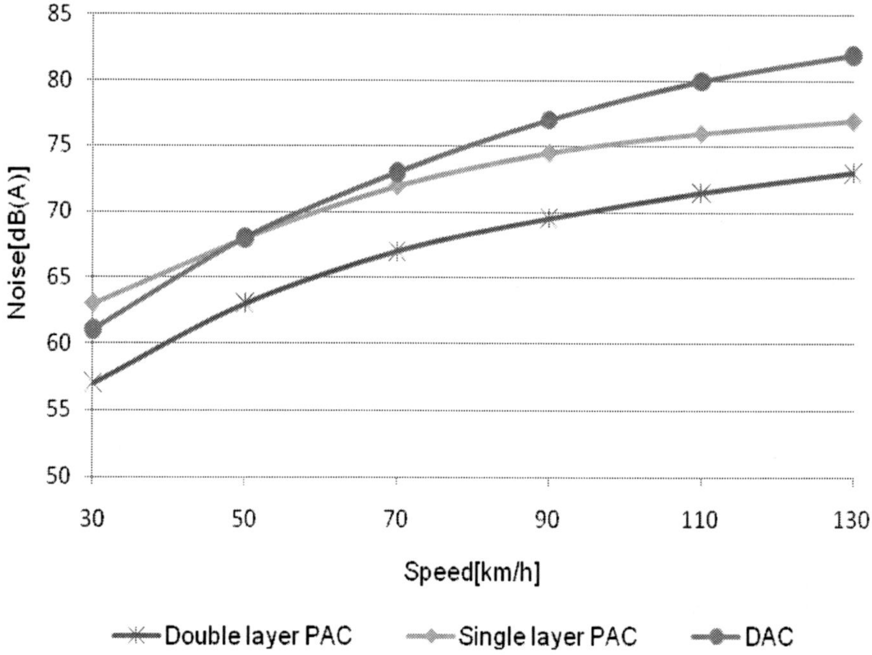

Fig. 10. Comparison of noise production by different types of top layers.

According to CROW (2005), the bitumen content should be at least 4.5% by mass on top of 100% aggregate. This means that 4.5 kg of bitumen should be added on top of 100 kg of aggregate. The traditional single layer PAC is a uniformly graded asphalt mixture with a minimum air void content of 20% after compaction. Such a high air voids content allows surface water to quickly penetrate into and drain through the PAC layer, offering considerably reduced splash and spray and improved visibility. The open structure of the surface also reduces

Table 5. The average construction and maintenance costs for DAC and PAC layer.

Type of top layer	The average construction costs (€per m^2)	The average maintenance costs (€per m^2)
DAC	19	1.18
PAC	23	2.16

the noise level produced by the tires rolling over the pavement surface and it is for this reason why PAC is so extensively used in the Netherlands. Figure 10 (Molenaar et al. 2006) shows the noise levels produced by different top layers.

In 2007, almost 75% of the Dutch motorways network has a PAC top layer. Another advantage of PAC is its high resistance to rutting (permanent deformation) due to its stone skeleton and its location as upper layer in the pavement structure. Porous asphalt has one major drawback, which is its limited lifespan. PAC is also more expensive than dense asphalt concrete (DAC). The construction costs of PAC are 21% higher than those of DAC and its maintenance costs are 83% higher than the DAC maintenance costs (See Table 5). The construction costs presented in Table 5 include the costs of preparation, administration, control and tax. The maintenance costs include the variable maintenance costs meaning the costs of (partly) replacing the top layer (Hofman et al. 2005).

5.1 Lifespan of Porous Asphalt Concrete

The lifespan of a PAC mixture depends on different variables like traffic loads, environmental effects, the composition of the mixture, the characteristics of the different mixture components and the production and laying process. Because of its high voids content, PAC is sensitive for damage due to mechanical (traffic) and environmental effects. The most dominant damage type is raveling which implies that aggregate particles get loose from the pavement surface and are whipped off. A more detailed description of raveling will be given later on. However, PAC is very resistant to permanent deformation (rutting).

Traffic loads on the slow lane are heavier than those on the fast lanes. As a result, the slow lane needs maintenance earlier. Often these lanes are the first where the PAC layer is replaced lane wide. On a later moment in time, the PAC layer needs to be replaced over the entire pavement width (slow and fast lanes). Furthermore, it should be mentioned that raveling occurs the earliest on locations where higher shear stresses occur (e.g. in curves).

In 2003 the Directorate-General for Public Works and Water Management has defined the following average lifespan of PAC (Molenaar and Miradi 2004):

Slow lane:
Average lifespan before repair equals 9.8 years;
Average lifespan after repair equals 7.5 years;
Other lanes:
Average lifespan before repair equals 15.4 years;
Average lifespan after repair equals 13.8 years;

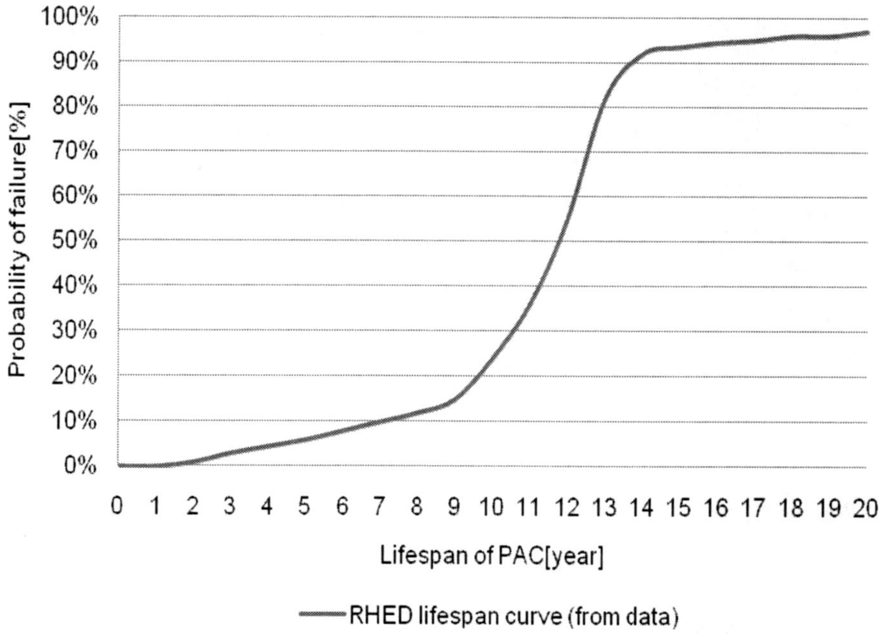

Fig. 11. Cumulative lifespan distribution as determined by the RHED.

The lifespan of PAC top layers is rather variable. Data from the Road and Hydraulic Engineering Division (RHED) of the Ministry of Transport, Public Works and Water Management show that the lifespan can be anywhere between 4 and 16 years. This is shown in Figure 11 as RHED lifespan curve.

Up to a reduction of 20% of the maintenance costs could be achieved as well as a reduction of 10% of the delay hours due to maintenance works, when the average lifespan of PAC could be extended with only a few years.

As has been stated before, PAC top layers show a significant amount of variation in lifespan. Not only the variation between road sections is large but also a significant variation within a section can occur. This latter has to do with the variation in quality within one section. Investigations by Meerkerk (2004) have shown that a significant variation in mixture composition can occur during construction. The result of his research showed a remarkable amount of variation in bitumen and voids content over a rather short transversal as well as longitudinal distance. The variation in voids content over the width of the paved lane seems to be as large as the variation in voids content in longitudinal direction. Meerkerk not only observed a striking variation in the bitumen content, he also found a significant variation in the properties of the recovered bitumen.

This leads to the conclusion that detailed information on the location, extent and severity of raveling as well as detailed information on the mixture composition and bitumen characteristics will be necessary in order to be able to capture the causes of raveling initiation and progression.

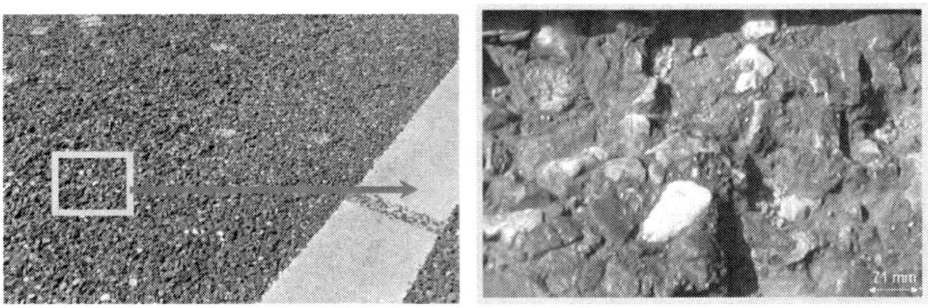

Fig. 12. Raveling of porous asphalt concrete (left) means that the pavement surface looses aggregate particles (right).

Fig. 13. Raveling of porous asphalt concrete on a Dutch motorway.

5.2 Raveling

As mentioned before, raveling is the most dominant type of damage of PAC top layers. Raveling means that the pavement surface looses aggregate particles (Figure 12) resulting in a rough texture and so in an increased noise level. Furthermore, raveling might result in windscreen damage (the loose particles on the road surface can hit the cars' windscreen) which may lead to dangerous traffic conditions.

The reason for raveling is the loss of bond between the aggregate particle and the bitumen coating. A large number of conditions can lead to raveling, varying from traffic loads and environmental effects to insufficient strength of the material.

The "strength" is influenced by the mixture composition, which can be rather variable. Furthermore the "strength" is affected in a negative way due to aging of the bituminous mortar making it brittle and sensitive to cracking.

Figure 13 shows raveling as observed on a specific motorway in the Netherlands. As one can observe, aggregate particles are being whipped off in the right-hand wheel track (indicated by means of the red arrow) and are swept towards the hard shoulder of the pavement (indicated by means of the yellow arrow).

6 Data

6.1 SHRP-NL Database

After evaluating different databases, it was concluded that SHRP-NL database is the only available database which contains all information this study needed for knowledge discovery about raveling of porous asphalt concrete.

The Strategic Highway Research Program in the Netherlands (SHRP-NL) has been performed between 1990 and 2000. It has been inspired by the SHRP program established in 1984 by the U.S government. The database provided by the SHRP-NL research program is called the SHRP-NL database.

The SHRP-NL database contains 34 PAC sections of 300 m. Because each section has three subsections of 100 m, a total of 102 subsections are available. It should be noted that although the specifications allow a minimum void content of 20%, lower void contents were observed on a number of sections.

As mentioned before, raveling severity was characterized by α% *L*(light), β% *M*(moderate) and γ% *S*(severe). The meaning of categories light, moderate and severe is explained in Table 6.

Table 6. The categories for severity of raveling.

Severity of raveling	Percentage of stone loss per m^2
Light	6 – 10
Moderate	11 - 20
Severe	>20

Some data management was needed in order to arrive to a logical input data set. For instance, it had to be decided whether the amount and severity of the raveling observed should be transformed into one single condition indicator or that the raveling should be treated as % light, % moderate and % severe. It was decided to use the *Meq* variable being defined as the "equivalent amount of moderate damage". This variable is calculated as follows

$$Meq(raveling) = \alpha\% L_{ra} + \beta\% M_{ra} + \gamma\% S_{ra} \tag{7}$$

where

$Meq(raveling)$ = equivalent amount of moderate raveling,

$\%L_{ra}, \%M_{ra}, \%S_{ra}$ = percentage of light, moderate and severe raveling,

α, β, γ = weighing factors.

Use of a single condition indicator allows easy comparisons and rankings to be made of the pavement condition. The most important reason to use the *Meq* variable to describe the amount and severity of the damage was the fact that available inspection and performance models are based on *Meq* (Sweere et al. 1996). After extensive investigation, the weighting factors used in the CROW system were chosen. It is recalled that in the CROW system, the following values are used:

$$\alpha = 0.25, \beta = 1, \gamma = 5 \tag{8}$$

These values were used to calculate *Meq* in this study.

6.2 Mixture Composition

For each test section, information was available on:

gradation,
density,
bitumen content,
void content,
type of aggregate used.

With respect to the gradation it was concluded that it doesn't make sense to include all the information available about the percentage passing the individual sieve sizes. It was concluded that the gradation could be characterized by means of the %fine and %coarse material and the D_{50} and the *Cu* of the coarse fraction. The following variables were taken as input.

D_{50} = sieve size through which 50% of the coarse material passes,
D_{60} = sieve size through which 60% of the coarse material passes,
D_{10} = sieve size through which 10% of the coarse material passes,
Cu = coefficient of uniformity = D_{60} / D_{10},
% fine = percentage of material passing the 2 mm sieve,
% coarse = percentage of material on the 2 mm sieve.

The %fine and %coarse are introduced as variables since it appeared from other investigations that the particles smaller than 2mm in diameter don't contribute to the formation of the stone skeleton in a porous asphalt mixture. This means that, together with the bitumen they form the mortar. The stone skeleton is characterized by taking the D_{50} and Cu of the coarse fraction.

The data of material composition was unfortunately not available for all data points. For some unclear reason the material data of 18 sections of 100 m was not present in the SHRP-NL database. At this stage the dataset for porous asphalt concrete contained 84 data points.

6.3 Traffic Data

An extensive search was done to find traffic information for all test sections at the library of the Ministry of Transport and Water Management. The information after 1986 was digitally available; traffic information from 1986 was taken from reports on traffic counts (Rijkswaterstaat, 1980, 1981, 1982, 1983, 1984, 1985).

In road engineering, it is a well-known fact that the number of the vehicles only gives a poor representation of the loads actually applied to the pavement. Axle load distributions should be available to quantify correctly the damaging effect of traffic. Such information however was not available. The only way to estimate the damaging effect of the truck traffic was to estimate the number of trucks and multiply this number by the damaging effect per truck. This later number can be retrieved from the Standard specifications CROW (2005). The last question to solve was to determine the amount of truck traffic from the total amount of traffic. Information on the relation between percentage of truck traffic and traffic intensity was not available for two lane roads. Most of the provincial roads are two lane roads. All in all, we had to deal with such a number of uncertainties that it had to be concluded that the predictions about the number of vehicles that have passed the various test sections is rather weak. Since the information on percentage of truck traffic as well as the information on the damaging effect per truck was weak, it was decided not to estimate the number of equivalent axle loads for each test section but to use the total traffic number.

To calculate the cumulative traffic intensity for each test section the following calculations were done. The cores in SHRP-NL project were taken from the right hand lane of the test section. The right hand lane also has the heaviest traffic intensity. Therefore, we were interested to calculate the cumulative amount of traffic on this lane. The traffic intensity in the dataset was the total traffic intensity in both directions. The right hand lane traffic intensity was not directly available and needed to be calculated. To do so, it had to be calculated which portion of the total traffic is traveling on the right hand lane. It was noticed that the right hand lane traffic intensity was available for the recent years (2005 and later). Therefore, it was decided to calculate the proportion for 2005 and generalize this to other years. For the cases that the traffic intensity of the right hand lane was not available, the average of proportion of other sections was used. Furthermore, the traffic measurements were not always done during the entire year. Therefore, in calculating the cumulative amount of traffic, the total intensity of each year was first divided by the number of days in which the measurements have been done. This gave the intensity for one day and could then be multiplied by 365 to calculate the intensity of the entire year. The next was to calculate the sum of intensities of different years. Finally, this sum was multiplied by the proportion of right hand lane to other lanes. This can be summarized as follows:

$$CITR = \frac{TIR}{TIOY} \sum_{i=CY}^{NCY} 365 \frac{TI_i}{D_i} \tag{9}$$

where

$CTIR$ = the cumulative traffic intensity on the right hand lane for the test section within N years from construction.
TIR = the traffic intensity on the right hand lane for year 2005,
$TIOY$ = the traffic intensity of all lanes for year 2005,
TI_i = traffic intensity of the ith year for a specific test section,
CY = the construction year (construction year) of the test section,
NCY = N year after the construction year of the test section,
D_i = the number of days the traffic intensity has measured during i^{th} year (in many cases the traffic intensity has been measured for less than 365 days).

6.4 Climate Data

The SHRP-NL database didn't contain information on annual rainfall, solar radiation etc. Furthermore, no information is available on the weather and working conditions during construction. Therefore, it was decided to gather climatic data for the test sections using other resources. The most reliable and complete resource is the Royal Netherlands Meteorological Institute (KNMI). KNMI has climate data digitally available from 1951 for all weather stations of the Netherlands, including minimum, maximum and mean temperature, duration of sunshine, average cloud cover, relative atmospheric humidity, precipitation in 24 hours and its duration, maximum and mean of wind speed, and mean air pressure. Furthermore, the mentioned climate data were available for almost each single day from 1951. This made it possible to calculate very accurate cumulative climate factors.

After thorough consideration, four climate variables were chosen as climate factors, being calculated as shown in Equations 10 to 15. These variables are cumulative number of cold days, cumulative number of warm days, sunshine duration in months May through September in hours, and cumulative amount of precipitation in mm.

$$CCDN = \sum_{i=1}^{N} \sum_{j=1}^{365} CD(year = i, day = j) \tag{10}$$

$$\begin{cases} CD(year = i, day = j) = 1 & if\,(Min(DT \leq 0)) \\ CD(year = i, day = j) = 0 & if\,(Min(DT > 0)) \end{cases} \tag{11}$$

$$CWDN = \sum_{i=1}^{N} \sum_{j=1}^{365} WD(year = i, day = j) \tag{12}$$

$$\begin{cases} WD(year = i, day = j) = 1 & if\,(Max(DT \geq 25) \\ WD(year = i, day = j) = 0 & if\,(Max(DT < 25) \end{cases} \tag{13}$$

$$CRN = \sum_{i=1}^{N} \sum_{j=1}^{365} R(year = i, day = j) \tag{14}$$

$$CUVN = \sum_{i=1}^{N} \sum_{j=1}^{365} UV(year = i, day_{May-September} = j) \tag{15}$$

where

$CCDN$	=	the cumulative number of cold days for N years after the construction year,
$Min(DT)$	=	the minimum daily temperature,
$Max(DT)$	=	the maximum daily temperature,
$CWDN$	=	the cumulative number of warm days for N years after the construction year,
CRN	=	the cumulative amount of precipitation (mm),
$CUVN$	=	the cumulative duration of sunshine for N years after the construction date,
$R(year = i, day = j)$	=	the amount of precipitation in j^{th} day of year i,
$UV(year = i, day_{M.-S.} = j)$	=	the duration of sunshine on j^{th} day of year i (only for days in months May through September).

6.5 Final Dataset

After explaining the background on knowledge discovery, machine learning techniques, and data, from this point on, it is possible to discuss the knowledge discovery process from pavement data raveling of porous asphalt concrete. In last section, it became clear that Meq of raveling is the output variable. the final dataset for raveling of porous asphalt concrete, contained 13 input variables and 84 data points. Table 7 gives a detailed list of all these 13 variables.

One important remark should be made before the results are being described. For this study, it was decided to let the data speaks for itself. This means that no qualitative knowledge from road experts was used for the selection of e.g., input parameters. The opinion of expert was only asked after completion of a certain step in the knowledge discovery process.

Concerning the input variables, the climate and traffic related input variables (the last five variables in Table 7) are all cumulative variables, being calculated for a certain number of years after construction (e.g. 5 years after construction). After ample consideration, it was decided to develop models that predict the amount of raveling five and eight years after construction. The reason for choosing five years was to perceive early appearance of raveling. Eight years after construction was considered to be important since in a number of contracts, contractors have to guarantee a proper performance of PAC top layers for at least 7 years. It should be

Table 7. 13 Input variables for raveling of PAC obtained from SHRP-NL dataset.

Index	Input variables	Unit/types
1	Mixture density	kg/m^3
2	Bitumen content	Mass percentage on 100% aggregate
3	Void content	Percentage
4	Type of stone	Four types: Crushed siliceous river gravel, Porphyry, Greywacke/ Greyquartsite, Greywacke
5	Percentage of fine aggregate	Mass percentage passing the 2 mm sieve
6	Percentage of coarse aggregate	Mass percentage on the 2 mm sieve
7	CU (Coefficient of uniformity)	D_{60}/D_{10}[5]
8	D_{50}	Sieve size through which 50% of the coarse material passes
9	Cumulative number of warm days	days
10	Cumulative number of cold days	days
11	Cumulative duration of sunshine	hours
12	Cumulative amount of rain	mm
13	Cumulative amount of traffic	-

noticed that the SHRP-NL dataset contained only 10 years of measurements. For 5 data points, the raveling five and eight years after construction was not available because these section were older than eight year at the beginning of the SHRP-NL project. As a result, 79 data points were available in the final dataset for raveling of PAC.

7 Data Preparation

As mentioned before, data preparation includes data cleaning, variable selection/reduction, and data scaling. This section discusses how the data of raveling of PAC is prepared to be used for the next step, data mining. Before data preparation, as mentioned before, the number of data points available in the dataset was 79.

7.1 Data Cleaning

To clean the data, the dataset is checked for missing values, wrong types, and outliers. Checking the SHRP-NL dataset showed that there were no wrong types or missing values in the final dataset for both *Meq raveling* five and eight years after construction.

An outlier is a data point that lies outside the overall pattern of a distribution. Using an statistical method (Renze 2008), it was investigated if the dataset

[5] D_x = Sieve size through which x% of the coarse material passes.

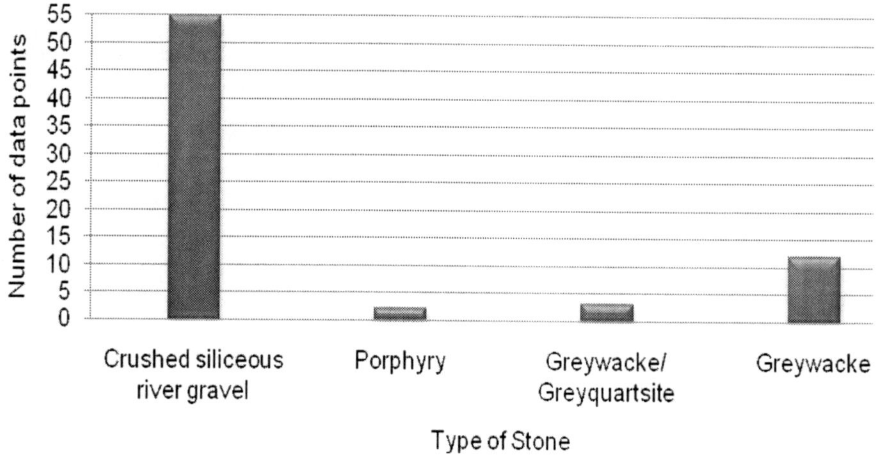

Fig. 14. The number of data points for each type of stone.

contained outliers. This method calculates two values called the inner fence and the outer fence. The data points falling outside of these fences are the outliers. The only difference is that the outer fence creates a larger window for non-outliers and as a result determines less data points as outliers.

The investigation showed that the input variable *Type of Stone* and the output variable *Meq of raveling* contained outliers. Hereafter, it is explained how these outliers are determined.

For *Type of Stone*, the number of data points for each type is visualized in Figure 14. As can be seen, the total number of data points with stone types *Prophyry* and *Greywacke/Greyquartsite* is five. The presence of these few data points in the training set will perhaps result in a less generalized model. In other words, use of these data points can result in a lower performance of the trained model and might very well confuse the learning process. For this reason, it was decided to delete the five data points containing the two mentioned types of stone to improve the quality of models. After deleting five data points from the dataset, 74 data points were left.

For *Meq of raveling*, the same statistical method was used to determine the outliers. It had to be decided if the inner fence or outer fence outliers should be taken into account. If the inner fence is used, a large number of data points are determined as outliers (For raveling eight years after construction, about 15% of data are above inner fence). If these data points would be eliminated, a low number of data points would stay in the dataset. This is less desirable because the SHRP-NL dataset is already a rather small dataset and each data point is a valuable one. Therefore, it was decided to choose the outer fence outliers, falling three times the interquartile above the third quartile For *Meq* raveling five years after construction, the outer fence was 4.62 + 3*4.62 = 18.5. There were no data points, having

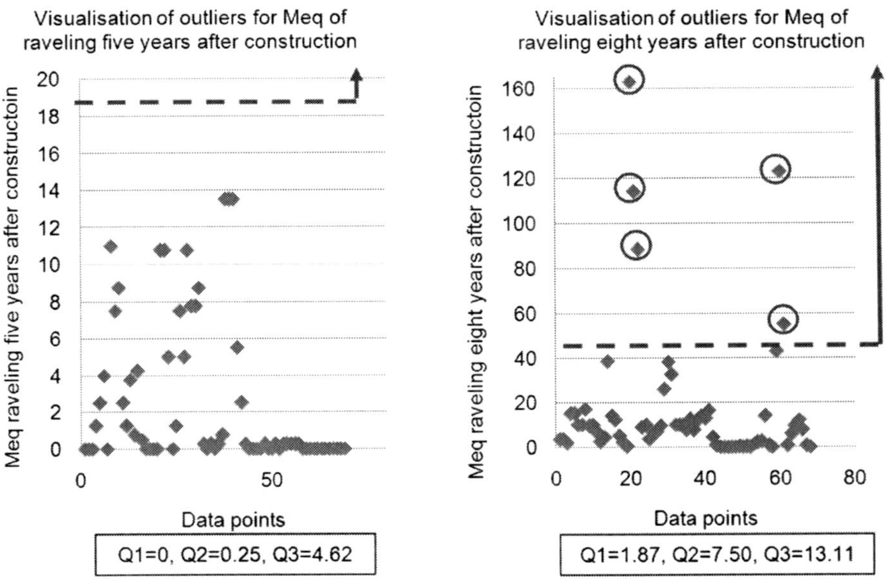

(a) (b)

Fig. 15. Determination of outliers for Meq of raveling five (a) and eight (b) years after construction.

a *Meq* raveling larger than 18.5 and therefore no outliers were determined for *Meq* five years after construction. The outer fence is shown in Figure 15(a) as a dotted line. Concerning *Meq* raveling eight years after construction, the value of outer fence was 13.11 + 3*11.24 = 46.83. As can be seen in Figure 15(b), five data points fall above this outer fence. Figure 15(b) shows the outer fence value with a dotted line and the outliers with circles around them. Now that the outliers have been determined, the question is if they should be eliminated from the dataset.

Doubting about deleting these outliers has two reasons. The most obvious reason is that the dataset at this point contained only 74 data points and after deleting another five data points as outliers, only 69 data points will be left. Another reason is that although the outliers have been determined with a statistical method, one is never certain whether these points are really measurement faults or if they contain important information which could make the problem distribution more complete.

To be able to decide about these five data points, more information was necessary, for instance the name and location of the road from where the data points were obtained and the mixture properties of the asphalt layer of those roads. The result of the search is shown in Table 8. The location of two of the outliers was on

Table 8. The information about the outliers including their location and their mixture properties.

SHRP-NL ID	Section	Meq raveling 8 years after construction	Location	Mixture density	Bitumen	Voids content	Type of stone
1107	1	114	A12	2063[6]	4.5[7]	16.4[8]	Crushed siliceous river gravel
1107	2	163	A12	2086	5.2	15	Crushed siliceous river gravel
1107	3	88	A12	2080	4.6	15.3	Crushed siliceous river gravel
5063	1	123	A1	2137[9]	4.4[10]	15.5[11]	Greywacke
5063	2	55	A1	2127	4.4	15.5	Greywacke

A1 highway close to the city of Apeldoorn with a moderate traffic intensity (see the map in Figure 16(a)) and the location of the other three data points is on A12 highway close to the city of Gouda with a high traffic intensity (see the map in Figure 16(b)). As can be seen in Figure 16(b), the test section number (1107) has been crossed off the map. This is because the road section was replaced in 1993 and therefore was partly present in the SHRP-NL project. However, in 1993 this section was eight years old and it was therefore included in the dataset for raveling eight years after construction.

Table 8 shows that the mixture density of the five identified outliers is much higher than the standard given for porous asphalt and the voids content is much lower on section 1107 the bitumen content was also high. However, the obvious deviation from standard properties of porous asphalt concrete cannot be seen as a reason to delete these outliers from the dataset because there are more data points in the dataset deviating from the standard mixture properties. Moreover, due to a higher bitumen content and mixture density in these sections, road engineers expect them to have a longer lifespan and less damage than has been observed. To decide about these data points, extra information about these sections could help. However, this additional information was not available.

---

[6] For test section with SHRP-NL ID 1107, average of *mixture density* is 2075 and the range is [2023, 2133].

[7] For test section with SHRP-NL ID 1107, average of *bitumen content* is 4.7 and the range is [4.4, 5.1].

[8] For test section with SHRP-NL ID 1107, average of *voids content* is 15.7 and the range is [13.2, 18.6].

[9] For test section with SHRP-NL ID 5063, average of *mixture density* is 2132 and the range is [2088, 2156].

[10] For test section with SHRP-NL ID 5063, average of *bitumen content* is 4.4 and the range is [4.3, 4.5].

[11] For test section with SHRP-NL ID 5063, average of *voids content* is 15.5 and the range is [14.7, 17.2].

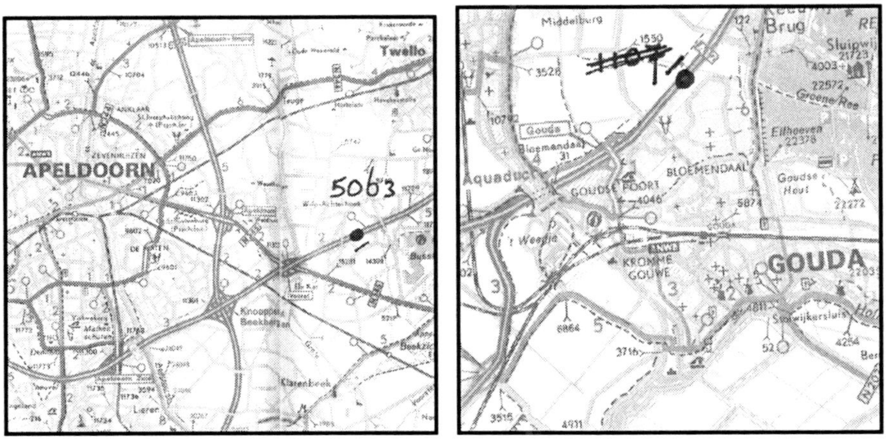

(a) (b)

Fig. 16. Location of SHRP-NL test section with ID number 5063 nearby the city of Apeldoorn (a) and the test section with ID number 1107 nearby the city of Gouda (b).

On one hand, the so called outliers are valuable for this study due to the lack of data. On the other hand, it is not known that if they are kept in the dataset, how they would influence the model performance. To solve this problem, an experiment was conducted. In the experiment, the outliers were deleted one for one, each time a model was developed using ANN technique and the model performance was compared with the previous one. This was continued until the outliers in the dataset did not negatively influence the performance of the model any more. One can wonder in which order the outliers should be deleted. The outlier which lies furthest from other points is expected to have the most negative influence on the model. Therefore, each time the outlier furthest from other points will be deleted. The results of this experiment are shown in Figures 17 and 18.

As Figure 17(a) shows, the R^2 of the ANN model trained with the dataset including all outliers is 0.53. This means that including all outliers in the model will result in a low performance. In the next step, the outlier with the largest *Meq* value (163) was eliminated from the dataset and the ANN model was trained again. This model had an R^2 of 0.58 (Figure 17(b)), which is still a poor performing model. In the next step, next to the first outlier, the outlier with the second largest *Meq* value (123) was deleted and the model was trained again. This time, the R^2 of the ANN model increased considerably (0.96) (see Figure 18(a)), showing that the model performs very well. It seemed that only the two mentioned outliers needed to be deleted from dataset. To be certain, an ANN model using the dataset excluding all five outliers was trained (see Figure 18(b)). The R^2 of this model is the same as the one excluding the two extreme outliers (0.96) (see Figure 18(a)). Therefore, it was decided to use the dataset excluding only the first two outliers with an *Meq* value of 163 (the road test section located on A12) and an *Meq* value of 123 (the road test section located on A1), resulting in a dataset with 72 data points.

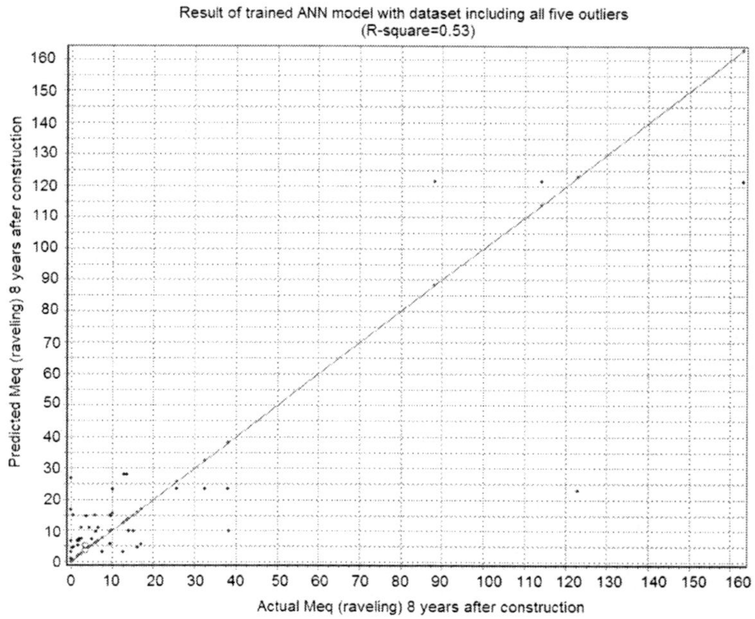

(a)

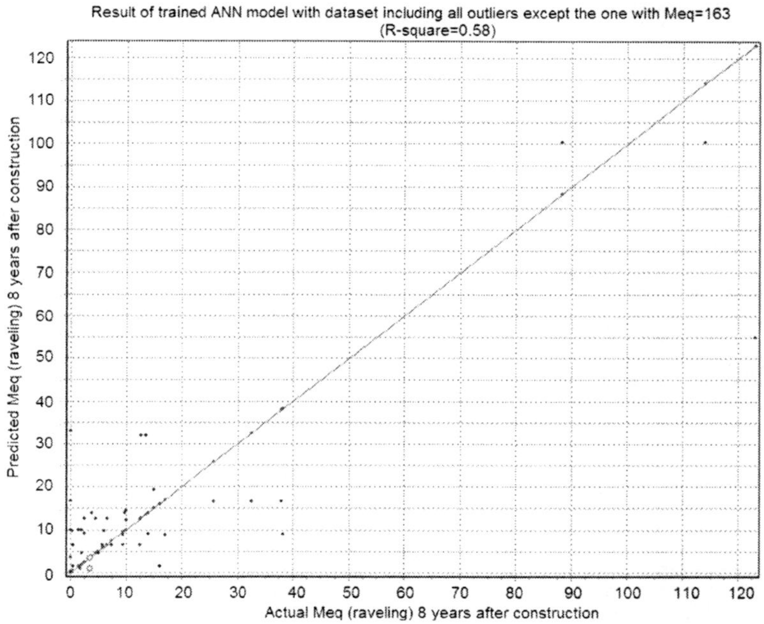

(b)

Fig. 17. The performance of trained ANN model using all data points (a) and using all data points except the one with Meq raveling value of 163 (b).

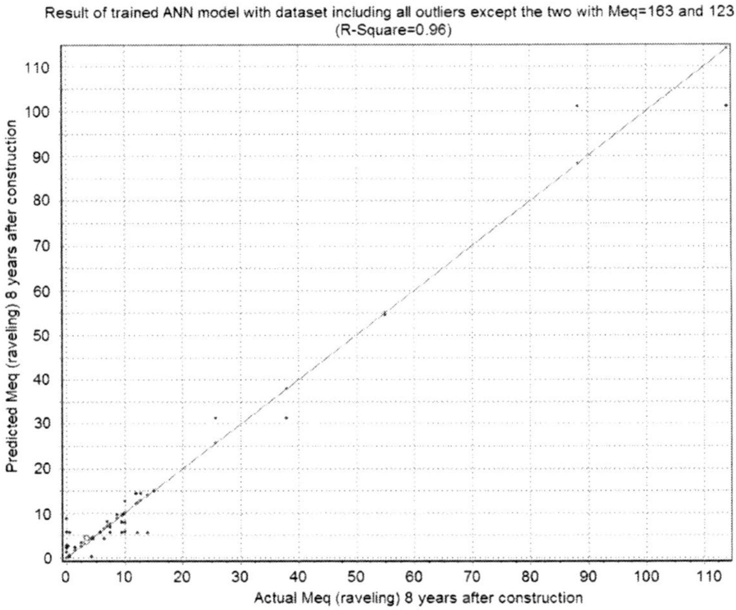

(a)

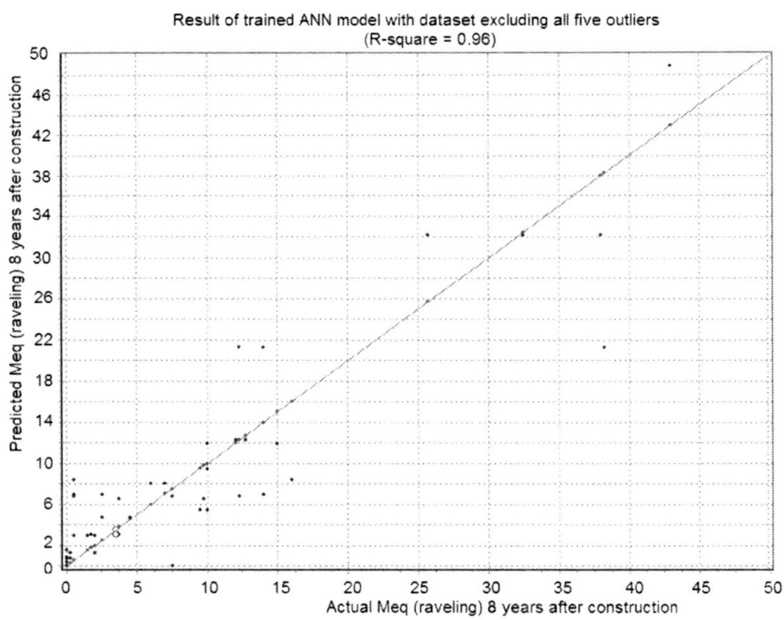

(b)

Fig. 18. The performance of trained ANN model ANN with all data points excluding the ones with M_{eq} of 163 and 123(a) and using the dataset excluding all five outliers (b).

7.2 Variable Selection

One important basic step in each type of problem is to determine the input variables that influence the performance of the model the most. As it was stated in the literature study, little attention has been given to variable selection/reduction methods by researchers working on pavement performance modeling and the few researchers, who employed these methods, used only one method at the same time. However, it should be noticed that using one type of variable selection method will not give strong evidence that the inputs selected are the most influential ones. A better approach would be to apply a number of variable selection algorithms to the dataset and then compare the variables selected by these algorithms. In case there are some input variables selected repeatedly by different algorithms, it can be concluded that those are the most influential input variables. This is a time-consuming and difficult approach. However, because it is believed that using this approach will result in a more reliable selection of input variables, it was employed in this study.

Eight different input selection methods were employed to select the most influential input variables:

regression trees,
genetic polynomial regression,
artificial neural network (weighted weight factor method),
rough set theory,
correlation based subset selection using bidirectional search,
correlation based subset selection using genetic search,
wrapper of artificial neural network using genetic search, and
relief ranking filter.

Before applying these methods, it was necessary to decide how many variables should maximally be selected. Because of the presence of variables representing material properties, climate, and traffic, a minimum of three variables was needed. However, eight of the input variables are related to the material properties, including two subgroups of mixture composition and gradation variables. As a result, more than one material related input variable was needed to model the problem. At the same time, because of the small number of data points, it was desirable to use the smallest number of input variables. Taking all these aspects into account, it was decided to choose a maximum of five most influential variables. Table 9 summarizes the result of all methods applied, to *Meq* (raveling) five years after construction, selecting a maximum of five input variables from the 13 variables listed in Table 7. In the setting column of Table 9, it can be seen that the cross validation method leave-one-out was used constantly. This method is suitable for small datasets (with less than 100 data points).

Table 9. The five most important input variables for Meq (raveling) five years after construction.

Method	Setting	Variable 1	Variable 2	Variable 3	Variable 4	Variable 5
Regression trees	Leave-one-out cross validation	Bitumen content	Traffic	Cold days	Voids content	%Coarse
Genetic polynomial	Polynomial degree = 3	Bitumen content	Traffic	Cold days	%Coarse	Voids content
Artificial neural network (WWF)	Leave-one-out cross validation	Bitumen content	Traffic	Voids content	Cold days	%Coarse
Rough sets	2-class output	Bitumen content				
Correlation-based subset selection (bidirectional search)	Greedy stepwise search Leave-one-out cross validation	Bitumen content	Traffic	Cold days		
Correlation-based subset selection (genetic search)	Genetic Search Leave-one-out cross validation	Bitumen content	Traffic	Voids content	Cold days	%Coarse
Wrappers of ANN (genetic search)	Genetic Search Leave-one-out cross validation	Bitumen content				
Relief ranking filter	K=20 Nearest neighbor (equal influence) Leave-one-out cross validation	Bitumen content	Traffic	Cold days		

Table 9 shows that for *Meq* 5 years after construction, *bitumen content* was selected by all methods as the most influential input variable and *Traffic* by all except one. The other three variables, determined by most methods, were *Cold days*, *Voids content*, and *percentage of Coarse.* The same approach was used for raveling eight years after construction, leading to the results shown in Table 10.

As can be seen in Table 10, *Voids content* and *Bitumen content* are the two most influential input variables for *Meq* 8 years after construction. Next to them, *Cold days*, *Coarse percentage*, and *Density* were selected by a majority of methods as influential input variables.

In the sake of completeness, it is recalled that the variables *type of stone, amount of rainfall, amount of sunshine,* and most of *gradation parameters* were not selected by any of the variable selection methods (see Tables 9 and 10).

Table 10. The five most important input variables for Meq(raveling) eight years after construction.

Method	Setting	Variable 1	Variable 2	Variable 3	Variable 4	Variable 5
Regression trees	Leave-one-out cross validation	Bitumen content	Voids content	Cold days	%Coarse	Density
Genetic polynomial	Polynomial degree = 3	Voids content	Bitumen content	Cold days	%Coarse	Density
Artificial neural network (WWF)	Leave-one-out cross validation	Voids content	Cold days	Bitumen content	%Coarse	Traffic
Rough Sets	3-class output	Bitumen content	Voids content			
Correlation-based subset selection (bidirectional search)	Greedy stepwise search Leave-one-out cross validation	Voids content	Bitumen content	Cold days	D50	Density
Correlation-based subset selection (genetic search)	Genetic Search Leave-one-out cross validation	Voids content	Bitumen con-tent	Density	Cold days	D50
Wrappers of ANN (genetic search)	Genetic Search Leave-one-out cross validation	Voids content				
Relief ranking filter	K=20 Nearest neighbor Leave-one-out cross validation	Cold days	Voids content	Bitumen content	Warm days	Density

Considering the order of variables importance/influence, it should be noticed that not all methods make it possible to determine the exact importance of ranking. Two of the methods, which give a rather clear ranking of input variables, are *genetic polynomial regression* and *artificial neural network (weighted weight factor)*. The ranking of these two methods are shown in Figure 19 (genetic polynomial) and 7 (artificial neural network) for both *Meq* raveling five and eight years after construction.

Figure 19 shows the number of iterations that each input variable is present in the model during 100 iterations. As can be seen for *Meq* raveling five years after construction, *Bitumen* is present in the model in all 100 iteration while for eight years after construction *Voids content* stayed in the model for about 90 iterations. The variable with the lowest presence in the model for raveling 8 years after construction is *Traffic*, which was present in the model only for five iterations.

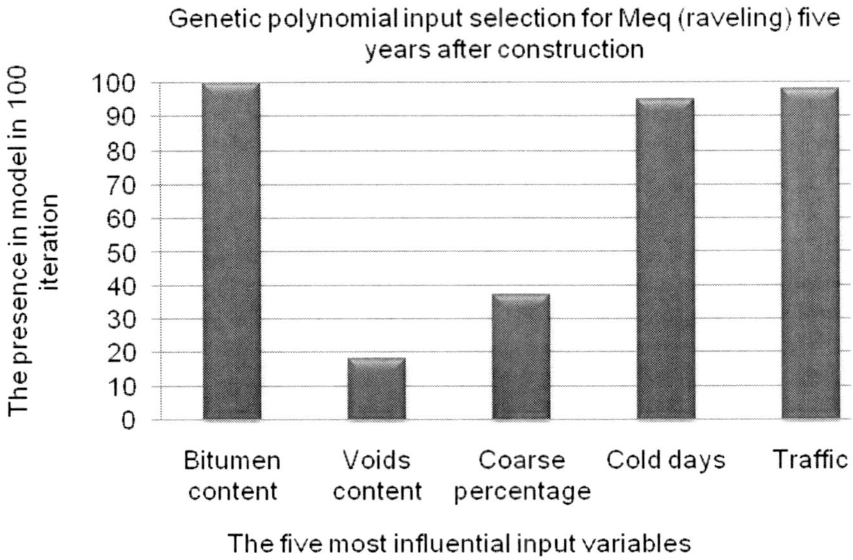

(a)

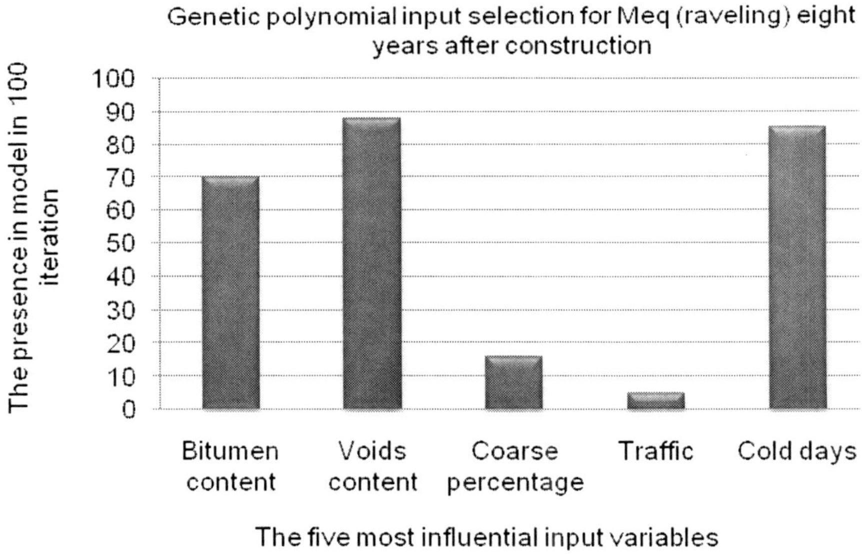

(b)

Fig. 19. The five most important input variables for raveling five years after construction (a) and raveling eight years after construction (b) determined by Genetic polynomial.

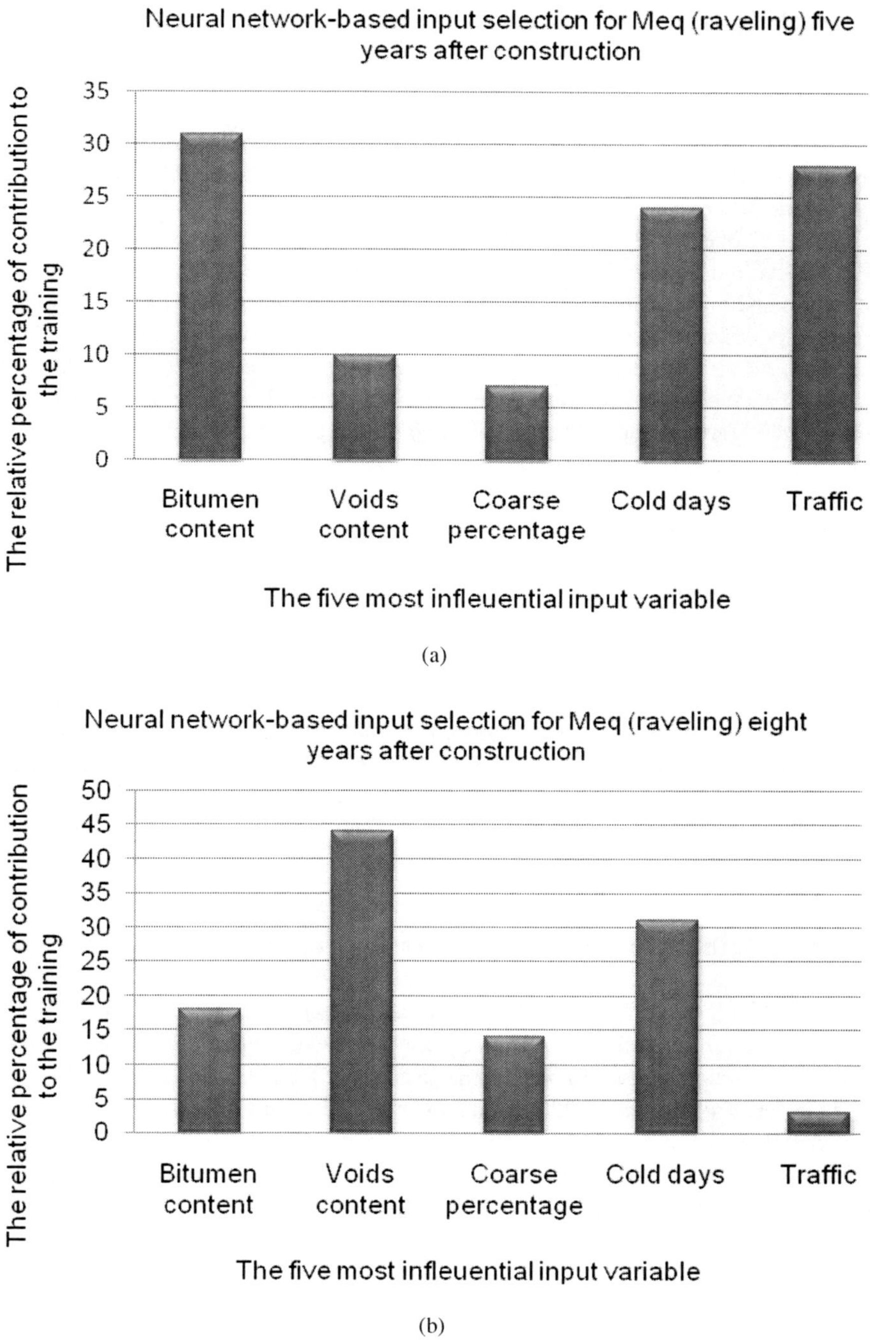

Fig. 20. The five most important input variables for raveling five years after construction (a) and raveling eight years after construction (b) determined by artificial neural network.

The ANN input selection is presented in Figure 20. This figure shows that for raveling 5 years after construction the input variables *Bitumen*, *Traffic* and *Cold days* are the most important ones. For raveling 8 years after construction, *Voids content*, *Bitumen*, and *Cold days* contribute most.

For road experts, it was very interesting to observe that *Cold days* is important for both *Meq* raveling five and eight years after construction while *Traffic* is not important for *Meq* raveling eight years after construction.

Based on the result of different input selection methods, listed in Table 9 and 10, the five most influential input variables for both raveling five years after construction and eight years after construction are easy to determine. For raveling five years after construction (results of Table 9), the five input variables *Bitumen*, *Traffic*, *Cold days*, *Voids content*, and *Coarse percentage* were selected for final modeling. For raveling eight years after construction, a close look into the results shown in Figures 6 and 7 learns that *Traffic* is not an important factor anymore. Therefore, it can be excluded from the input variables. It means that the modeling can be done using four input variables. These four variables are *Voids content*, *Bitumen*, *Coarse percentage*, and *Cold days*.

7.3 Data Scaling

As mentioned before, all variables are numerically continuous except for the input variable *Type of stone*, which is a categorical one. Initially database included four types of stones (see Table 7) but when the outliers were deleted only two types were left. In the previous section, the results of variable selection (Tables 9 and 10) showed that *Type of stone* was not included in the five most influential input variables. As a result, all variables that need to be scaled are numerical one. The numerical input variables and the output variable were scaled to the range of [-1..1].

8 Data Mining and Evaluation/ Interpretation of Models

In the last two steps of the knowledge discovery process, being data mining and evaluation/interpretation of the mined pattern (model), a specific technique with certain parameters is used to develop a model from the data (find a pattern in data) and the result of the model is examined (Figure 1). From the discussion so far, it became clear that data mining for raveling of porous asphalt concrete will be performed on a dataset with 72 data points, with the output variable *Meq* raveling five or eight years after construction and the selected four respectively five input variables. This can be summarized as follows

$$Meq5 = f(Bitumen, Voids\ content, Coarse\ percentage, Cold\ days, Traffic) \tag{16}$$

$$Meq8 = f(Bitumen, Voids\ content, Coarse\ percentage, Cold\ days) \tag{17}$$

where $Meq5_{Rav}$ and $Meq8_{Rav}$ are the *Meq* raveling five and eight years after construction, respectively. As explained before, four machine learning techniques are employed in the data mining step: *artificial neural networks, support vector machines, decision trees*, and *rough set theory*. The next four sections discuss the last two steps of data mining for raveling of porous asphalt concrete using each of the four mentioned techniques.

9 Data Mining Using Artificial Neural Network

In this section, the models developed for raveling five and eight years after construction are called $Meq5_{Rav}$*_ANN* and $Meq8_{Rav}$*_ANN*, respectively. Before starting with data mining, the dataset was partitioned into two subsets: a training set (85% of data points) and a test set (15% of data points). A part of the training set is used for the cross validation. The size of this part depends on the type of cross validation method being used.

9.1 Parameter Determination for ANN

The first step in data mining is to determine the parameters needed for the techniques applied. The parameters necessary to develop an ANN model are *type of activation function, number of hidden neurons, type of learning algorithm, learning rate*, and *momentum* (Haykin 1999).

According to the universal approximation theorem, one hidden layer is sufficient to solve many problems (Haykin 1999). Therefore, the number of hidden layers was set to one. To estimate the number of hidden neurons, a 10-fold cross validation was employed. The calculated training and cross-validation errors of 12 neural networks with 1 to 12 hidden neurons are shown in Figure 21. The number of hidden neurons resulting in the lowest validation error is the optimal number of hidden neurons, which is in this case three for both $Meq5_{Rav}$*_ANN* and $Meq8_{Rav}$*_ANN* models. Consequently, the optimal architecture for both models is one hidden layer containing three hidden neurons. Using the mentioned architecture, different types of activation functions were tried. The *hyperbolic tangent* gave the lowest prediction error and therefore it was chosen as the activation function for both hidden and output layers.

Concerning the other three parameters, the investigation showed that the learning algorithm *batch backpropagation* with a learning rate of *0.1* and a momentum of *0.3* for $Meq5_{Rav}$*_ANN* model resulted in the best performance. For $Meq8_{Rav}$*_ANN* model, a *batch backpropagation* algorithm performed the best with a learning rate of *0.1* and a momentum of *0.2*.

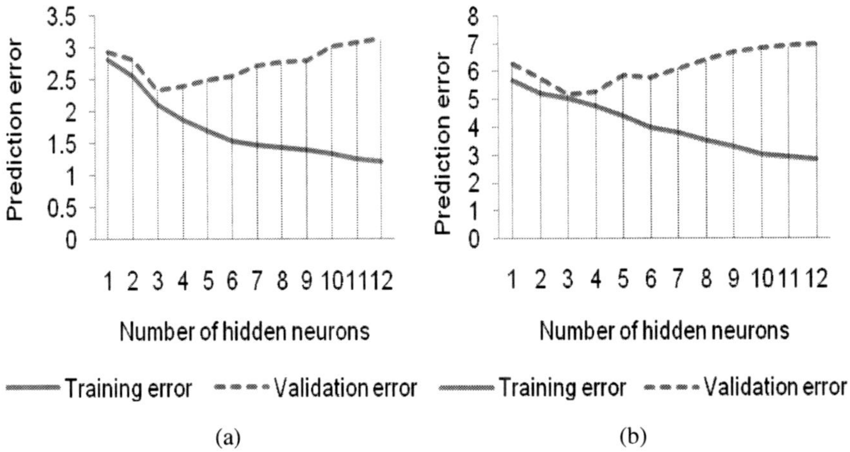

Fig. 21. Determination of the optimal number of hidden neurons for model $Meq5_{Rav}$_ANN (a) and $Meq8_{Rav}$_ANN (b).

9.2 Modeling Using ANN

After parameter determination, *$Meq5_{Rav}$_ANN* and *$Meq8_{Rav}$_ANN* models were only trained using the parameters mentioned above. They were tested using the test set. The training, cross validation, and testing errors for both models are shown in Table 11. The cross validation method *leave-one-out* was used to calculate the cross validation error. The reason for using leave-one-out was that the dataset is small (number of data points less than 100). As mentioned before, the number of data points after data cleaning was 72. For the leave-one-out method, 72 data points formed the training set and one data point the validation set, repeating this 71 times each time using another single data point as validation set.

From the results given in Table 11 shows that *$Meq5_{Rav}$_ANN* and *$Meq8_{Rav}$_ANN* models with four/five input parameters perform better than the models which were developed earlier using all input variables (Miradi and Molenaar, 2005). This is most likely the result of reducing the input dimension. The prediction plot of the training and test set for these models are shown in Figures 22 and 23, respectively. In these figures, the x-axis of the plots shows the actual Meq of raveling while the y-axis shows the predicted Meq. The line on the plot is called the line of equality. The closer the points are located to the line of equality, the better the prediction.

Table 11. The result of model $Meq5_{Rav}$_ANN and $Meq8_{Rav}$_ANN.

Model	Training error	Cross validation error	Testing error	R-square
$Meq5_{Rav}$_ANN	0.55	0.61	0.24	0.95
$Meq8_{Rav}$_ANN	2.70	2.88	4.01	0.94

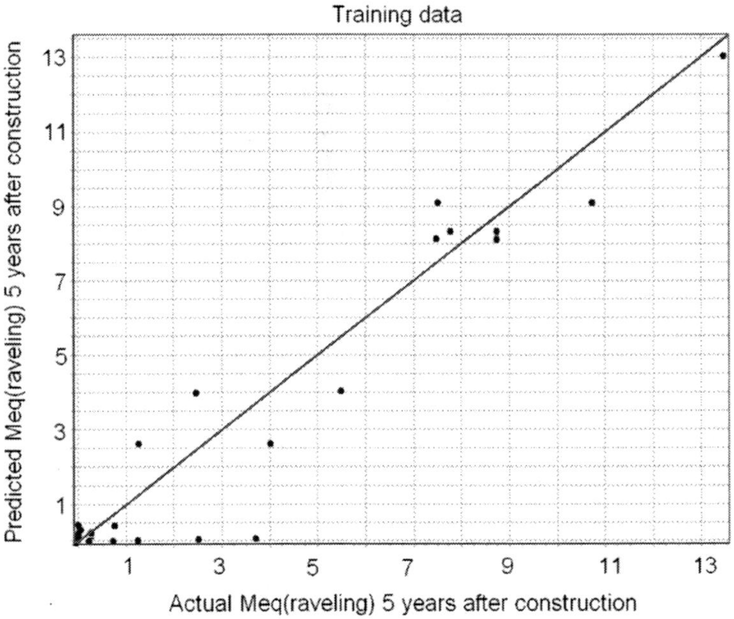

(a)

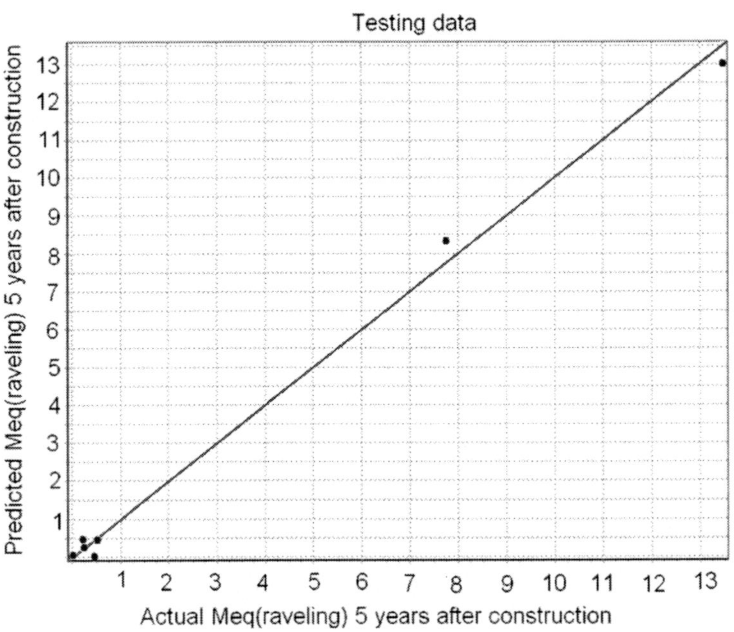

(b)

Fig. 22. Prediction of Meq(raveling) five years after construction by model $Meq5_{Rav}$_ANN for the training set (a) and the test set (b).

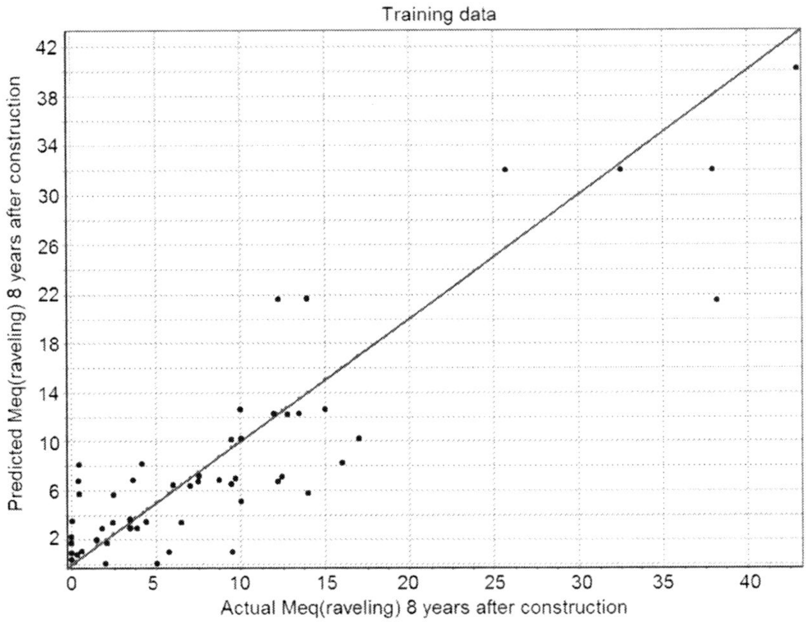

(a)

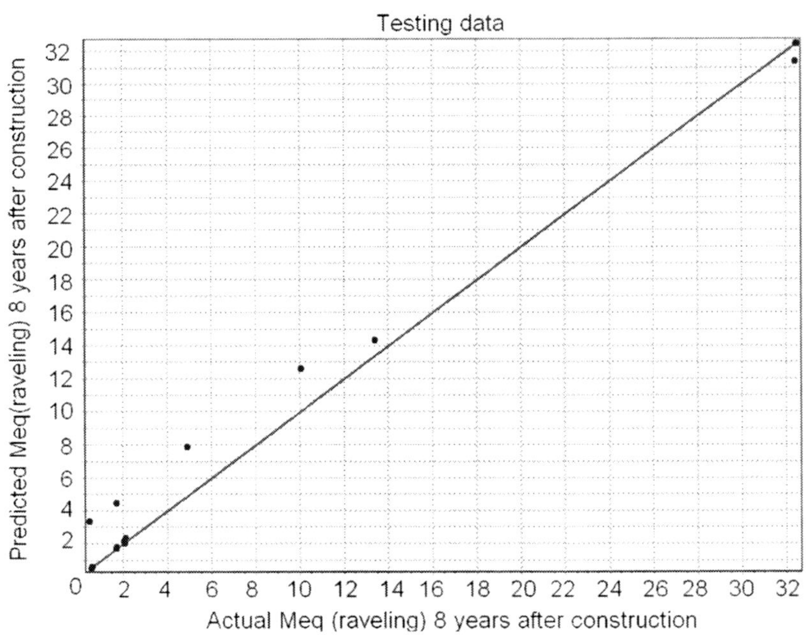

(b)

Fig. 23. Prediction of Meq(raveling) eight years after construction by model $Meq8_{Rav}$_ANN for training set (a) and testing set (b).

9.3 Evaluation/Interpretation of ANN Models

Figure 22 (a) shows that the prediction made by the *Meq5*$_{Rav}$*_ANN* model in the range of [0, 4] is less accurate for the training set. This is because of the discernibility of some data points. Discernibility of data points means that although the input variables of those data points are identical, their output (Meq raveling) is not the same. This is due to the fact that although some road samples are taken from the same test section, the raveling observed on the three subsections within that section. This means that their material properties, climate circumstances, and the traffic load are the same but the raveling observed on the subsections is not the same. Since the data contain this variability, some error tolerance should be allowed. As can be seen in Figure 23 (b), this also applies to model *Meq8*$_{Rav}$*_ANN* for the range [0, 15].

One of the tools used for the interpretation of the ANN result is the response graph. A response graph displays the response of the model output as one input variable is varied while other input variables are held constant. The constant value for each variable is the average value of that variable in the dataset. The average value for *Bitumen content* was 4.3, for *Voids content* 18.8, for *Coarse percentage* was 83.1, for *Cold days* was 329, and for *Traffic* was 22,538,978. This graph is called a response graph because it is the response to the different values of the selected input variable. Figure 11 shows the result of this investigation into all five input variables of model *Meq5*$_{Rav}$*_ANN.*

As can be seen in Figure 24(a), if *Bitumen content* < *4%*, the Meq raveling 5 years after construction is between 1 and 10. *Bitumen content* ≥ *4%* causes no raveling. Figures 24(b) and 24(c) show that by increasing *Voids content* or *Coarse percentage* the *Meq* raveling 5 years after construction increases. Figure 24 (d) shows that if *Cold days* > 310, then the *Meq* raveling 5 years after construction is between 0 and 3. Finally Figure 24(e) shows that if traffic increases, the raveling increases as well. It should be noticed that the above if-then rules are valid only when the input variable present in the rule is varied and other variables are held constant.

The response graphs shown in Figure 11 are in agreement with practical experience. As can be seen in Figure 24(b), the value of *Voids content* is between 13% and 24%. The void content of PAC is around 20% and a road section with a void content less than 17% cannot be rated as PAC anymore. The presence of these low void contents clearly indicates that something must have gone wrong during construction.

When one compares the response graphs presented in Figure 24, one should pay attention to the y-axis. It can be seen that for input variables *Bitumen content* and *Traffic*, changes in their value cause larger changes in the *Meq* of raveling (between 0 and 10) comparing to other variables. For the other four input variables, the *Meq* raveling increases up to a maximum of 6.

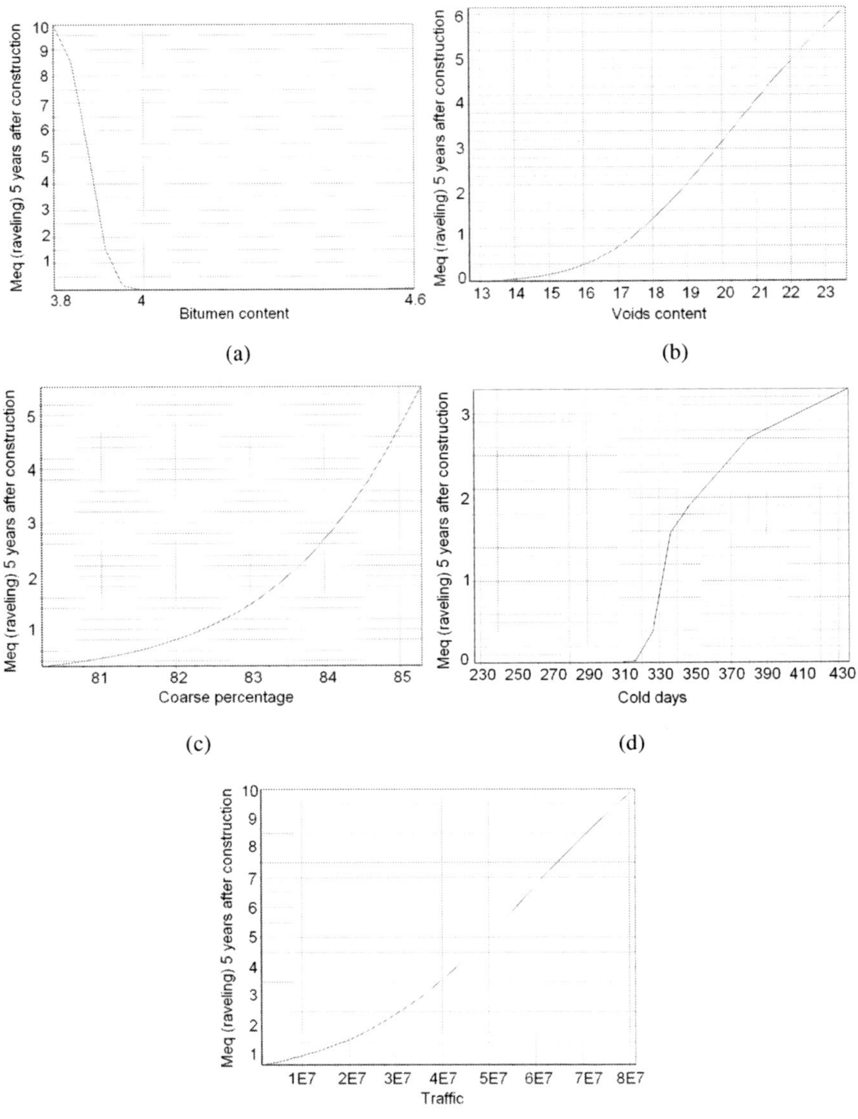

Fig. 24. Response graph of the input variables bitumen content (a), voids content (b), coarse percentage (c), cold days (d), and traffic (e) for model $Meq5_{Rav}$_ANN.

The response of model *Meq8$_{Rav}$_ANN* to its four input variables is shown in Figure 25. As was the case for raveling five years after construction, when response graph deals with one input, it holds other variables constant. The constant

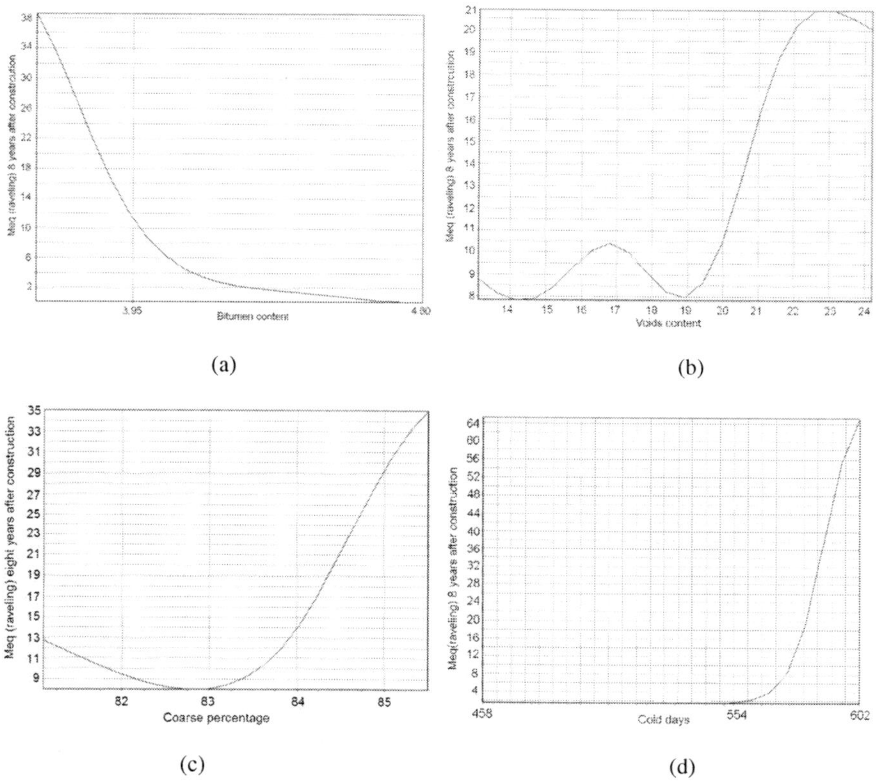

Fig. 25. Response graph of the three input variables Bitumen content (a), Voids content (b), and Cold days (c) for model $PM_{eq}8$_ANN.

value for each variable is the average value of that variable in the dataset. The average value for *Bitumen content* was 4.3, for *Voids content* 18.8, for *Coarse percentage* was 83.1, for *Cold days* was 515.

Figure 25 (a) clearly shows that *Bitumen content < 3.95%* causes *Meq* of raveling between 11 and 38. If *3.95% ≤ Bitumen content between ≤ 4.60%*, then *Meq* of raveling is between 0 and 11. Figure 25(b) shows that when *Voids content is between 13%* and *20%* raveling does not vary that much (between 8 and 10). However, if *Voids content > 20%*, the amount of *Meq* of raveling is between 10 and 21. Figure 25(c) shows that if the *82.5% ≤ Coarse percentage ≤ 83.5%*, the *Meq* value will be about 8. In the case *Coarse percentage > 83.5%*, *Meq* raveling is between 13 and 35. From Figure 25(d), it can be concluded that if the eight year cumulative number of *Cold days > 554* days, the raveling will increase fast to a maximum of 64 but if *Cold days ≤ 554, Meq* raveling eight years after construction stays very low (around 2). As in the $Meq5_{Rav}$*_ANN* model, the if-then rule for each input variable is valid under the condition that that variable is varied and the rest are held constant.

Next to ANN, support vector regression is also employed for data mining in this dissertation. The next section will describe the development process of data mining using SVR for *Meq* of raveling five and eight years after construction.

10 Data Mining Using Support Vector Regression

In this section, the models extracted from the data using support vector regression for raveling five and eight years after construction are called $Meq5_{Rav}$*_SVR* and $Meq8_{Rav}$*_SVR*, respectively.

10.1 Parameter Determination for SVR

The first step in SVR modeling is to determine the optimal modeling parameters. Concerning the kernel type, pre-investigation showed that the radial basis kernel function showed the highest performance.

Parameter C is one of the necessary parameters for SVR modeling. Due to the use of a radial basis kernel function, its parameter, γ, should also be determined. Using a 10-fold cross validation grid search, the optimal value of parameter γ was searched between 1 and 20 for models $Meq5_{Rav}$*_SVR* and $Meq8_{Rav}$*_SVR*.

As can be seen in Figure 26(a) and 26(b), for both $Meq5_{Rav}$*_SVR* and $Meq8_{Rav}$*_SVR*, $\gamma = 18$ showed the lowest error (lowest point in the graph) and as a result, 18 is the optimal value for γ ($\gamma = 3$ is also optimal but results in slightly more error than $\gamma = 18$). The determination of parameters C and γ is done in parallel. It means that for each value of γ, parameter C is calculated for the whole range. This explains the many dots on Figure 26. A better way of presenting the result is a 3D plot. However, due to presence of many combinations of C and γ (many dots on the plot), it would be difficult for the reader to observe which value results in the lowest performance error. Therefore it was decided to plot the results as 2D plots.

As was done for γ, a 10-fold cross validation grid search was performed to determine the optimal value for parameter C. Looking at values between 1 and 250, the value $C = 30$ showed the lowest error for model $Meq5_{Rav}$*_SVR* and was therefore chosen as the optimal value (see Figure 27 (a)). As can be seen in Figure 27(b), for model $Meq8_{Rav}$*_SVR*, the optimal value for C was 40. The final parameters used in SVR modeling are summarized in Table 12.

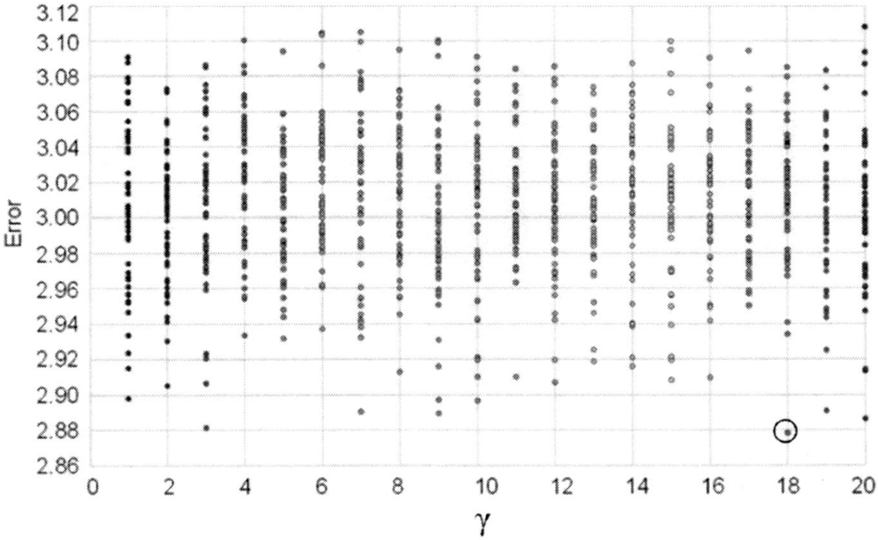

(a)

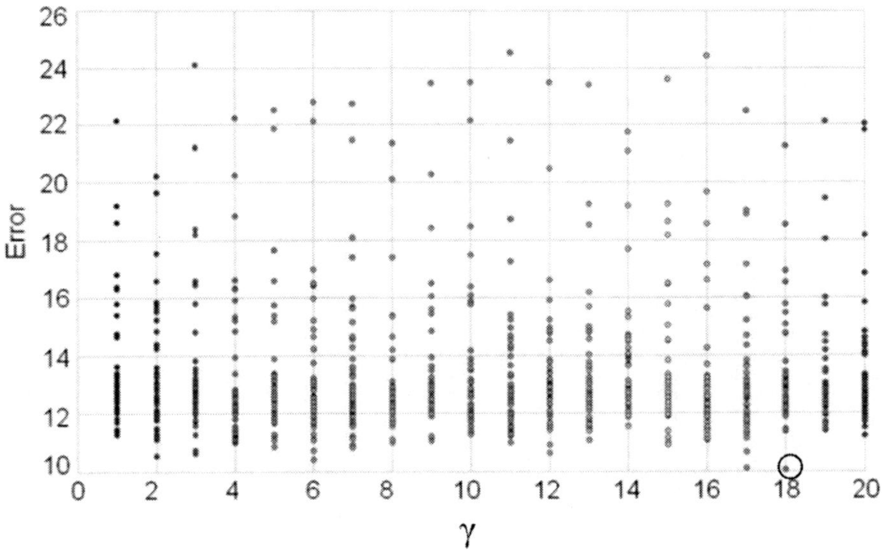

(b)

Fig. 26. Cross validation grid search for selection of optimal value of parameter γ of radial basis kernel function for models $Meq5_{Rav}$_SVR (a) and $Meq8_{Rav}$_SVR (b).

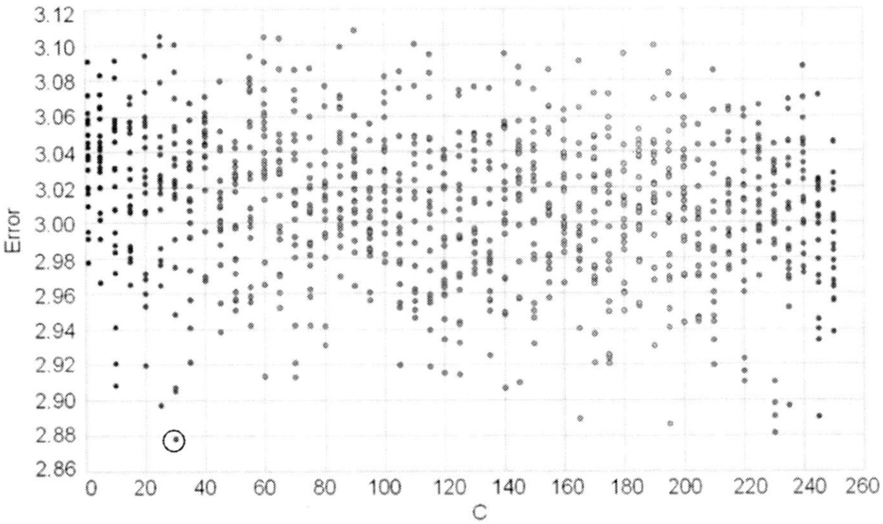

(a)

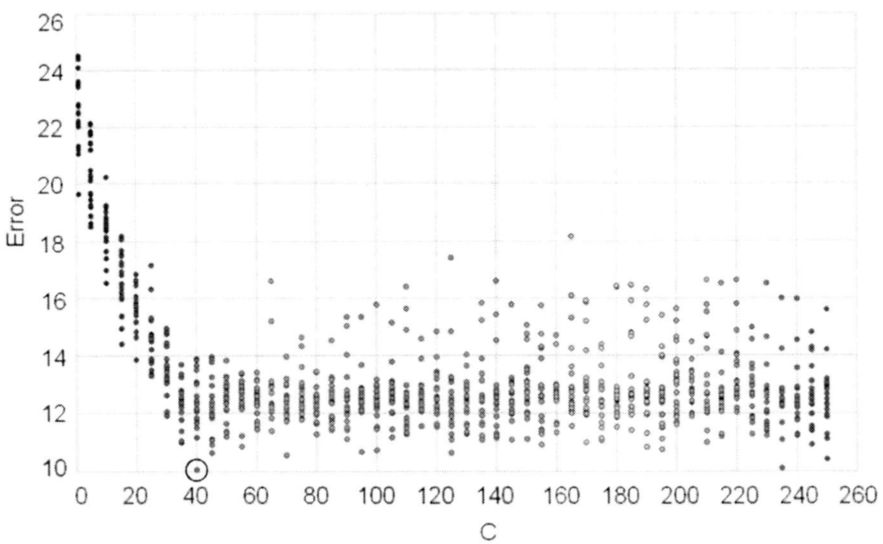

(b)

Fig. 27. Cross validation grid search for selection of optimal value of parameter C for models $Meq5_{Rav}$_SVR (a) and $Meq8_{Rav}$_SVR (b).

Table 12. The setting for SVR models $Meq5_{Rav}$_SVR and $Meq8_{Rav}$_SVR.

Parameter	Value for model $Meq5_{Rav}$_SVR	Value for model $Meq8_{Rav}$_SVR
SVM type	Epsilon SVR	Epsilon SVR
Kernel type	Radial basis	Radial basis
	18	18
C	30	40

Table 13. The number of support vectors, weights of the inputs, and the bias of the models $Meq5_{Rav}$_SVR and $Meq8_{Rav}$_SVR.

Parameter	Value for model $Meq5_{Rav}$_SVR	Value for model $Meq8_{Rav}$_SVR
Number of support vectors	49	58
Weights	W(Bitumen) = 832.48 W(Voids content) = 1,179.3 W(%Coarse) = 1,313.2 W(Cold days) = 829.1 W(Traffic) = 357.1	W(Bitumen) = -5,471.8 W(Voids content) = 1,132.0 W(%Coarse) = 970.3 W(Cold days) = 2,226.6
Bias	-3.1	-7.9

Table 14. The quality measures for SVR models $Meq5_{Rav}$_SVR and $Meq8_{Rav}$_SVR.

Measure	Value for model $Meq5_{Rav}$_SVR	Value for model $Meq8_{Rav}$_SVR
RMSE of test set	2.9	6.4
R-square	0.97	0.87

10.2 Modeling Using SVR

Using the parameters given in Table 12, the SVR models *$Meq5_{Rav}$_SVR* and *$Meq8_{Rav}$_SVR* were trained using LibSVMLearner. Developing an SVR model results in finding some parameters: support vectors, weights, and bias. Table 13 reports these parameters for the *$Meq5_{Rav}$_SVR* and *$Meq8_{Rav}$_SVR* models.

10.3 Evaluation/Interpretation of SVR Models

To evaluate the SVR models, the trained models were tested using the test set. As shown in Table 14, the RMSE of the test set for models *$Meq5_{Rav}$_SVR* and *$Meq8_{Rav}$_SVR* were 2.9 and 6.4, respectively. The R-square of *$Meq5_{Rav}$_SVR* model is higher than *$Meq8_{Rav}$_SVR* (0.97 against 0.87). Comparing the results of Tables 11 and 14, it can be seen that ANN has a higher prediction performance for this specific problem. The prediction plot of SVR models, *$Meq5_{Rav}$_SVR* and *$Meq8_{Rav}$_SVR*

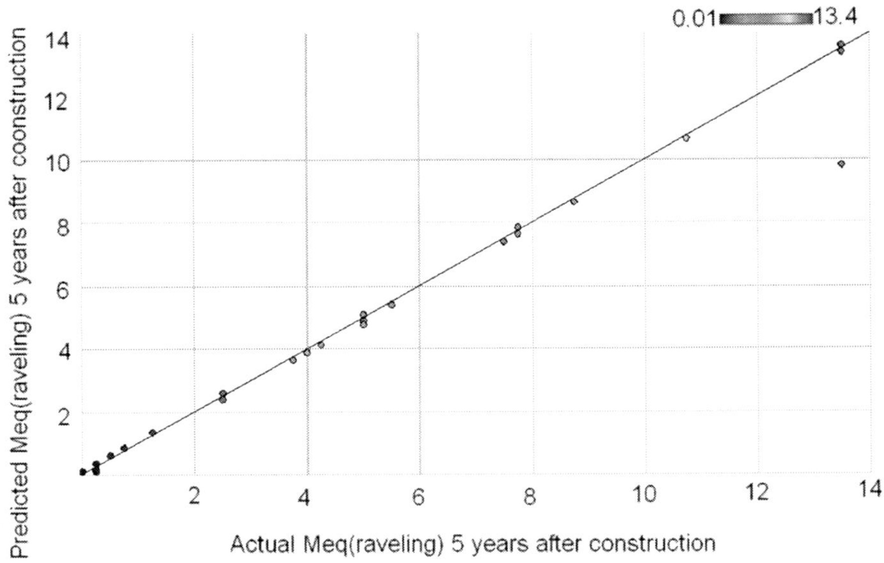

(a)

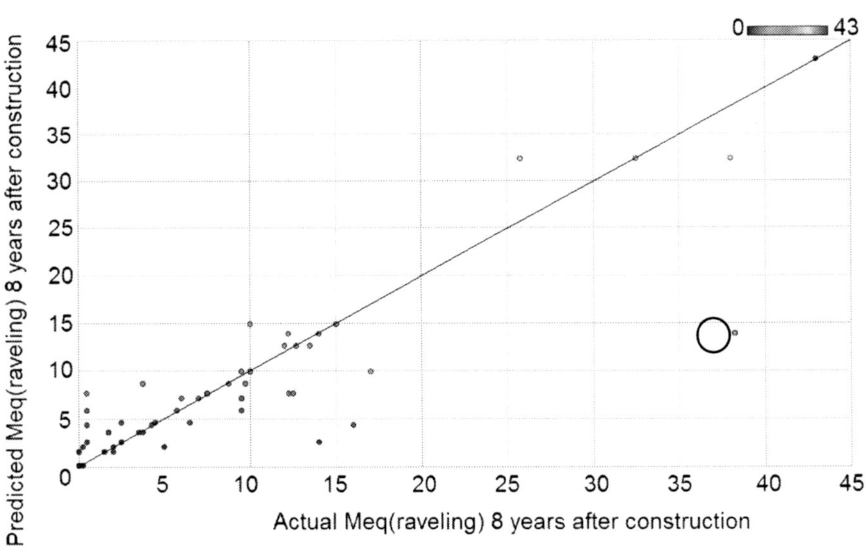

(b)

Fig. 28. Prediction of Meq(raveling) five years after construction by model $Meq5_{Rav}$_SVR (a) and (raveling) eight years after construction by model $Meq8_{Rav}$_SVR (b).

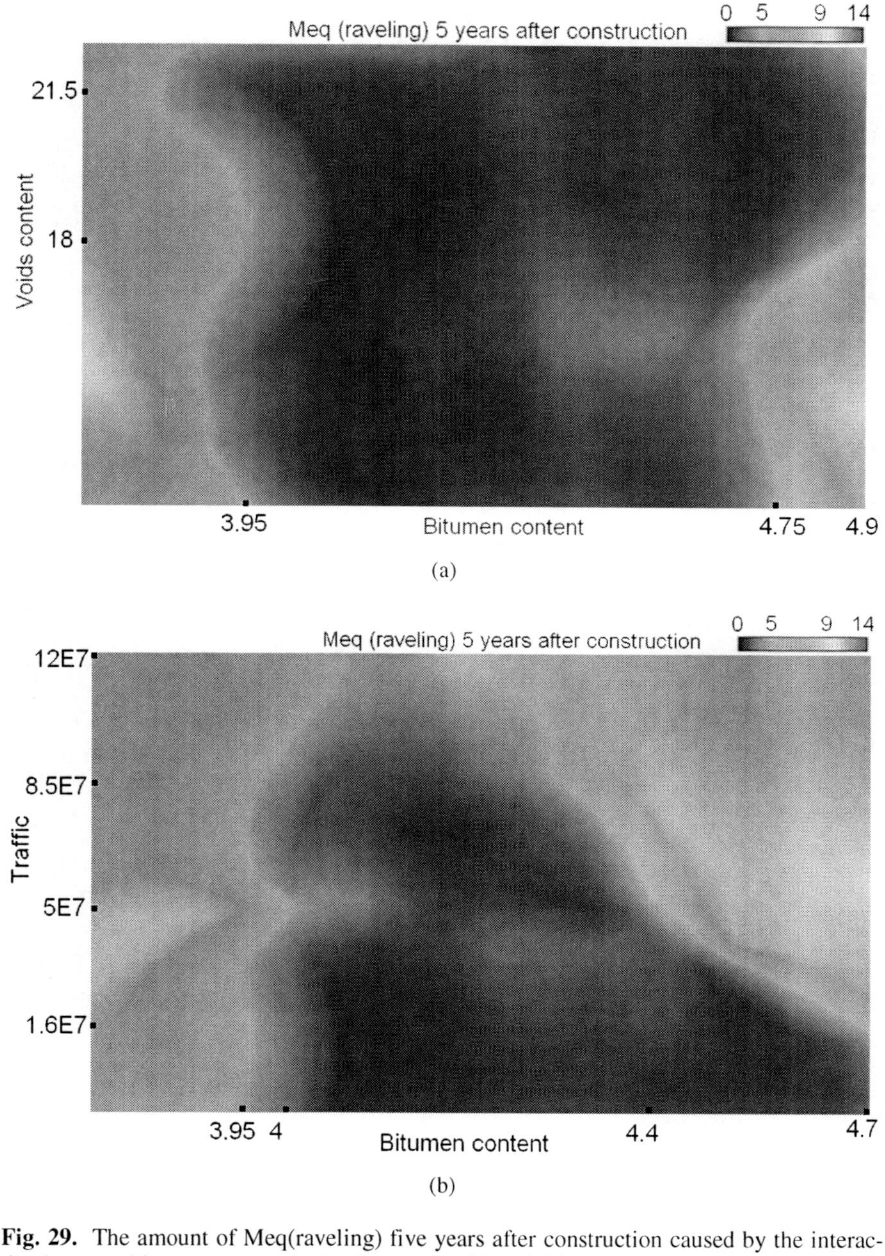

Fig. 29. The amount of Meq(raveling) five years after construction caused by the interaction between bitumen content and voids content (a) and bitumen content and traffic (b).

are shown in Figures 28(a) and 28(b). As was the case in the ANN plots, the x-axis gives the actual output (either M_{eq} five years after construction or M_{eq} eight years after construction) and the y-axis gives the output, predicted by the SVR models.

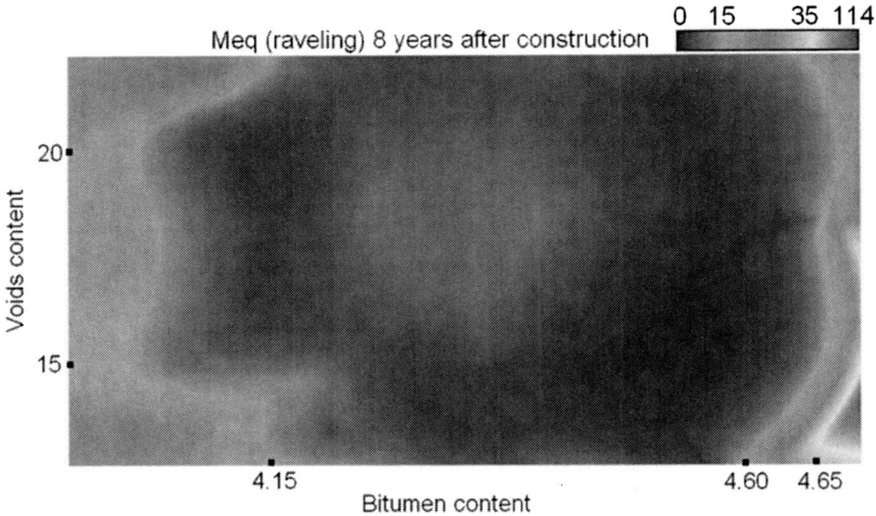

(a)

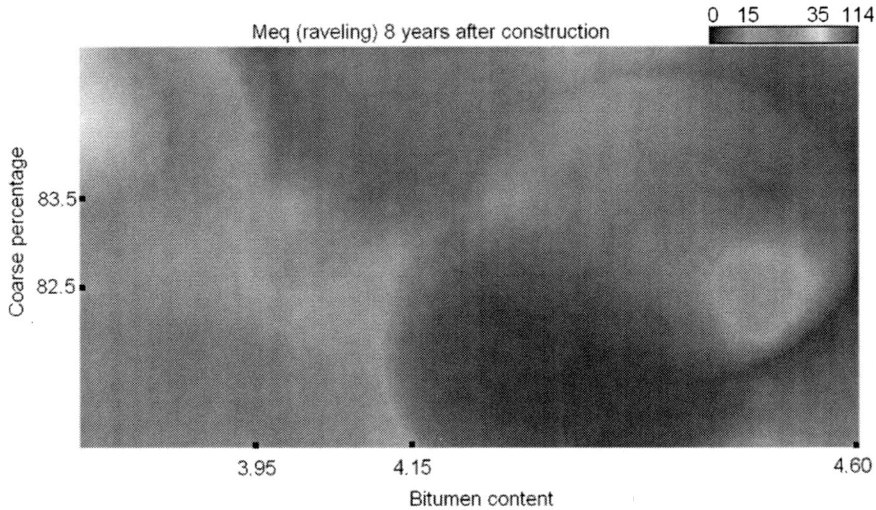

(b)

Fig. 30. The amount of Meq(raveling) eight years after construction caused by the interaction between bitumen content and voids content (a) and bitumen content and percentage of coarse (b).

In Figure 28(b), one data point is predicted very poorly. This data point has been marked with a black circle. It is interesting to know why this data point lies so far from the general pattern of prediction. Looking into the material properties of this data point, it became clear that the data point has a *Bitumen content* of 3.8, a *Void content* of 19.3, a *Coarse percentage* of 84.1, and a number of *Cold days* of

555. The data point has the lowest bitumen content in the whole dataset and therefore this data point is a rather unique one in the dataset. If the model does not have similar examples to learn from, the new combination of input variables is less easy to predict. This is most likely the explanation for the poor prediction of this specific data point.

One of the tools used for interpretation of the results is the color contour, which shows how the interaction between two input variables influences the output variable while other input variables are held constant (the average of that variable in the dataset). By using color contours, it is possible to investigate how much raveling is caused by the interaction between each two input variables. In this way, the values of input variables which cause a large amount of raveling can be identified. For model $Meq5_{Rav}\_SVR$ with five input variables 10 interactions are possible. The number of interactions is 6 for the $Meq8_{Rav}\_SVR$ model with four input variables. Examples of these interactions are given in Figures 29 and 30. The figures are self explaining.

11 Data Mining Using Regression Trees

11.1 Parameter Determination for Regression Trees

Another technique which is used in this dissertation for data mining is regression trees (RT). Regression trees are decision trees generated for regression purposes. The tree structure of these models, especially of the binary trees, is directly interpretable by users in the form of if-then rules. The models developed using regression trees for *Meq* raveling five and eight years after construction are called $Meq5_{Rav}\_RT$ and $Meq8_{Rav}\_RT$. The modeling with RT includes two stages, being the generation of the tree and pruning the tree. It should be determined how far the generated tree should be pruned. This is done using a 10-fold cross validation method for both $Meq5_{Rav}\_RT$ and $Meq8_{Rav}\_RT$ (Figure 31).

11.2 Modeling Using RT

As shown in Figure 31, for raveling five years after construction the optimal tree has six terminal nodes (Figure 31(a)) and the one for raveling eight years after construction has two terminal nodes (Figure 31(b)). As a result, the tree of model $Meq5_{Rav}\_RT$ was pruned until six terminal nodes were present in the tree. This is shown in Figure 32(a). The pruned tree of model $Meq8_{Rav}\_RT$ with three terminal nodes can be seen in Figure 32(b). In regression trees, the input variable on top of the tree is the most important input variable. Figure 19 shows that *Bitumen content* is the most important input variable for raveling of PAC.

11.3 Evaluation/Interpretation of RT Models

Although the structure of the tree is clear and the rules can be discovered by the reader, the generated rules for Meq raveling five years after construction ($Meq5_{Rav}\_RT$) and Meq raveling eight years after construction ($Meq8_{Rav}\_RT$) are given hereafter:

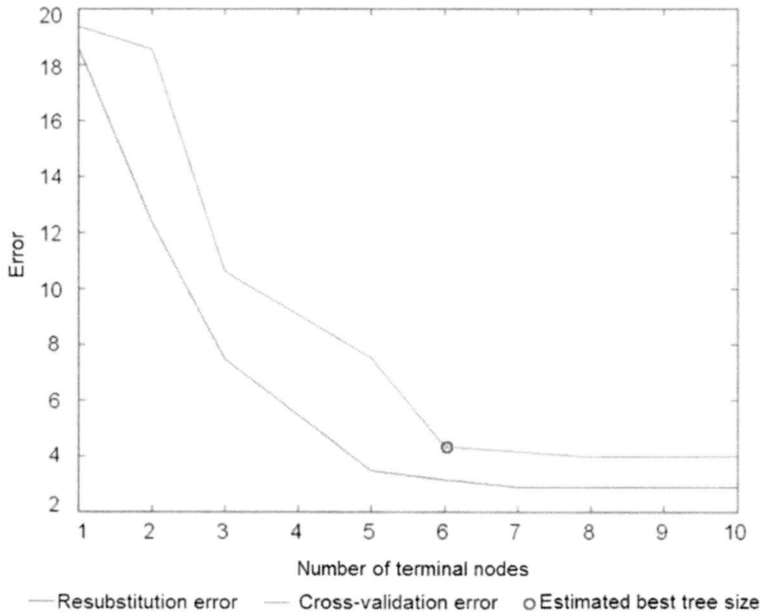

(a)

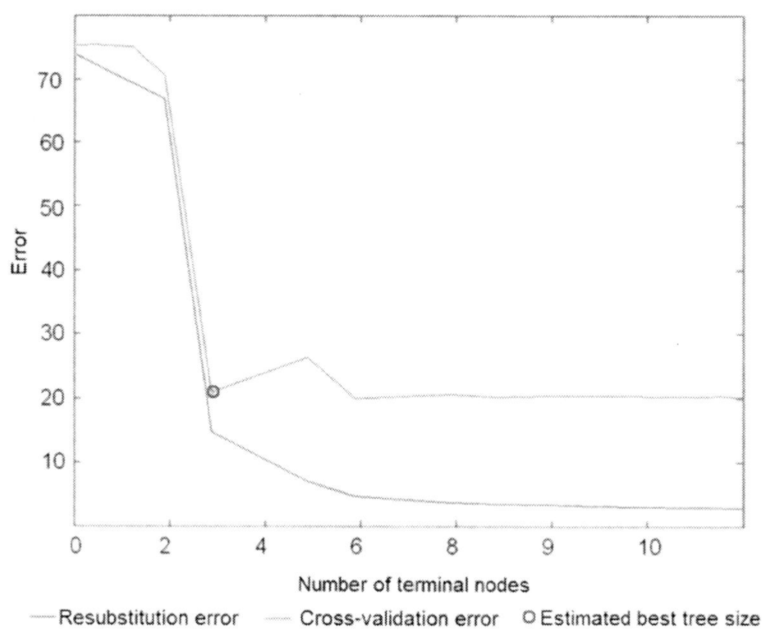

(b)

Fig. 31. The optimal number of terminal nodes for pruning of models $Meq5_{Rav}$_RT (a) and $Meq8_{Rav}$_RT (b).

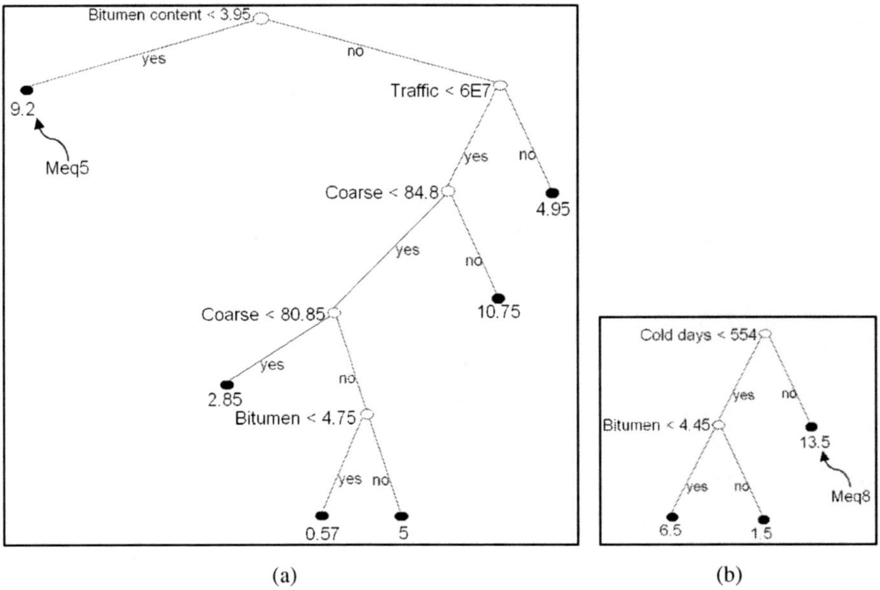

Fig. 32. The optimal pruned tree for models $Meq5_{Rav}$_RT (a) and $Meq8_{Rav}$_RT (b). Note that the terminal nodes (the fold black circles) show the value of Meq.

$Meq5_{Rav}$_RT:

IF *Bitumen content < 3.95* THEN *Meq raveling 5 years after construction = 9.2*
IF *Bitumen content ≥ 3.95* AND *Traffic ≥ 6E7*
THEN *Meq raveling 5 years after construction = 4.95*
IF *Bitumen content ≥ 3.95* AND *Traffic < 6E7* AND *Coarse percentage ≥ 84.8*
THEN *Meq raveling 5 years after construction = 10.75*
IF *Bitumen content ≥ 3.95* AND *Traffic < 6E7* AND *Coarse percentage < 80.85*
THEN *Meq raveling 5 years after construction = 2.85*
IF 3.95 ≤ *Bitumen content < 4.75* AND *Traffic < 6E7* AND *80.85 ≤ Coarse percentage < 84.8*
THEN *Meq raveling 5 years after construction = 0.57*
IF *Bitumen content ≥ 4.75* AND *Traffic < 6E7* AND *80.85 ≤ Coarse percentage < 84.8*
THEN *Meq raveling 5 years after construction = 5*

$Meq8_{Rav}$_RT:

IF *Cold days <554* AND *Bitumen content < 4.45*
THEN *Meq raveling 8 years after construction = 6.5*
IF *Cold days < 554* AND *Bitumen content ≥ 4.45*
THEN *Meq raveling 8 years after construction = 1.5*
IF *Cold days ≥ 554* THEN *Meq raveling 8 years after construction = 13.5*

Table 15. Accuracy of RST classification, upper and lower approximation for model $Meq5_{Rav}$_RST.

Class	Number of data points	Number of lower approximation	Number of higher approximation	Accuracy (Leave-one-out)
NoneLow	51	40	57	74.51%
LowModerate	21	21	15	71.34%

12 Data Mining Using Rough Sets Theory

12.1 Parameter Determination for Rough Sets Theory

This section applies the rough sets theory method to develop models *$Meq5_{Rav}$_RST* and *$Meq8_{Rav}$_RST*. The first step in applying RST is to classify the output variable to discrete classes. Because of the low number of data points for *$Meq5_{Rav}$_RST*, it was decided to classify the output variable to only two classes: *NoneLow* ($0 \leq Meq5 \leq 5$) and *LowModerate* ($5 < Meq5 \leq 13.5$). The reason for choosing these specific classes is that the output needs to be classified into the classes which show the severity of the damage and, at the same time, contain enough data points. This is due to the fact that a class with a low number of data points will not perform well. The upper limit of *Meq* of raveling five years after construction is 13.5 because the maximum value of this variable in the dataset is 13.5.

Due to the large range of the output variable, *Meq* raveling eight years after construction was classified into three discrete calsses, being *NoneLow* ($0 \leq M_{eq}8 \leq 14$), *LowModerate* ($14 < Meq8 \leq 34$), and *ModerateSevere* ($34 < Meq8 \leq 114$).

12.2 Modeling Using Rough Sets Theory

The second step in RST is to calculate the lower and upper approximation for each class. The result of this calculation for *Meq* of raveling five years after construction is shown in Table 15. Next to that, as can be seen in Table 15, the accuracy of classes *NoneLow* and *LowModerate* have been calculated using leave-one-out cross validation. The classification accuracy of class *LowModerate* is lower. This is perhaps because of the low number of data points in this class (21 data points).

Table 16 gives the lower and upper approximation for *Meq* of raveling eight years after construction. As can be seen, the accuracy of classification of all classes was also determined using leave-one-out cross validation.

RST is well suited to identify the most significant input variable by computing *Reducts* and *Core*. Thus, given the data, six *Reducts* were calculated for *Meq* of raveling five years after construction:

R1 = { Bitumen content, %Coarse}
R2 = {Bitumen content, Voids content, %Coarse}
R3 = { Bitumen content, Voids content}
R4 = {Bitumen content, %Coarse, Cold days}
R5 = {Bitumen content, %Coarse, Traffic}
R6 = {Bitumen content, Voids content, Traffic}

Intersecting all *Reducts* leads us to *Core*, which is in this case *Bitumen content.* Then the classification rate using only the Core variable (Bitumen content) was calculated, being 18% (total of both classes). This means although *Bitumen content* is the most important input variable for *Meq* of raveling five years after construction, the other four input variables are still significant for a reasonable quality of the models. The *Reducts* were also generated for *Meq* of raveling eight years after construction, resulting in three *Reducts*:
R1={ Voids content, Bitumen content, Cold days }
R2={Voids content, Coarse}
R3={Voids content, Coarse, Cold days}
The intersection of the Reducts, the *Core*, was *Voids content* for *Meq* of raveling eight years after construction.

12.3 Evaluation/Interpretation of RST Models

In the next step, MODLEM2 algorithm was used to generate a set of if-then rules, i.e., the set does not contain any redundant rules. For *Meq* of raveling five years after construction, the induced set contained 6 rules, where four rules correspond to class *NoneLow* and two rules to class *LowModerate*. All rules were supported by at least four data points. The number of data points supporting a rule is also called the strength of that rule. Rules related to class *NoneLow*, have a minimum strength of six and a maximum of 36. From the two rules related to class *LowModerate*, one has the strength of six and the other one the strength of four. This shows that the rules belonging to class *LowModerate* are less strong rules (supported by less data points). Table 17 shows the rules and their strength.

Table 18 shows the RST rules generated for *Meq* of raveling eight years after construction. As can be seen the maximum and minimum strength of the rules is lower than the one from *Meq* of raveling five years after construction (26 and three comparing to 36 and four). In total, nine rules were generated, five related class

Table 16. Accuracy of RST classification for model $Meq8_{Rav}$_RST.

Class	Number of data points	Number of lower approximation	Number of higher approximation	Accuracy (Leave-one-out)
NoneLow	25	14	36	60.00%
LowModerate	31	18	42	61.29%
ModerateSevere	12	9	18	83.33%

Table 17. RST rules generated for $Meq5_{Rav}$_RST using MODLEM2 algorithm and their strength.

RST rule	Strength
IF *(Bitumen content ≥ 3.95)* AND *(Cold days < 310)* THEN *(Meq5 = NoneLow)*	36
IF *(Bitumen content ≥ 3.95)* AND *(Voids content < 18.3%)* AND *(%Coarse < 84.8)* THEN *(Meq5 = NoneLow)*	33
IF *(Cold days < 310)* THEN *(Meq5 = NoneLow)*	12
IF *(Voids content < 20.6%)* THEN *(Meq5 = NoneLow)*	11
IF*(Traffic < 7.5E7)* THEN *(Meq5 = NoneLow)*	6
IF *(Bitumen content < 3.95)* AND *(Voids content > 20.6%)* THEN *(Meq5 = LowModerate)*	6
IF *(Voids content > 20.6%)* AND *(Cold days > 310)* THEN *(Meq5 = LowModerate)*	4

Table 18. RST rules generated for $Meq8_{Rav}$_RST using MODLEM2 algorithm and their strength.

RST rule	Strength
IF *(Voids content < 20.6)* AND *(Cold days < 554)* THEN *(Meq8 = NoneLow)*	26
IF *(82.5 ≤ Coarse < 83.5)* AND *(Cold days < 554)* THEN *(Meq8 = NoneLow)*	24
IF *(Voids content < 20.6)* THEN *(Meq8 = NoneLow)*	16
IF *(Coarse ≥ 80.7)* AND *(Cold days < 474)* THEN *(Meq8 = NoneLow)*	14
IF *(Coarse < 84.7)* THEN *(Meq8 = NoneLow)*	9
IF *(Bitumen content < 3.95)* AND *(Cold days ≥ 554)* THEN *(Meq8 = LowModerate)*	4
IF *(Voids content ≥ 20.6)* AND *(Coarse < 80.7)* AND *(Cold days ≥ 554)* THEN *(Meq8 = LowModerate)*	3
IF *(Bitumen content < 4.1)* AND *(Coarse < 80.7)* THEN *(Meq8 = LowModerate)*	3
IF *(Voids content ≥ 22)* THEN *(Meq8 = ModerateSevere)*	3

NoneLow, three to *LowModerate,* and one to class *ModerateSevere*. The maximum strength of rules for class *NoneLow* was 26, for class *LowModerate* 4 and for class *ModerateSevere* 3.

13 Summary and Conclusions

This study carried out the process of knowledge discovery for raveling of porous asphalt concrete in the Netherlands. The results were demonstrated in the form of graphs and plots of the mined models for raveling five and eight years after construction.

Table 19. Comparison of results of ANN and SVR models.

Model	Testing error	R-square
$Meq5_{Rav}\_ANN$	0.24	0.95
$Meq5_{Rav}\_SVR$	2.9	0.97
$Meq8_{Rav}\_ANN$	3.26	0.94
$Meq8_{Rav}\_SVR$	6.4	0.87

Table 20. Interpretation of results of ML techniques for raveling five years after construction.

IF	THEN	Method
Bitumen < 4	0 < Meq5 ≤ 10	ANN
Bitumen ≥ 4	Meq5 = 0	ANN
Cold days < 310	Meq5 = 0	ANN
Cold days ≥ 310	0 < Meq5 ≤ 3	ANN
3.95 ≤ Bitumen ≤ 4.75	0 ≤ Meq5 ≤ 5	SVR
4 ≤ Bitumen < 4.4 AND Traffic < 5E7	0 ≤ Meq5 ≤ 5	SVR
3.95 ≤ Bitumen < 4.7 AND Traffic < 1.6E7	0 ≤ Meq5 ≤ 5	SVR
Bitumen < 3.95	6 ≤ Meq5 ≤ 9	SVR
Bitumen ≥ 4.7	6 ≤ Meq5 ≤ 9	SVR
Bitumen < 3.95	Meq5 = 9.2	RT
Bitumen ≥ 3.95 AND Traffic ≥ 6E7	Meq5 = 4.95	RT
Bitumen ≥ 3.95 AND Traffic < 6E7 AND %Coarse ≥ 84.8	Meq5 = 10.75	RT
Bitumen ≥ 3.95 AND Traffic < 6E7 AND %Coarse < 80.85	Meq5 = 2.85	RT
3.95 ≤ Bitumen < 4.75 AND Traffic < 6E7 AND 80.85 ≤ %Coarse < 84.8	Meq5 = 0.57	RT
Bitumen ≥ 4.75 AND Traffic < 6E7 AND 80.85 ≤ %Coarse < 84.8	Meq5 = 5	RT
Cold days < 310	0 ≤ Meq5 ≤ 5	RST
Voids content < 20.6%	0 ≤ Meq5 ≤ 5	RST
Traffic < 7.5E7	0 ≤ Meq5 ≤ 5	RST
Bitumen ≥ 3.95 AND Cold days < 310	0 ≤ Meq5 ≤ 5	RST
Bitumen ≥ 3.95 AND Voids content < 18.3% AND %Coarse < 84.8	0 ≤ Meq5 ≤ 5	RST
Bitumen < 3.95 AND Voids content > 20.6	5 < Meq5 ≤ 13.5	RST
Voids content > 20.6% AND Cold days > 310	5 < Meq5 ≤ 13.5	RST

A detailed explanation of knowledge discovery steps, being data preparation, data mining, and evaluation/interpretation of the results is given. In the data preparation, an extended variable selection was performed to choose a maximum of five input variables. Reduction of the input dimension was needed due to the low number of data points available (after preparation 72 data points). For the data mining step of knowledge discovery, four ML based techniques were used: artificial neural network, support vector machine, regression trees, and rough set theory. The prediction power of ANN and SVR were tested on a small part of the dataset, which is called the test set. The test results of ANN and SVR are summarized in Table 19.

As can be seen in the table, there is not much difference between the ANN and SVR models for *Meq* raveling five years after construction. However, ANN performs better than SVR for *Meq* raveling eight years after construction. The results of the other two techniques (regression tree and rough set theory) were in the form of if-then rules. For evaluation of the models, different tools were employed such as scatter plots, color contours, and response graphs. A summary of the interpretation of the results of all four techniques is given in Table 20 for raveling five years after construction.

Having the results of different techniques reported in Table 20, the question now is "can some common conclusions be drawn by the different methods about any of the input variables?" To answer this, graphs can be made from the results of all techniques about a specific variable. For instance, Figure 33 shows the results of ANN, SVM, RT, and RST for the input variable *Bitumen content*.

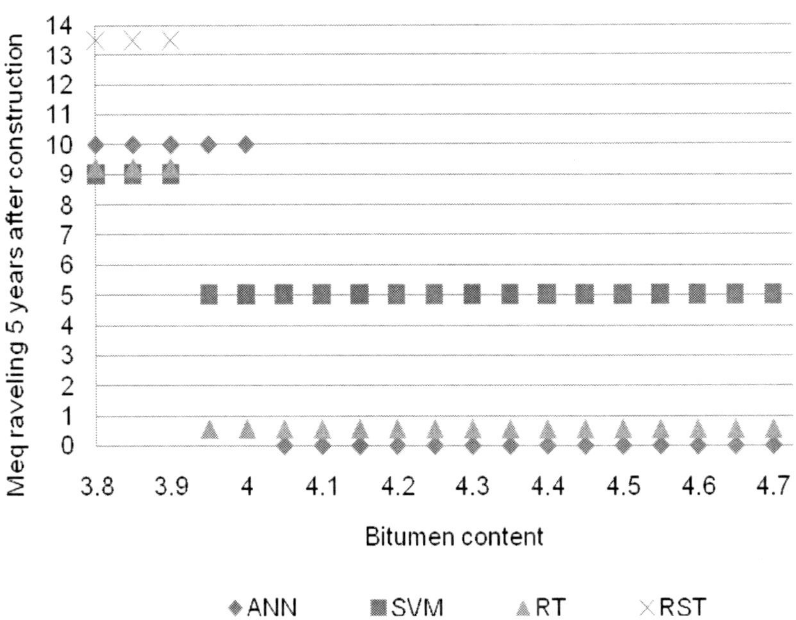

Fig. 33. The result of different methods for the input variable "Bitumen content". The output of all techniques is Meq (raveling) 5 years after construction.

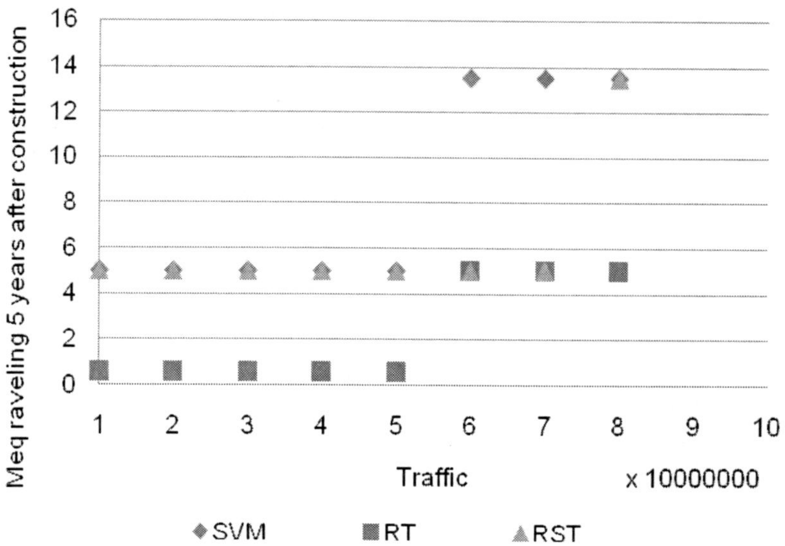

Fig. 34. The result of different methods for the input variable "Traffic for Meq (raveling) 5 years after construction (this values are valid under condition that 3.95 ≤ Bitumen < 4.75).

Figure 33 shows that all techniques agreed on the fact that a *bitumen content lower than 3.95% causes a high amount of raveling during the first five years after construction*. It also can be concluded *that* a *bitumen content between 3.95% and 4.75 causes an amount of raveling that is limited to Meq = 5. Therefore a bitumen content in that range is recommended.*

Figure 34 shows the result of methods SVM, RT, and RST for input variable Traffic intensity. The figure shows that RST and SVM agree that if the bitumen content is between 3.95 and 4.75%, a cumulative traffic intensity five years after construction less than 5E7 will results in a maximum raveling of Meq = 5.

Furthermore, the results of both ANN and RST show that if the cumulative number of cold days < 310, then low amount of raveling five years after construction will occur (Meq between 0 and 5).

It is not possible to give similar graphs for all input variables. These variables appear in combination with other variables (Table 20) and not individually. Table 21 shows the results of models for raveling eight years after construction in the form of if-then rules.

Also for *Meq* raveling eight years after construction (Table 21), the question is again if the results of different techniques imply the same conclusions about any of the input variables.

ANN and SVR show again that *Bitumen content ≤ 3.95 is not recommended.* Next to that, ANN and RST declare that *using a maximum of 20 to 22% of voids content in the PAC mixture can avoid high amount of raveling.* Further, taking the ANN and RT

Table 21. Rules generated by different methods for raveling eight years after construction.

IF part of the rule	THEN part of the rule	Method
Bitumen < 3.95%	$12 \leq$ Meq8 < 38	ANN
3.95% ≤ Bitumen < 4.60%	$0 \leq$ Meq8 ≤ 12	ANN
Voids content > 20%	$10 \leq$ Meq8 ≤ 21	ANN
82.5 ≤ %Coarse < 83.5	Meq8 ≈ 8	ANN
%Coarse > 83.5	$13 \leq$ Meq8 ≤ 35	ANN
Cold days ≥ 554	$2 \leq$ Meq8 ≤ 64	ANN
Cold days < 554	Meq8 ≈ 2	ANN
4.15% ≤ Bitumen ≤ 4.60%	$0 \leq$ Meq8 ≤ 14	SVR
82.5 ≤ %Coarse < 83.5 AND Bitumen > 3.95	$0 \leq$ Meq8 ≤ 14	SVR
Cold days ≥ 554 AND Bitumen < 4.10	$15 \leq$ Meq8 ≤ 34	SVR
Cold days < 554 AND Bitumen > 3.95	$0 \leq$ Meq8 ≤ 14	
Bitumen < 3.95%	$15 \leq$ Meq8 ≤ 34	SVR
Bitumen > 4.65%	$35 \leq$ Meq8 ≤ 114	SVR
Cold days < 554 AND Bitumen < 4.45	Meq8 = 6.5	RT
Cold days < 554 AND Bitumen ≥ 4.45	Meq8 = 1.5	RT
Cold days ≥ 554	Meq8 = 13.5	
Voids content < 20.6 AND Cold days < 554	$0 \leq$ Meq8 ≤ 14	RST
82.5 ≤ %Coarse < 83.5 AND Cold days < 554	$0 \leq$ Meq8 ≤ 14	RST
Voids content < 20.6	$0 \leq$ Meq8 ≤ 14	RST
%Coarse ≥ 80.7 AND Cold days < 474	$0 \leq$ Meq8 ≤ 14	RST
%Coarse < 84.7	$0 \leq$ Meq8 ≤ 14	RST
Bitumen content < 3.95 AND Cold days ≥ 554	$14 <$ Meq8 ≤ 34	RST
Voids content ≥ 20.6 AND %Coarse < 80.7 AND Cold days ≥ 554	$14 <$ Meq8 ≤ 34	RST
Bitumen content < 4.1 AND %Coarse < 80.7	$14 <$ Meq8 ≤ 34	RST
Voids content ≥ 22	$34 <$ Meq8 ≤ 114	RST

rules into account, it can be concluded that *Cold days ≥ 554 results in moderate raveling*. Finally, it seems that ANN, SVR, and RST recommend *the percentage of coarse material to be between 82.5 and 83.5 in order to keep raveling low.*

In general, the presence of the input variable *Bitumen content* in the results of all techniques for both raveling five and eight years after construction shows the importance of this input variable for raveling. Another general point to notice is that *Traffic* is strongly present in the first five years after construction but is less important for eight years after construction. This may imply that heavy traffic causes raveling mainly in the first few years of the lifespan of the porous asphalt. It was also noticeable that a high number of *Cold days* after both models of five and eight years after construction of PAC layers cause raveling. Finally, the results showed that an optimum *Coarse percentage* will avoid an excessive amount of raveling.

Acknowledgment

Ir. Marc Eijbersen from the national Information and Technology Platform for Transport, Infrastructure and Public space (CROW) made the SHRP-NL database available to us. His kind contribution is highly appreciated. we also want to extend our gratitude to the Ministry of Transport, Public Works, and Water Management, Dienst Verkeer en Scheepvaart (DVS) (formerly DWW) for their financial support.

References

Abdallah, I., Ferregut, C., Melchor-Lucero, O., Nazarian, S.: Stiffness properties of composite pavements using artificial neural network-based methodologies. 0-1711, Centre for Highway Material research, The University of Texas at El Paso, El Paso (2001)

Abonyi, J., Feil, B., Abraham, A.: Computational Intelligence in Data Mining. Informatica 29, 3–12 (2005)

Apte, C., Hong, S.J.: Predicting Equity Returns from Securities Data with Minimal Rule Generation. Advances in Knowledge Discovery and Data mining, 514–560 (1996)

Attoh-Okine, N.O.: Combining Use of Rough Set and Artificial Neural Networks in Doweled-Pavement-Performance Modeling—A Hybrid Approach. Journal of Transportation Engineering 28(3) (2002)

Aultman-Hall, L., Jackson, E., Dougan, C.E., Choi, S.-N.: Models relating pavement quality measures. Transportation research record, 119–125 (2004)

Avila, C., Shiraishi, Y., Tsuji, Y.: Crack width prediction of reinforced concrete structures by artificial neural networks. In: 7th Seminar on Neural Network Applications in Electrical Engineering, Belgrade, Serbia and Montenegro, pp. 39–44 (2004)

Bayrak, M.B., Teomete, E., Agarwal, M.: Use of Artificial neural network for predicting rigid pavement roughness. In: Midwest Transportation Consortium, Fall Student Conference, Ames, Iowa (2004)

Bishop, C.M., Tipping, M.E.: Bayesian regression and classification. NATO Science Series III: Computer & Systems Sciences. IOS Press, Amsterdam (2003)

Bosurgi, G., Trifirò, F., Xibilia, M.G.: Artificial Neural Network for Predicting Road Pavement Conditions. In: 4th International SIIV Congress, Palermo, Italy (2007)

Brachman, R.J., Anand, T.: The Process of Knowledge Discovery in Databases: A First Sketch. In: KDD Workshop (1994)

Bray, J., Verma, B., Li, X., He, W.: A Neural Network based Technique for Automatic Classification of Road Cracks. In: International Joint Conference on Neural Networks, Vancouver, BC, Canada, pp. 907–912 (2006)

Bredenhann, S.J., van de Ven, M.F.C.: Application of Artificial Neural Networks in the Back-calculation of Flexible Pavement Layer Moduli from Deflection Measurements. In: Proceedings of the 8th Conference on Asphalt Pavements for Southern Africa (CAPSA 2004), Sun City, South Africa (2004)

Butt, A.A., Shahin, M.Y., Carpenter, S.H., Carnahan, J.V.: Application of Markov Process to Pavement Management System at Network Level. In: Third International Conference on Managing Pavements, San Antonio, Texas, pp. 159–172 (1994)

Carey, W.N., Irick, P.E.: The pavement servicibility-performance cencept, Bulletin 250, Washington DC, pp. 40–58 (1960)

Ceylan, H., Guclu, A., Tutumluer, E., Thompson, M.R.: Use of Artificial Neural Networks for Analyzing Full Depth Asphalt Pavements. In: TRB 2005, Annual Meeting, Washington DC, USA (2005a)

Ceylan, H., Guclu, A., Tutumluer, E., Thompson, M.R.: Backcalculation of full-depth asphalt pavement layer moduli considering nonlinear stress-dependent subgrade behavior. International Journal of pavement Engineering 6(3), 171–182 (2005b)

Ceylan, H., Gopalakrishnan, K., Guclu, A.: Nonlinear Pavement Analysis Using Artificial Neural Network Based Stress-Dependent Models. In: 86th Annual Meeting of Transportation Research Board, Washington DC (2007)

Chang, J.-R., Tzeng, G.-H., Hung, C.-T., Lin, H.-H.: Non-Additive Fuzzy Regression Applied to Establish Flexible Pavement Present Serviceability Index. In: The IEEE International Conference on Fuzzy Systems, pp. 1020–1025 (2003)

Chang, J.-R., Hung, C.-T.u., Tzeng, G.-W., Lin, J.-D.: Non-additive Grey Relational Model: Case Study on. Evaluation of Flexible Pavement. In: FUZZ-IEEE Budapest, Hungary (2004)

Chang, J., Hung, C., Chen, D.: Application of An Artificial Neural Network on Depth to Bedrock Prediction. International Journal of Computational Intelligence Research 2(1), 33–39 (2006)

Choi, J., Adams, T.M., Bahia, H.U.: Pavement roughness modeling using back-propagation neural network. Computer-Aided Civil and Infrastructure Engineering 19, 295–303 (2004)

Chou, J., O'Neill, W.A., Cheng, H.D.: Pavement Distress Classification Using Neural Networks. In: IEEE International Conference on Systems, Man, and Cybernetics, pp. 397–401 (1994)

Cios, K.J., Pedrycz, W., Swiniarski, R.W., Kurgan, L.A.: Data Mining. A Knowledge Discovery Approach. Springer, New York (2007)

CROW, Manual global visual inspection (in Dutch), National Information and Technology Platform for Transport, Infrastructure and Public space, Ede (2005)

Demir, F.: Prediction of elastic modulus of normal and high strength concrete by artificial neural networks. Construction and Building Materials (2007) (in Press)

DWW. PAC. Ministry of Transport, Public Works and Water management, Road and Hydraulic Engineering Division, Delft, Website (2005) (in Dutch)

Dy, J.G., Brodley, C.E.: Feature Selection for Unsupervised Learning. Journal of Machine Learning Research 5, 845–889 (2004)

Eldin, N.N., Senouci, A.B.: Use of Neural Network for Condition rating of Jointed Concrete Pavements. Advances in Engineering Software 23, 133–141 (1996)

Engelbrecht, A.P.: Computational Intelligence: An Introduction. Wiley, Chichester (2007)

Fayyad, U., Piatetsky-Shapiro, G., Smyth, P.: From Data Mining to Knowledge Discovery in Databases. American Association for Artificial Intelligence (1996)

Ferregut, C., Abdallah, I., Melchor-Lucero, O., Nazarian, S.: Artificial Neural Networks Based Methodologies for Rational Assessment of Remaining Life of Existing Pavements. Texas Department of Transportation, Austin, TX (1999)

Geipot, Research on the Interrelationships Between Costs of Highway Construction, Final Report, 12 volumes. Maintenance and Utilisation (PICR), Brasilia, Brazil (1982)

Goktepe, A.B., Agar, E., Lav, A.H.: Comparison of Multilayer Perceptron and Adaptive Neuro-Fuzzy System on Backcalculating the Mechanical Properties of Flexible Pavements. ARI The Bulletin of the Istanbul Technical University 54(3) (2004)

Goktepe, A.B., Altun, S.: Artificial intelligence application in the backcalculation of the mechanical properties of flexible pavements (2006)

Gopalakrishnan, K., Thompson, M.R., Manik, A.: Rapid Finite Element Based Airport Pavement Moduli Solutions Using Neural Networks. Int. J. of Computational Intelligence 3(1), 63–71 (2006)

Guclu, A., Ceylan, H.: Condition Assessment of Composite Pavement Systems Using Neural-Network-Based Rapid Backcalculation Algorithms. In: Transportation Research Board 86th Annual Meeting (2007)

Haykin, S.: Neural Networks: A Comprehensive Foundation, 2nd edn. Prentice-Hall, Englewood Cliffs (1999)

Hoffman, P.C., Chou, K.C.: Infrastructure assessment: fuzzy regression with neural networks. In: Proceedings of the First International Joint Conference of the North American Fuzzy Information Processing Society Biannual Conference. The Industrial Fuzzy Control and Intelligent Systems Conference, and the NASA Joint Technolo, San Antonio, TX, USA, pp. 273–274 (1994)

Hofman, R., Fafié, J.J., Sule, M.S., Hooghwerff, J., Kegel, J.C., Langebach, W.J., Hermens, P.: IPG-advice, application of two layer porous asphalt concrete on Dutch highway network - Part II. In: DWW-2005-031, Ministry of Transport, Public Works and Water management, Delft (2005) (in Dutch)

Hong, H.P., Wang, S.S.: Stochastic Modeling of Pavement Performance. International Journal of Pavement Engineering 4(4), 235–243 (2003)

Hsu, D.S., Tsai, C.H.: Reinforced concrete structural damage diagnosis by using artificial neural network. In: IASTED International Conference on Intelligent Information Systems (IIS 1997), vol. 149 (1997)

Huang, C.C.: Development of Freeway Pavement Performance Prediction Model Using Markov Chain, Tamkang University (1997)

Huang, C., Najjar, Y.M., Romanoschi, S.: Predicting the Asphalt Concrete Fatigue Life Using Artificial Neural Network Approach. In: TRB 2007 Annual Meeting, Washington DC, USA (2007)

Jang, J.-S.R., Sun, C.-T., Mizutani, E.: Neuro-Fuzzy and Soft Computing - A Computational Approach to Learning and Machine Intelligence. Prentice-Hall, NJ (1997)

Karan, M.A., Haas, R.: Determining Investment Priorities for Urbana Pavement Improvements. Journal of Assoc. of Asphalt Paving Technology 45 (1976)

Karlaftis, M., Loizos, A.: Neural Networks and Nonparametric Statistical Models: Comparative Analysis in Pavement Condition Assessment. In: Proceedings of the 85th Transportation Research Board Annual Meeting, Washington D.C., U.S.A (2006)

Kaur, D., Chou, E.: Applying Neuro-Fuzzy Techniques for Intelligent Highway Pavement Performance Prediction Modlel. In: 42nd Midwest Symposium on Circuits and Systems, Las Cruces, NM, USA (1999)

Kaur, D., Tekkedil, D.: Fuzzy Expert System for Asphalt Pavement Performance Prediction. In: IEEE Intelligent Transportation Systems, Dearborn (MI), USA (2000)

Kaur, D., Pulugurta, H.: Fuzzy decision tree based approach to predict the type of pavement repair. In: Proceedings of the 7th Conference on 7th WSEAS International Conference on Applied Informatics and Communications, Vouliagmeni, Athens, Greece (2007)
Kim, Y.R., Lee, Y.-C., Ranjithan, S.: Flexible Pavement Condition Evaluation Using Deflection Basin Parameters and Dynamic Finite Element Analysis Implemented by Artificial Neural Networks. In: Tayabji, S.D., Lukanen, E.O. (eds.) Nondestructive Testing of Pavements and Backcalculation of Modili, vol. 17, American Society for Testing and Materials, West Conshohocken (2000)
Kira, K., Rendell, L.: A practical approach to feature selection. In: International Conference on Machine Learning, Aberdeen, pp. 368–377 (1992)
Kononenko, I., Kukar, M.: Machine learning and data mining: introduction to principles and algorithms. Horwood publishing, Chichester (2007)
Lea, J., Harvey, J.T.: Data Mining of the Caltrans Pavement Management System (PMS) Database. California Department of Transportation, Richmond, CA (2004)
Lee, B.J., Lee, H.D.: A Robust Position Invariant Artificial Neural Network for Digital Pavement Crack Analysis. In: TRB 2003 Annual Meeting, Washington, D.C., USA (2003)
Lee, Y., Liu, Y., Ker, H.: Application of Modern Regression Techniques and Artifcial Neural Networks To Pavement Prediction Modeling. In: 86th Annual Meeting of the Transportation Research Board, 2007, Washington D.C (2007)
Li, N., Xie, W.C., Haas, R.: Reliability-based processing of Markov chains for modeling pavement network deterioration. Transportation research record, 203–213 (1996)
Loia, V., Sessa, S., Staiano, A., Tagliaferri, R.: Merging Fuzzy Logic, Neural Networks, and Genetic Computation in the Design of a Decision-Support System. International Journal of Intelligent Systems 15, 575–594 (2000)
Loizos, A., Georgiou, P., Plati, C.: Assessment of Asphalt Pavement Remaining Life using Artificial Neural Network Modelling. In: Advanced Characterisation of Pavement and Soil Engineering Materials, pp. 993–1002 (2007)
Lou, Z., Lu, J.J., Gunaratne, M.: Road surface crack condition forecasting using neural network models College of Engineering, University of South Florida (1999)
Lytton, R.L., Michalak, C.H., Scullion, T.: The Texas flexible pavement system. In: Fifth International Conference on Structural Design of Asphalt Pavements, The University of Michigan and the Delft University of Technology (1982)
Maertens, K., Baerdemaeker, J.D., Babuska, R.: Genetic polynomial regression as input selection algorithm for non-linear identification. Soft Computing 10(9), 785–795 (2006)
Meerkerk, A.J.J.: Variation in Quality during the Construction of PAC, Master Thesis, Delft University of Technology, Delft (2004) (in Dutch)
Mei, X., Gunaratne, M., Lu, J.J., Dietrich, B.: Neural Network for Rapid Depth Evaluation of Shallow Cracks in Asphalt Pavements. Computer-Aided Civil and Infrastructure Engineering 19(3), 223–230 (2004)
Meier, R.W., Rix, G.J.: Backcalculation of Flexible Pavement Moduli from Dynamic Deflection Basins Using Artificial Neural Networks. Transportation Research Record No. 1473, 72-81 (1995)
Meignen, D., Bernadet, M., Briand, H.: One application of neural networks for detection of defects using video data bases: identification of road distresses. In: Eighth International Workshop on Database and Expert Systems Applications, Toulouse, France, pp. 459–464 (1997)

Molenaar, A.A.A., Meerkerk, A.J.J., Miradi, M., van der Steen, T.: Performance of Porous Asphalt Concrete. Journal of the association of asphalt paving technologists 75, 1053–1094 (2006)

Molenaar, A.A.A., Miradi, M.: Development of a Maintenance Planning Model for Motorways Based on an Artificial Neural Network. Delft University of Technology (2004)

Mu-yu, L., Shao-yi, W.: Genetic optimization method of asphalt pavement based on rutting and cracking control. Journal of Wuhan University of Technology–Materials Science Edition 18(1), 72–75 (2003)

Nakatsuji, T., Miyasaka, J., Kawamura, A., Shirakawa, T.: Discriminant Analyses of Winter Road Surface Conditions Using Vehicular Motion Data Based on Artificial Intelligence Techniques. In: TRB 2005 Annual Meeting, Washington DC, USA (2005)

Ozbay, K., Laub, R.: Models for Pavement Deterioration Using LTPP. New Jersey Department of Transportation Division of Research and Technology and U.S. Department of Transportation Federal Highway Administration, New Jersy (2001)

Ozsahin, T.S., Oruc, S.: Neural network model for resilient modulus of emulsified asphalt mixtures. Construction and Building Materials (2007) (in press)

Parsley, L.L., Robinson, R.: The TRRL, road investment model for developing countries (RTIM2), TRRL Laboratory Report 1057. Transport and Road Research Laboratory, Crowthorne, UK (1982)

Pawlak, Z.: Rough Sets: Theoretical Aspects of Reasoning About Data. Kluwer Academic Publishing, Dordrecht (1991)

Pekcan, O., Tutumluer, E., Thompson, M.R.: Analyzing pavements on lime-stabilized soils with artificial neural networks. In: 2007 Advanced Characterisation of Pavement and Soil Engineering Materials, London (2007)

Rababaah, H., Vrajitoru, D., Wolfer, J.: Asphalt Pavement Crack Classification: A Comparison of GA, MLP, and SOM. In: Genetic and Evolutionary Computation Conference (2005)

Rakesh, N., Jain, A.K., Reddy, M.A., Reddy, K.S.: Artificial neural networks - genetic algorithm based model for backcalculation of pavement layer moduli. International Journal of Pavement Engineering 7(3), 221–230 (2006)

Reddy, M.A., Reddy, K.S., Pandey, B.B.: Selection of Genetic Algorithm Parameters for Backcalculation of Pavement Moduli. International Journal of Pavement Engineering 5(2) (2004)

Renze, J.: Outlier. In: Weisstein, E.W. (ed.) MathWorld–A Wolfram Web Resource (2008), `http://mathworld.wolfram.com/Outlier.html`

Rijkswterstaat, Verkeersgegevens 1980, Ministerie van Verkeer en Waterstaat Directoraat-Generaal Rijkswaterstaat Adviesdienst Verkeer en Vervoer (AVV), Den Haag (1980)

Rijkswterstaat, Verkeersgegevens 1981, Ministerie van Verkeer en Waterstaat Directoraat-Generaal Rijkswaterstaat Adviesdienst Verkeer en Vervoer (AVV), Den Haag (1981)

Rijkswterstaat, Verkeersgegevens 1982, Ministerie van Verkeer en Waterstaat Directoraat-Generaal Rijkswaterstaat Adviesdienst Verkeer en Vervoer (AVV), Den Haag (1982)

Rijkswterstaat, Verkeersgegevens 1983, Ministerie van Verkeer en Waterstaat Directoraat-Generaal Rijkswaterstaat Adviesdienst Verkeer en Vervoer (AVV), Den Haag (1983)

Rijkswterstaat, Verkeersgegevens 1984, Ministerie van Verkeer en Waterstaat Directoraat-Generaal Rijkswaterstaat Adviesdienst Verkeer en Vervoer (AVV), Den Haag (1984)

Rijkswterstaat, Verkeersgegevens 1985, Ministerie van Verkeer en Waterstaat Directoraat-Generaal Rijkswaterstaat Adviesdienst Verkeer en Vervoer (AVV), Den Haag (1985)

Roberts, C.A., Attoh-Okine, N.O.: Comparative Analysis of Two Artificial Neural Networks using Pavement Performance Prediction. Computer Aided Civil and Infrastructure Engineering 13(5), 339–348 (1998)

Saltan, M., Tigdemir, M., Karashin, M.: Artificial Neural Network Application for Flexible Pavement Thickness Modeling. Turkish J. Eng. Env. Sci. 26, 243–248 (2002)

Saltan, M., Sezgin, H.: Hybrid neural network and finite element modeling of sub-base layer material properties in flexible pavements. Materials & Design 28(5), 1725–1730 (2006)

Saltan, M., Terzi, S.: Modeling deflection basin using artificial neural networks with cross-validation technique in backcalculating flexible pavement layer moduli. Advances in Engineering Software (2007) (in Press)

Sweere, G.T.H., Zwieten, J., Eijbersen, M.J., Huipen, H.: Wegverhardingen op termijn bekeken - Technische Verslag SHRP-NL peiode 1990-1995, CROW, Ede (1996)

Tarefder, R.A., White, L., Zaman, M.: Development and Application of A Rut prediction model for flexible Pavement. TRB 2005, USA (2005)

Terzi, S., Saltan, M., Yildirim, T.: Optimization of The Deflection Basin By Genetic Algorithm And Neural Network Approach. In: Kaynak, O., Alpaydın, E., Oja, E., Xu, L. (eds.) ICANN 2003 and ICONIP 2003. LNCS, vol. 2714. Springer, Heidelberg (2003)

Terzi, S.: Modeling the Pavement Present Serviceability Index of Flexible Highway Pavements Using Data Mining. Journal of Applied Science 6(1), 193–197 (2006)

Terzi, S.: Modeling the pavement serviceability ratio of flexible highway pavements by artificial neural networks. Construction and Building Materials 21, 590–593 (2007)

Thube, D.T., Thube, A.D.: An alternative approach for modelling and simulation of pavement deterioration models: Artificial neural networks. In: TRB 2007 Annual Meeting, Washington DC, USA (2007)

Way, G.B., Eisenberg, J.: Pavement Management System for Arzona Phase II: Verificiation of Performance Prediction Models and Developement of Database, Arizona Department of Transportation (1980)

Weiss, S.I., Kulikowski, C.: Computer systems that learn: Classification and prediction methods from statistics, neural network, machine learning and expert systems. Morgan Kaufmann, San Francisco (1991)

Xiao, W., Yan, X., Zhang, X.: Pavement Distress Image Automatic Classification Based on DENSITY-Based Neural Network. Rough Sets and Knowledge Technology. In: First International Conference, RSKT 2006, Chongquing, China, pp. 685–692 (2006)

Yang, J., Lu, J.J., Gunaratne, M.: Application of neural network models for forecasting of pavement crack index and pavement condition rating. Florida Deprtment of Transportation, Tampa, Florida (2003)

Backcalculation of Pavement Layer Thickness and Moduli Using Adaptive Neuro-fuzzy Inference System

Mehmet Saltan[1] and Serdal Terzi[2]

[1] S.Demirel University, Engineering and Architectural Faculty, Civil Engineering Department, 32260 Isparta/Turkey
msaltan@mmf.sdu.edu.tr

[2] S. Demirel University, Technical Education Faculty, Structural Education Department, 32260 Isparta/Turkey
sterzi@tef.sdu.edu.tr

Abstract. Efficient and economical methods are important in determination of the structural properties of the existing flexible pavements. An important pavement monitoring activity performed by most highway agencies is the collection and analysis of deflection data. Pavement deflection data are often used to evaluate a pavement's structural condition non-destructively. It is essential not only to evaluate the structural integrity of an existing pavement but also to have accurate information on pavement structural condition in order to establish a reasonable pavement rehabilitation design system. Pavement structural adequacy is often evaluated by calculating elastic modulus of each layer using the so-called "backcalculation". Backcalculating the pavement layer properties is a well-accepted procedure for the evaluation of the structural capacity of pavements. The ultimate aim of the backcalculation process from Nondestructive Testing (NDT) results is to estimate the pavement material properties. Using backcalculation analysis, flexible pavement layer thicknesses together with *in-situ* material properties can be backcalculated from the measured field data through appropriate analysis techniques. In this study, adaptive neural based fuzzy inference system (ANFIS) is used in backcalculating the pavement layer thickness and moduli from deflections measured on the surface of the flexible pavements. Experimental deflection data groups from NDT are used to show the capability of the ANFIS approaches in backcalculating the pavement layer thickness and moduli, and compared each other.

Keywords: Backcalculation, Nondestructive Testing, Adaptive neural based fuzzy inference system, Flexible pavements.

1 Introduction

Highway pavements are generally constructed in the form of flexible pavements. Flexible pavements are layered systems with better materials on top and inferior materials at the bottom. Starting from the top, the pavement consists wearing course, base and sub-base layers. The base material may be a bituminous mix or a granular material, depending on the number of heavy vehicles on the considered

K. Gopalakrishnan et al. (Eds.): Intel. & Soft Comp. in Infra. Sys. Eng., SCI 259, pp. 177–204.
springerlink.com

section of the road. However, local and cheaper materials can be used as a subbase layer on top of the subgrade (Huang, 1993). Repeated applications of vehicle loads, weather conditions and other factors decrease the serviceability of the pavement. For this reason, a maintenance program should be set up to decide when and where to carry out maintenance works. The most difficult aspect is to determine the remaining life of the pavement. In order to determine the remaining life, the pavement should be analyzed structurally with material properties for each layer being elastic modulus, Poisson's ratio and thickness of layer.

The response of a multilayered flexible pavement to the load of a vehicle is complex. The variety of possible geometries, materials, layer interfaces, loads, weather effects, etc. make the prediction of pavement behavior a challenging task. In order to study the response of the entire sites are usually instrumented with deflection, strain, pressure, temperature and humidity sensors. Data obtained from *in situ* measurements are used to validate and develop theoretical models. Using the models it is possible to predict the response of pavements under different conditions, thus helping to improve design, extended pavement life and reduce maintenance costs (Arraigada et al., 2008). Due to the combined influence of traffic loading and environment, the structural condition of pavement deteriorates with time. Studies on structural damage, which usually result in functional deterioration are important for the highway engineer to plan the maintenance/strengthening strategies to extend the life of pavements, and provides valuable information to a highway agency for proper work planning and budged allocation (Reddy and Veeraragavan, 1997; Madanat et al., 2002). One of the most important tasks of highway officials and engineers is the maintenance of deteriorating existing highway system. Deflection has the virtue of being much simpler to measure and would be expected to show a broad correlation with performance. Due to increased magnitude of wheel load, tire pressure, and traffic load repetitions, the pavement deterioration starts taking place much earlier than the anticipated design life and steps have to be undertaken the existing pavement with suitable overlay material and thickness so as to extend the service life to the required period. The appropriate timing to carry out maintenance/rehabilitation is crucial; if delayed, the road structure may fail beyond any scope of restoration (Reddy and Veeraragavan, 1997). There are two major modes of flexible pavement structural failure – fatigue cracking and rutting. These failure modes are a result of tensile strain cycles in the asphalt concrete layer and stress cycles on the subgrade surface. The pavement structural capacity deteriorates with time (or traffic) due to the fatigue of pavement materials and other types of pavement distress (Hassan et al., 2003).

Nowadays, applied pavement is the most popular area of engineering science. During these days, guesswork has been replaced by reliable methods for evaluating, designing, managing, and maintaining pavements. Precise Nondestructive Testing (NDT) equipments have been developed for determining pavement structural condition. Hence, we can say that pavement construction and maintenance practices based on trial and error or tradition have been replaced by science. Pavement design and management can now be based on factual information gathered with specialized equipment.

Among nondestructive deflection measurement methods, commercially available devices are the Benkelman Beam, Dynaflect, Road Rater, Falling Weight Deflectometer (FWD), Rolling Weight Deflectometer (RWD) and Rolling Dynamic Deflectometer (RDD). In the last 15-20 years, the Benkelman Beam and Dynaflect have been applied successfully to many projects all over the world (Chang et al., 2002).

In recent years, one of the most important and promising research field has been "Heuristics from Nature", an area utilizing some analogies with natural or social systems and using them to derive non-deterministic heuristic methods and to obtain very good results. Artificial Neural Network (ANN) and Fuzzy Logic Approach (FLA) methods are among the heuristic methods.

Artificial neural networks (ANN) are valuable computational tools that are increasingly being used to solve resource-intensive complex problems as an alternative to using more traditional techniques (Ceylan *et al.,* 2004). Ceylan employed ANN in the analysis of concrete pavement systems and developed ANN-based design tools that incorporated the state-of-the-art finite element solutions into routine practical design at several orders of magnitude faster than those sophisticated finite element programs (Ceylan, 2002). Meier and Rix (1994 and 1995) and Meier *et al.* (1999) firstly attempted to backcalculate the pavement layer properties using ANN. Also, the first author (Saltan et al., 2002) and the authors of the paper (Saltan and Terzi, 2004; Saltan and Terzi, 2005) used the ANN approach in backcalculating pavement layer properties. From these studies, it can be said that there are several advantages to using ANN for NDT evaluation of highway flexible pavements. The mathematical simplicity of ANN makes them computationally efficient (Meier and Rix, 1994).

Jang (1993) first proposed the adaptive neural based fuzzy inference system (ANFIS) method and applied its principles successfully to many problems. It identifies a set of parameters through a hybrid learning rule combining the backpropagation gradient descent error digestion and a least squares method. It can be used as a basis for constructing a set of fuzzy IF-THEN rules with appropriate membership functions in order to generate the preliminary stipulated input-output pairs.

Kaur and Chou (1999) applied the Neuro-Fuzzy techniques for modeling the highway pavement performance prediction. Also, Göktepe *et al* (2005) used the ANFIS methodology for backcalculating the mechanical properties of flexible pavements.

The main purpose of this paper is to develop an ANFIS methodology for estimating the layer thickness and moduli values by considering seven deflection data.

2 Backcalculation of Pavement Layer Thickness

The highway maintenance engineer wishes to determine the structural integrity of a road pavement by nondestructive means (McMullen *et al.,* 1986). It is essential not only to evaluate a structural integrity of existing pavement but also to have accurate information on pavement surface conditions in order to establish a reasonable pavement rehabilitation design system (Inoue and Matsui, 1990).

Nondestructive Testing (NDT) of asphalt concrete pavements is one of the most useful and cost-effective methods developed by engineers to assist in the management of pavements (Zhou *et al.,* 1990). NDT enables the use of a mechanistic approach for pavement design and rehabilitation because *in-situ* material properties may be backcalculated from the measured field data through appropriate analysis techniques (Kang, 1998). NDT conducted on pavement surfaces are also used in the design of pavement overlays. Pavement is not destroyed and the test is relatively quick. If the FWD is used, then stress levels comparable to those anticipated for truck loadings can be obtained. The backcalculation procedure is an inverse operation of knowing the pavement characteristics and estimating the deflections due to an applied load (Hassan *et al.,* 2003). Estimation of layer elastic mechanical characteristics by the use of pavement surface deflections and backcalculation computer programs has been rapid (Mahoney *et al.*, 1993). In order to backcalculate reliable values, it is essential to accomplish several deflection tests at different locations along a highway section having relating uniform layer thicknesses (Uzan *et al.*, 1998). Pavement structural properties may be generally stated in terms of moduli values which is key element in pavement analysis and evaluation procedures (Zhou *et al.,* 1990). The backcalculation of layer properties from surface deflection measurements is of considerable importance for the accurate evaluation and design of overlays and the management of existing pavements (Harichandran *et al.*, 1993). Deflections obtained from the FWD are used to backcalculate the layer material properties, which are elastic modulus, Poisson's ratio and layer thicknesses. Flexible pavement layer thicknesses must also be known to get realistic results. Layer thicknesses can be obtained by coring the flexible pavement. Although laboratory testing of core samples yields much valuable information to assess pavement layer conditions, the required traffic control and time delay lowers the value of the service to the general public (Chang *et al.*, 2002). But it is important that nondestructive tests are carried out on flexible pavements for preventing to be damaged.

In order to analyze flexible pavements, individual layers are characterized by their characteristic parameters (elasticity moduli – E_1, E_2, ..., E_n, Poisson's Ratios - μ_1, μ_2, ..., μ_n and the layer thicknesses- h_1, h_2, ..., h_n of the layers above subgrade) (Rakesh *et al.*, 2006).

Elastic modulus is one of the most important mechanical properties of asphalt concrete mixes because it is related to the strength of asphalt concrete and thus the pavement distress resistance (Li *et al.*, 1999). The thickness of pavement layers is an important parameter used in Pavement Management System (PMS). Thickness data are used for pavement condition assessment, performance predictions, selection of maintenance strategies and rehabilitation treatments, basic quality assessment, and as input to overlay thickness design. Pavement thickness is usually determined from direct testing such core samples, nondestructive testing such as radar, or historical records such as pavement network database (Attoh-okine and Roddis, 1998). Due to limitations in the backcalculation software, and the limited time available to perform backcalculation activities in a production environment, pavement layer thicknesses are generally assumed to be constant over the pavement section under test. This is seldom the case. Pavement layer thickness variations result from various construction and maintenance details, even under specially controlled conditions. By considering the layer thicknesses as unknown parameters, layer thicknesses can also be backcalculated as well as layer moduli (Wang and Lytton, 1993).

Pavement structural evaluation analysis has long been a problem for pavement engineers. Previous approaches concentrated on statistical formulae mostly based on regression analysis to predict the performance. These equations illustrate the effects of various factors on the performance of pavements. These equations are valid only under certain conditions and should not be used if the actual conditions are different. Besides, this approach is very cumbersome and time consuming in terms of the calculation and in terms of the acquisition of the data required for doing the calculations (Kaur and Chou, 1999).

"Backcalculation" is a mechanistic evaluation of pavement surface deflection basins generated by various pavement deflection devices. Backcalculation takes a measured surface deflection and attempts to match it (to within some tolerable error) with a calculated surface deflection generated from an identical pavement structure using assumed layer stiffnesses (moduli). The assumed layer moduli in the calculated model are adjusted until they produce a surface deflection that closely matches the measured one. The combination of assumed layer stiffnesses that results in this match is then assumed to be near the actual *in situ* moduli for the various pavement layers. The backcalculation process generally refers to an iterative procedure and normally done with computer software. NDT and backcalculation processes are well-accepted procedures for the evaluation of structural capacity of the flexible pavements.

The backcalculation of pavement moduli from surface deflection measurements using nondestructive tests has been used for more than four decades to assess and manage existing pavements and to design overlays. Unfortunately, the backcalculated pavement moduli lack the accuracy in spite of the existence of many backcalculation programs employing different backcalculation procedures and algorithms (Alkasawneh, 2007).

A basic flowchart (patterned after Lytton, 1989) that represents the fundamental elements in all known backcalculation programs is shown as Figure 1. Briefly, these elements include (http://training.ce.washington.edu/wsdot/):

- *Measured deflections.* Includes the measured pavement surface deflections and associated distances from the load.
- *Layer thicknesses and loads.* Includes all layer thicknesses and load levels for a specific test location.
- *Seed moduli.* The seed moduli are the initial moduli used in the computer program to calculate surface deflections. These moduli are usually estimated from user experience or various equations.
- *Deflection calculation.* Layered elastic computer programs are generally used to calculate a deflection basin.
- *Error check.* This element simply compares the measured and calculated basins. There are various error measures which can be used to make such comparisons (more on this in a subsequent paragraph in this section).

- *Search for new moduli.* Various methods have been employed within the various backcalculation programs to converge on a set of layer moduli which produces an acceptable error between the measured and calculated deflection basins.
- *Controls on the range of moduli.* In some backcalculation programs, a range (minimum and maximum) of moduli are selected or calculated to prevent program convergence to unreasonable moduli levels (either high or low).

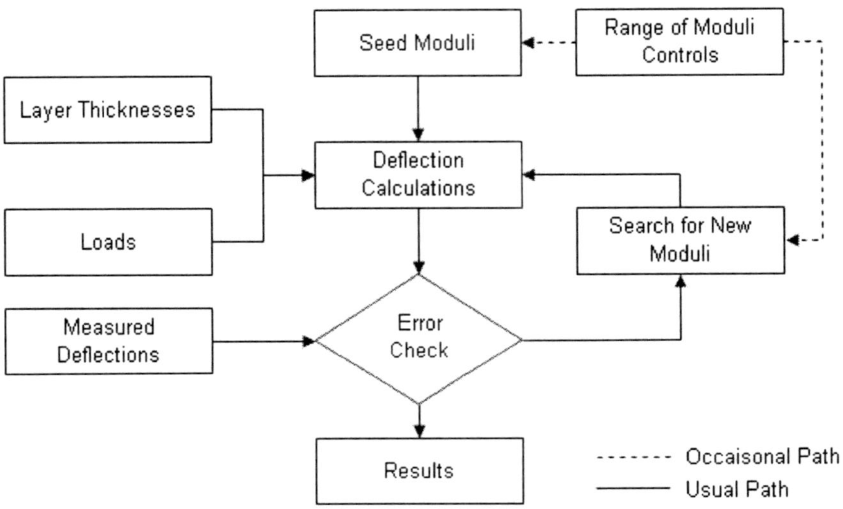

Fig. 1. Backcalculation Flowchart (http://training.ce.washington.edu/wsdot/)

In general, backcalculation is a laborious process, requiring a high degree of skill, and the results are known to be moderately to highly dependent on the individual doing the backcalculation. This comes about for a number of reasons, including the lack of a consensus standard addressing all aspects of the backcalculation process.

Measurement of an impulse deflection wave by the FWD appears to have emerged as the coming method of structural pavement evaluation. A weight of known magnitude is dropped from different heights, creating various levels of impulse loads. The pavement structure responds by a dynamic wave of deflections which spreads outward from the centre under the load. The peaks of this deflection wave are measured at several points by sensors called geophones. One of the sensors is placed in the centre, accessible through a hole in the disk, and the others at various distances outside the disk. The outer sensors are placed on the pavement surface by lowering a boom. The measured deflections generated by the FWD test load represent a deflection bowl or basin such as it may occur under a passing wheel load of corresponding magnitude and speed and of similar distribution area of tire contact pressure (Inoue and Matsui, 1990; Jung, 1990).

The ultimate aim of the backcalculation process from NDT results is to estimate the pavement material properties and layer thicknesses. The backcalculation procedure finds the set of parameters corresponding to the best fit to the measured deflection bowls. It is important to obtain the layer thicknesses through *in-situ* deflection test data equally non-destructively.

Maximum precision is needed from the backcalculation procedures, and more realistic models will reduce the size of systematic errors. This will make it possible to predict the remaining life of a pavement realistically in the field immediately after it has been tested.

Elastic layered programs used in asphalt pavement analysis assume linear elasticity. Pavement geo-materials do not, however, follow linear type stress-strain behaviour under repeated traffic loading. In effect, nonlinear stress sensitive response of unbound aggregate materials and fine-grained subgrade soils has been well established. Unbound aggregates exhibit *stress hardening* and fine-grained soils show *stress softening* type behaviour. When these geo-materials are used as pavement layers, the layer stiffnesses, i.e., moduli are no longer constant but functions of the applied stress state. Pavement structural analysis programs that take into account nonlinear geo-material characterization need to be employed to more realistically predict pavement responses needed for mechanistic based pavement design (Ceylan *et al.*, 2004).

Current basin-matching programs fall into two broad groups. Most programs employ gradient search techniques to adjust the pavement layer moduli iteratively until the theoretical and experimental deflection basins agree within a specified tolerance. Required inputs include experimental deflection measurements and pavement layer thicknesses. The iterative solution technique also requires an initial estimate of the solution (seed moduli) and a range of moduli to constrain the solution. A second approach is to interpolate within a database of theoretical basins. A database of theoretical basins is generated for prescribed pavement layer thicknesses by parametrically varying the pavement layer moduli within expected ranges (Meier and Rix, 1994; Meier and Rix, 1995).

The main problems that any classical backcalculation procedure faces are convergence, accuracy, and the number of layers in the backcalculation program. The selection of the seed moduli controls the convergence of the backcalculation procedure to pavement moduli that minimizes the mean square error (objective function) between the measured deflection basin and the backcalculated deflection basin using the backcalculated moduli. It is known that more than one solution can satisfy the objective function criterion in backcalculating the pavement moduli due to the multimodal nature of the backcalculation search space where many local optima exist. In turn, the arrival at local optima will lead to "inaccurate" pavement moduli that can be as twice as the "accurate" value. On the other hand, the maximum number of layers than can be used in any backcalculation program does not exceed 5 layers with recommendations to use layers to reduce the error associated with the backcalculation process. In some cases, increasing the number of the layers in the backcalculation process is desirable to obtain more representative variation of the moduli with depth (Alkasawneh, 2007).

3 Nondestructive Test Devices for Pavement Structural Evaluation

Structural evaluation of pavement deflection response using Non Destructive Test (NDT) data has been growing since the introduction of the Benkelman Beam at the Western Association of State Highway Organizations (WASHO) Road Test in 1952 is a simple device that operates on the lever arm principle. The Benkelman Beam is used with a loaded truck - typically 80 kN (18,000 lb) on a single axle with dual tires inflated to 480 to 550 kPa (70 to 80 psi). Measurement is made by placing the tip of the beam between the dual tires and measuring the pavement surface rebound as the truck is moved away. The Benkelman Beam is low cost but is also slow, labor intensive and does not provide a deflection basin (Figures 2 and 3) (http://training.ce.washington.edu/wsdot/). Developments in analytical techniques, coupled with improved deflection measurement capabilities, have resulted in the current so-called backcalculation techniques widely employed in pavement evaluation.

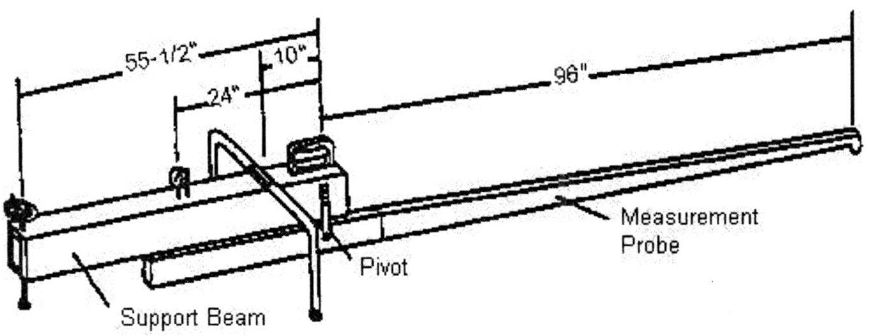

Fig. 2. Benkelman Beam Schematic (http://training.ce.washington.edu/wsdot/)

Calculation of load-related pavement surface deflections of specific points, using material properties of pavement layers (modulus, Poisson's ratio, and thickness), is well established (Noureldin, 1993). Of the different load responses (stress, strain, and deflection), only surface deflections are measured easily. Pavement deflection is the basic response of the structural system (surface-base-subgrade) to the applied load. It is used frequently as an indicator of pavement structural capability and performance potential. Surface deflection measurements are rapid, relatively cheap, and nondestructive (Garg and Thompson, 1999). Pavement deflection is measured through a series of velocity transducers at various distances from the baseplate, and the data can be used to backcalculate *in situ* pavement layer properties. This information can in turn be used in pavement

Fig. 3. Benkelman Beam in Use (http://training.ce.washington.edu/wsdot/)

structural analysis to determine the bearing capacity, estimate the remaining life, and calculate an overlay requirement over a desired design life (Wang and Lytton, 1993). The Falling Weight Deflectometer (FWD) was first introduced in France in 1970s to test the flexible road networks. It has since gained increasing acceptance as one of the most effective methods for evaluating flexible roads (Karadelis, 2000). In order to simulate the truck loading on the pavement, a circular mass is dropped from a certain height on the pavement. The height is adjusted according to the desired load level. Underneath the circular plate a rubber pad is mounted to prevent shock loading. Seven geophones are generally mounted on the trailer (the number of geophones can change). When the vertical load is applied on the pavement, the geophones collect the deflection data. The duration and magnitude of the force applied is representative of the load pulse induced by a truck moving at moderate speeds (Garg and Thompson, 1999). FWD is commonly used in many countries. Structural evaluation of road pavements using the Falling Weight Deflectometer (FWD) is essential tool of Pavement Management System (PMS) all over the world.

Benkelman beam and Dynaflect (Figure 4) which are most commonly used devices in the developing countries, give the information about underneath the centre of circular mass (i.e. these devices give one deflection data in each measurement) whereas the FWD (Figure 5) gives the information about other six points (or more points) which are away from the circular plate. Therefore, the effect of the wheel loading can also be seen in other points.

A FWD is a device that applies an impulsive load to a pavement surface and the deflection response is recorded at a series of radial points. The level of impact load, loading duration, and area are adjusted in such a way that it corresponds to

Fig. 4. Dynaflect in use (http://training.ce.washington.edu/wsdot/)

Fig. 5. FWD Application (http://training.ce.washington.edu/wsdot/)

the actual loading by a standard truck moving on an in-service road. The impulsive load is sensed by load cells and the shape of the deflection profile is captured by a number of geophones. The basic aim of Nondestructive Evaluation by FWD measurements is to estimate the *in situ* layer characteristic parameters when the deflection profile is given – this is called a backcalculation problem. The problem of backcalculation of layer moduli of asphalt pavement from FWD data is indeed a complex one (Sharma and Das, 2007).

Except these equipments, Rolling Weight Deflectometer (RWD) and Rolling Dynamic Deflectometer (RDD) are newly developed testing devices. RWD is a trailer-mounted device that continuously measures maximum pavement deflections under a load wheel under moving conditions. The RWD collects deflections at highway speeds of 70 to 80 kph under a 40 kN wheel loads (Briggs *et al.*, 2000). On the other hand, the RDD measures continuous deflection profiles rather than deflections at discrete points. The RDD is a truck-mounted deflectometer, which applies large cyclic loads to the pavement and measures cyclic deflections as it moves along the pavement. Thus, rather than measuring deflections at discrete locations, the RDD continuously measures the deflection profile along the entire pavement (Bay and Stokoe, 2000).

4 Adaptive Neural-Based Fuzzy Inference System

Various fuzzy inference system (FIS) types are studied in the literature and each one is characterized by consequent parameters. In this section a brief description of adaptive neural-based fuzzy inference system (ANFIS) model principles are presented.

ANFIS applications and properties are investigated, and number of methods is proposed for partitioning the input space and hence addresses the structure identification problem. Fundamentally, ANFIS is a graphical network representation of Sugeno-type fuzzy systems, endowed by neural learning capabilities. The network is comprised of nodes and with specific functions, or duties, collected in layers with specific functions (Tsoukalas and Uhrig, 1997).

In order to illustrate ANFIS's representational strength, the neural fuzzy control systems are considered based on the Tagaki-Sugeno-Kang (TSK) fuzzy rules whose consequent parts are linear combinations of their preconditions. The TSK fuzzy rules are in the following forms:

$$R^j : IF\ x_1\ is\ A_1^j\ AND\ x_2\ is\ A_2^j\ AND...AND\ x_n\ is\ A_n^j,$$
$$THEN\ y = f_j = a_0^j + a_1^j x_1 + a_2^j x_2 + ... + a_n^j x_n \quad (4.1)$$

where x_i's (i = 1,2,...n) are input variables, y is the output variable (pavement serviceability ratio), A_i^j are linguistic terms of the precondition part with membership functions $\mu A_1^j (x_i)$, (j = 1, 2,..., n) a_1^j ЄR are coefficients of linear equations f_i (x_1, x_2,..., x_n). To simplify the discussion it is necessary to focus on a specific Neuro-Fuzzy Controller (NFC) of this type called ANFIS.

Let us assume that the fuzzy control system under consideration of two inputs x_1 and x_2 and one output y and that the rule base contains two TSK fuzzy rules as follows:

$$R^1 : IF\ x_1\ is\ A_1^1\ AND\ x_2\ is\ A_2^1\ , THEN\ \ y = f_1 = a_0^1 + a_1^1 x_1 + a_2^1 x_2 \quad (4.2)$$

$$R^2 : IF\ x_1\ is\ A_1^2\ AND\ x_2\ is\ A_2^2\ , THEN\ \ y = f_2 = a_0^2 + a_1^2 x_1 + a_2^2 x_2 \quad (4.3)$$

In TSK fuzzy system, for given input values x_1 and x_2, the inferred output y^* is calculated by the following formula

$$y^* = (\mu_1 f_1 + \mu_2 f)_2 / (\mu_1 + \mu_2) \quad (4.4)$$

where μ_j are firing strengths of R^j, j=1, 2, and they are given by the equation below,

$$\mu_j = \mu_{A_1^j}(x_1) + \mu_{A_2^j}(x_2), \qquad j=1, 2 \quad (4.5)$$

If product inference is used, the corresponding ANFIS architecture is shown in Figure 6, where node functions in the same layers are of the type described below.

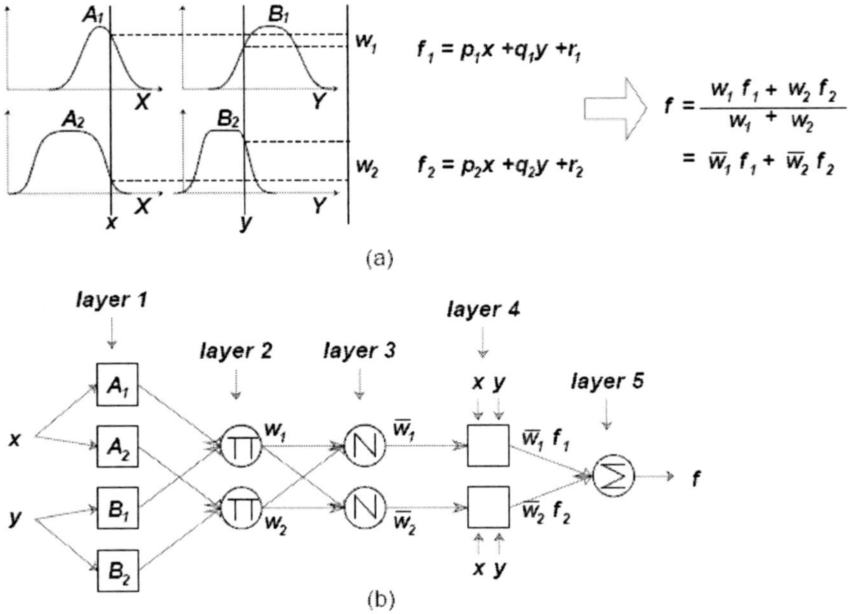

Fig. 6. Structure of ANFIS (a) Fuzzy inference system (b) Equivalent ANFIS (Jang, 1993)

This is an ANFIS architecture where the following meanings can be attached to each layer.

Firstly, every node in this layer implies an input and it just passes external signals to the next layer,

Layer 1. Every node in this layer acts as a membership function $\mu_{A_i^j}(x_i)$, and its output specifies the degree to which the given x_i satisfies the quantifier A_i^j. Generally, $\mu_{A_i^j}(x_i)$ is selected as bell-shaped with a maximum membership degree equal to 1 and minimum equal to zero, such as

$$\mu_{A_i^j}(x_i) = 1/\left(1 + \{[(x_i - m_i^j)/\sigma_i^j]^2\}^{b_i^j}\right) \tag{4.6}$$

or

$$\mu_{A_i^j}(x_i) = \exp\{-[((x_i - \mu_i^j)/(\sigma_i^j))^2]^{b_i^j}\} \tag{4.7}$$

where $\{m_I^j, \sigma_I^j, b_i^j\}$ is the parameter set to be tuned. In fact, continuous and piecewise differentiable functions, such as commonly used trapezoidal or triangular membership functions, are also qualified candidates for node functions in this layer. Parameters in this layer are referred to as precondition parameters,

Layer 2. Every node in this layer is labeled Π and multiplies the incoming signals $\mu_j = \mu_{A_1^j}(x_1) + \mu_{A_2^j}(x_2)$ and sends the product out. Each node output represents the firing strength of a rule,

Layer 3. Every node in this layer is labeled by N and calculates the normalized firing strength of a rule. That is the jth node calculates the ratio of the jth rule's firing strength of all the rules' firing strengths as,

$$\overline{\mu}_j = \mu_j / \left(\mu_{A_1^j}(x_1) + \mu_{A_2^j}(x_2)\right) \tag{4.8}$$

Layer 4. Every node j in this layer calculates the weighted consequent value as,

$$\overline{\mu_j}\left(a_0^j + a_1^j x_1 + a_2^j x_2\right) \tag{4.9}$$

where $\overline{\mu_j}$ is the output of layer 4 and $\{a_0^j, a_1^j, a_2^j\}$ is the set to be tuned. Parameters in this layer are referred to as consequent parameters,

Layer 5. The only node in this layer is labeled as Σ, and it sums all incoming signals to obtain the final inferred result for the whole system (Lin and Lee, 1995).

5 Development of the ANFIS Model

Fuzzy modeling is a system identification task, which involves two phases: structure identification and parameter estimation. Structure identification includes the issues such as selecting relevant input variables, choosing a specific type of fuzzy inference system, determining the number of fuzzy rules, their antecedents and consequents, and determining the type and number of membership functions. Determining the optimum number and form of fuzzy rules is the most crucial step and various algorithms have been developed to automate this process, such as k-means clustering, fuzzy C-means clustering, and subtractive clustering. The subtractive clustering method was used in here. This method assumes each data point as a potential cluster center and calculates a measure of the likelihood that each data point would define the cluster center, based on the density of surrounding data points. The steps of the fuzzy model algorithm can be summarized as: (1) select the data point with the highest potential to be the first cluster center, (2) remove all data points in the vicinity of the first cluster center as determined by the range of influence (radius), and (3) iterate on this process until all the data are within the radii of a cluster center.

Range of influence indicates the radius of a cluster when the data space is considered as a unit hypercube. Acceptable values for radii are usually between 0.2 and 0.5. A small cluster radius will usually yield many small clusters in the data, resulting in many rules and vice versa. In multi-dimensional data, different radii may be specified for each dimension. If the same value is applied to all data dimensions, each cluster center will have a spherical neighborhood of influence with the given radius.

Squash factor is the factor used to multiply the radii values that determine the neighborhood of a cluster center, so as to squash the potential for outlying points to be considered as part of that cluster. High values, e.g. 20, are used to find clusters that are far from each other.

Accept ratio sets the potential, as a fraction of the potential of the first cluster center, above which another data point will be accepted as a cluster center. High values are used to accept data points that have a very strong potential for being cluster centers.

Reject ratio sets the potential, as a fraction of the potential of the first cluster center, below which a data point will be rejected as a cluster center. High value, like 0.7 is used to reject all data points without a strong potential. Another important step is parameter optimization, or fine-tuning of parameter values to best fit the input–output data set. Neural network learning techniques can automate this tuning process and this is a strong motivation for the development of neuro-fuzzy system. ANFIS is a new inference system that incorporates Sugeno type fuzzy inference system into adaptive neural networks structure (Akbulut *et al.*, 2004). Table 1 shows the used parameters in the model. These parameters were found in the results of various applications.

Table 1. Model parameters

Model parameters	Value
Range of influence	0.05
Squash factor	1.25
Accept ratio	0.5
Reject ratio	0.15

Highway flexible pavements have some characteristics such as layer elastic modulus, layer Poisson ratio, layer thickness, and surface deflections against dynamical traffic loading. Among these characteristics, surface deflection values are especially important for evaluating the structural properties of the highway flexible pavements. A typical flexible pavement in which wearing course, base layer and subgrade exist was chosen for the study (Figure 7).

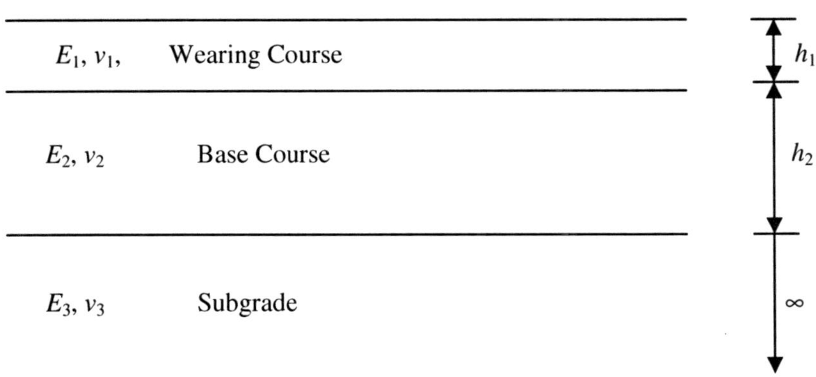

Fig. 7. A typical flexible pavement used in the analysis

In this study, two models were established which estimate the layer thickness and moduli from surface deflection values of a flexible pavement from using ANFIS. Seven surface deflection values at different locations were used as input. Layer thickness and moduli were used as output variables in first and second model respectively. Data sets included 114 different data configurations. These data sets are obtained from the earlier studies of the first author by using axially symmetric finite element software, SDUFEM (Saltan, 1999).

The models are incorporating the effects of seven input parameters used to simulate the pavement layer thickness and moduli. The schematic of the architectures of ANFIS used in here are shown in Figures 8 and 9.

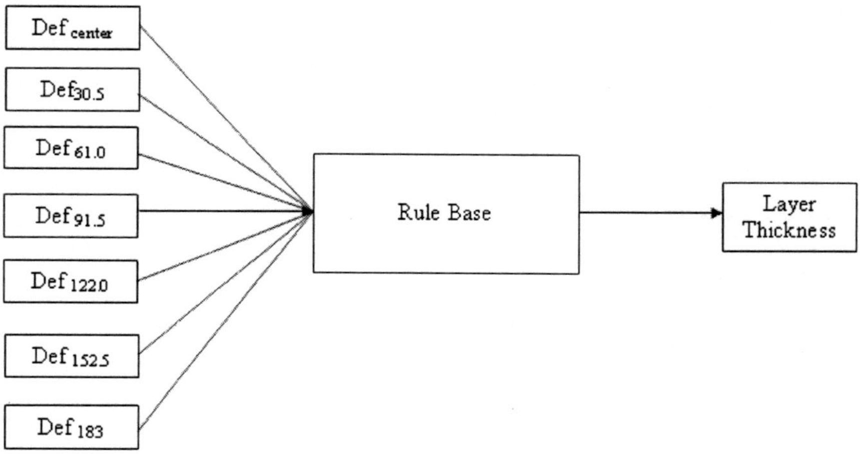

Fig. 8. The schematic of the architectures of ANFIS for layer moduli

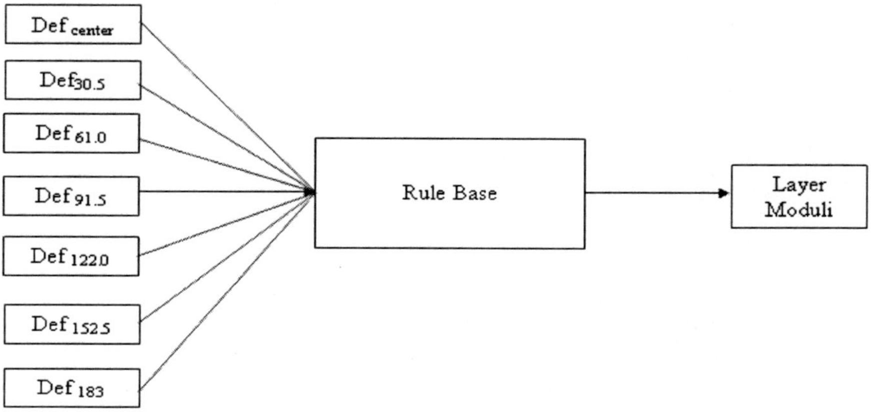

Fig. 9. The schematic of the architectures of ANFIS for layer thickness

ANFIS models are trained by using approximately 80% of the data and tested using the rest of the data. Prior to execution of the model, standardization, x_i^1, on the data, X_i (i = 1, 2, …, n) is done according to the following expression such that all data values fall between zero and 1.

$$x_i^1 = \left(X_i - X_{\min}\right)/\left(X_{\max} - X_{\min}\right) \tag{5.1}$$

where the X_i is the actual value and the X_{max} and X_{min} are the maximum and the minimum of the measurement values. Such standardization procedure renders the data also into dimensionless form. Furthermore, standardization removes the arbitrary effects of similarity between objects or variables. Membership function types for inputs are selected as Gauss-bell whereas it is linear for the output.

5.1 Layer Thickness Model Properties

The membership functions plots of deflections are shown in Figures 10, 11, 12, 13, 14, 15, and 16. The membership functions plot of layer thickness is shown in Figure 17.

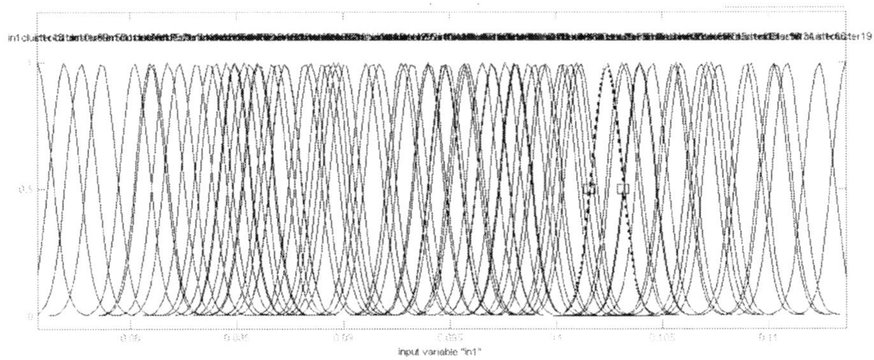

Fig. 10. The membership functions plots of deflection 1 (loading center)

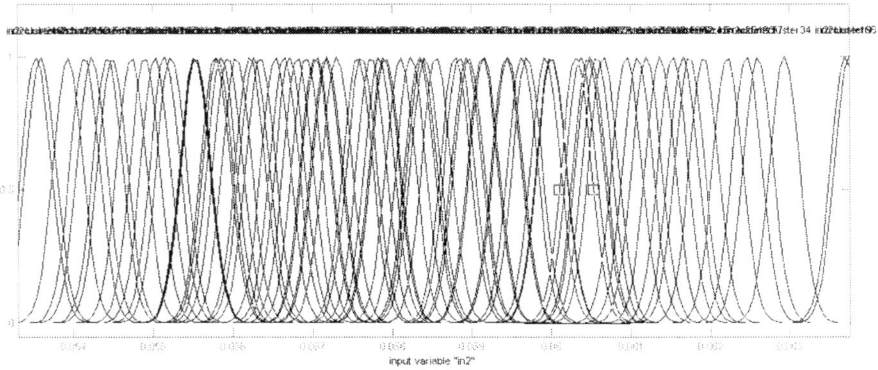

Fig. 11. The membership functions plots of deflection 2 (30.5 cm distance from center)

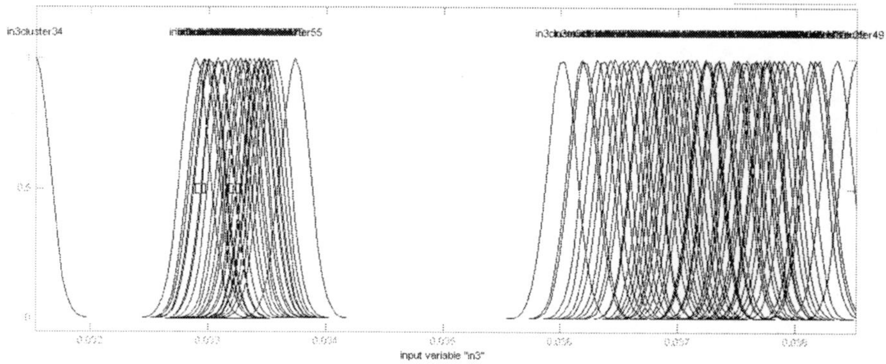

Fig. 12. The membership functions plots of deflection 3 (61 cm distance from center)

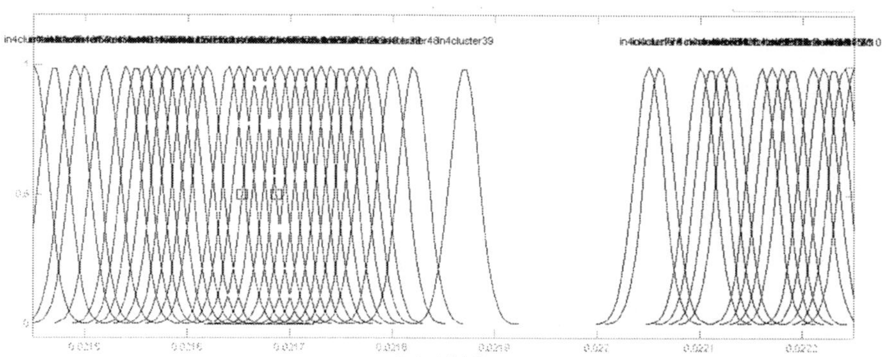

Fig. 13. The membership functions plots of deflection 4 (91.5 cm distance from center)

Fig. 14. The membership functions plots of deflection 5 (122 cm distance from center)

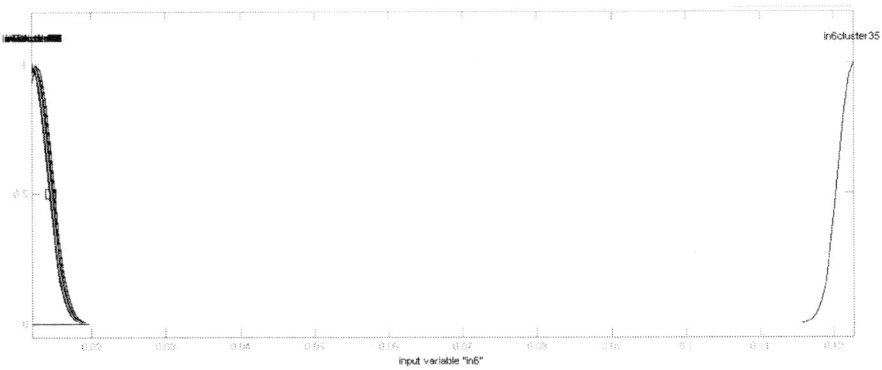

Fig. 15. The membership functions plots of deflection 6 (152.5 cm distance from center)

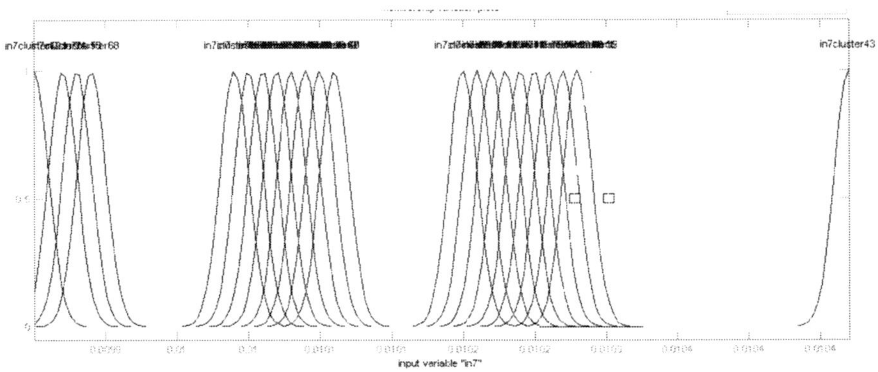

Fig. 16. The membership functions plots of deflection 7 (183 cm distance from center)

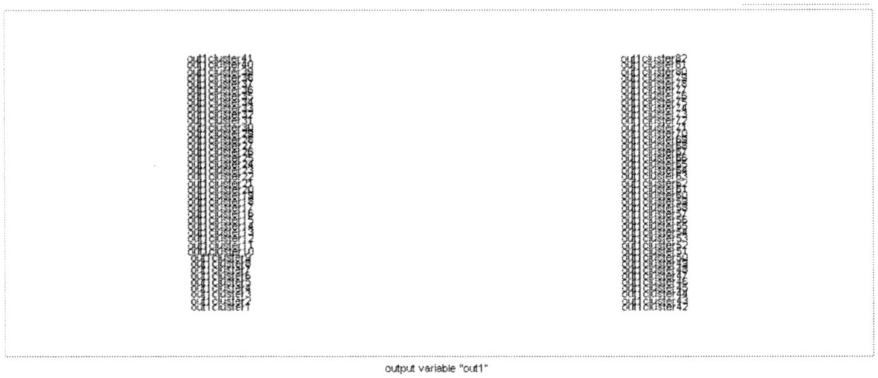

Fig. 17. The membership functions plots of layer thickness

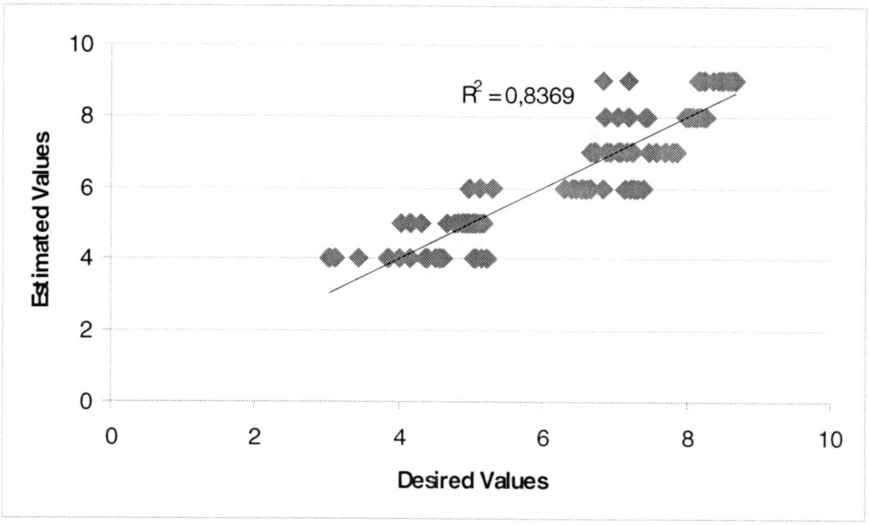

Fig. 18. Comparison of the estimated values (MLR) and desired values for training set for layer thickness

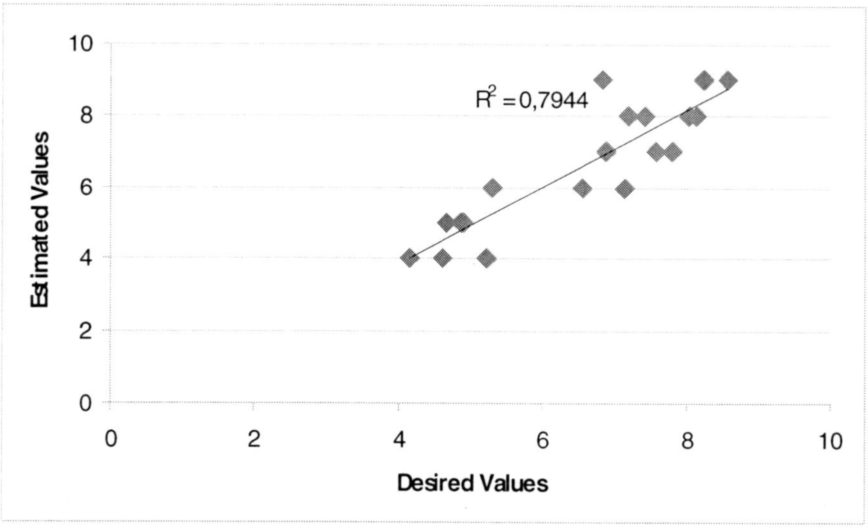

Fig. 19. Comparison of the estimated values (MLR) and desired values for testing set for layer thickness

Multiple linear regression model (MLR) is also developed to estimate layer thickness for the same training and testing data set. The results are given in Figures 18 and 19.

Comparison of between obtained and ANFIS model for layer thickness are given in Figures 20 and 21 for training and testing sets, respectively. As can be

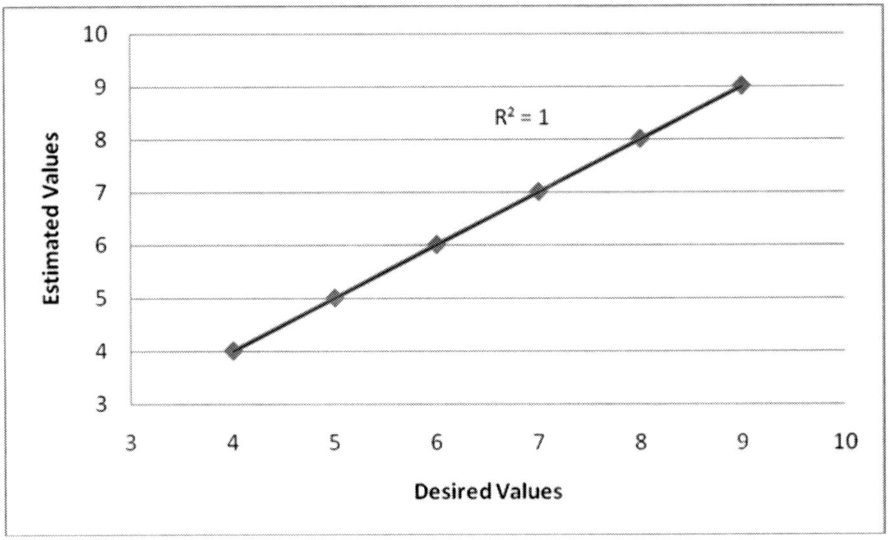

Fig. 20. Comparison of the estimated values (ANFIS) and desired values for training set for layer thickness

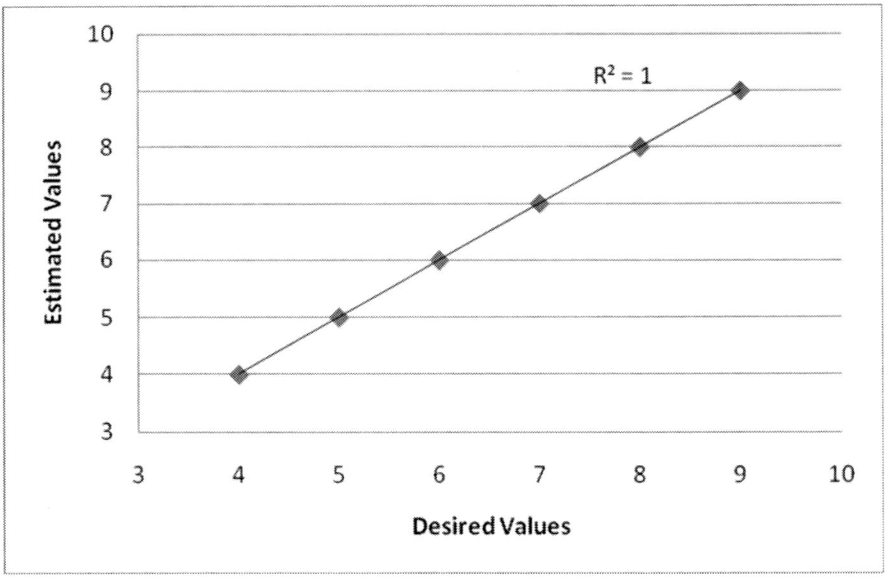

Fig. 21. Comparison of the estimated values (ANFIS) and desired values for testing set for layer thickness

seen from the figures, ANFIS model gives the satisfactory pavement surface layer thickness prediction.

As seen from the figures, ANFIS gives excellent results than MLR values for layer thickness estimation.

5.2 Layer Moduli Model Properties

The membership functions plots of deflections are shown in Figures 22, 23, 24, 25, 26, 27 and 28. The membership functions plot of layer moduli are shown in Figure 29.

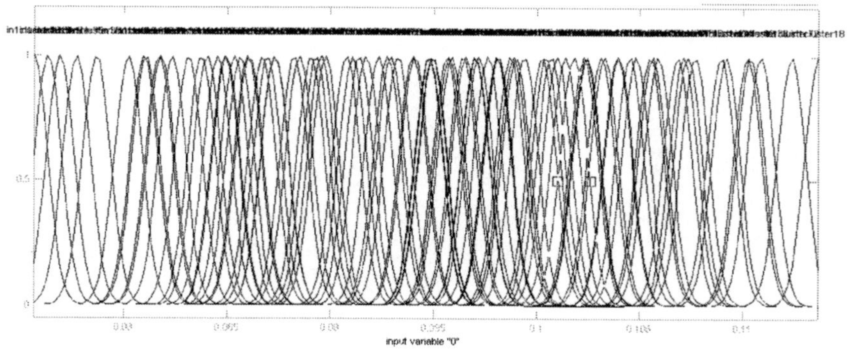

Fig. 22. The membership functions plots of deflection 1 (loading center)

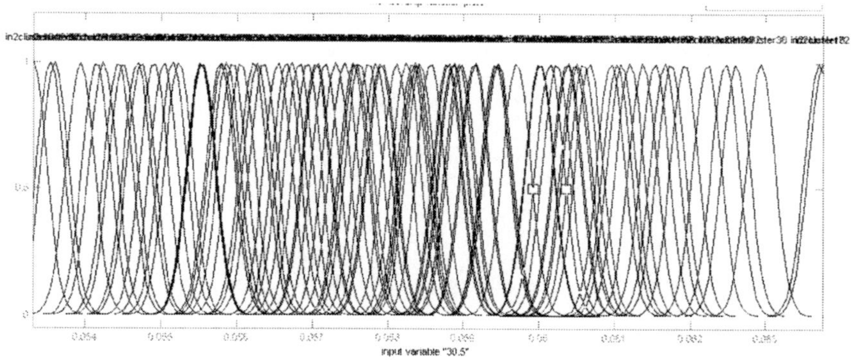

Fig. 23. The membership functions plots of deflection 2 (30.5 cm distance from center)

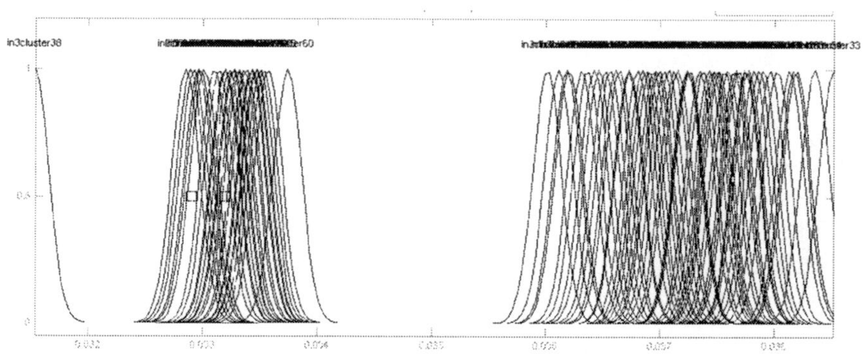

Fig. 24. The membership functions plots of deflection 3 (61 cm distance from center)

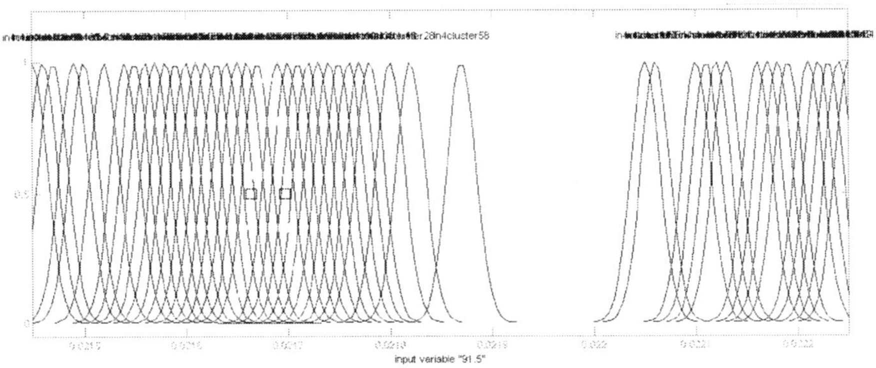

Fig. 25. The membership functions plots of deflection 4 (91.5 cm distance from center)

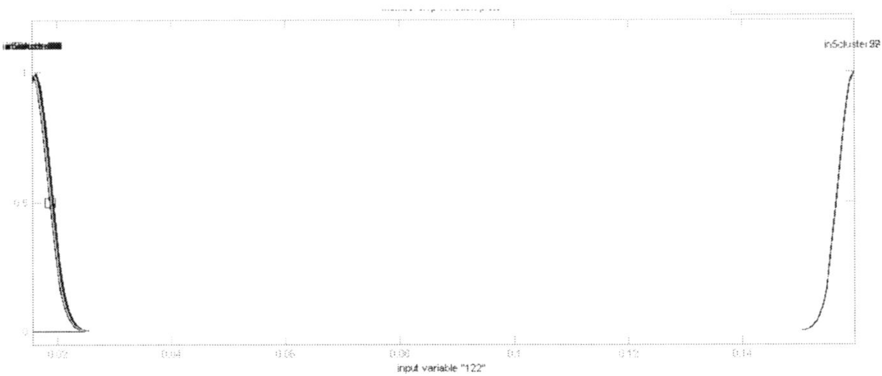

Fig. 26. The membership functions plots of deflection 5 (122 cm distance from center)

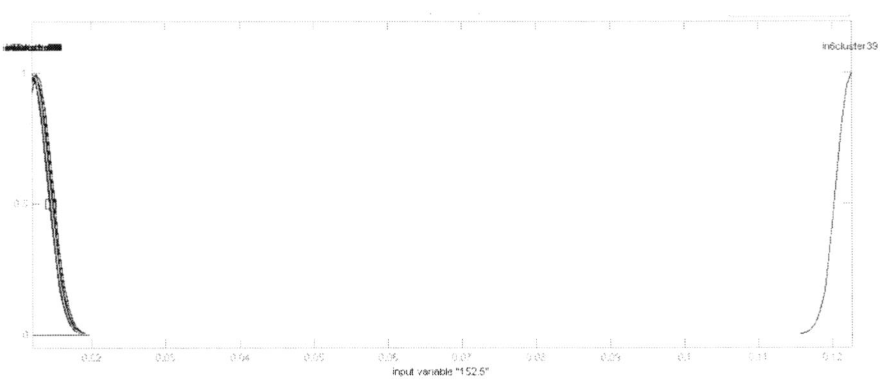

Fig. 27. The membership functions plots of deflection 6 (152.5 cm distance from center)

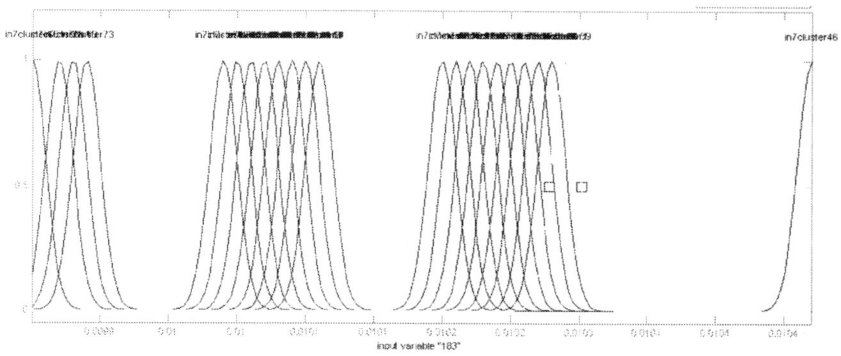

Fig. 28. The membership functions plots of deflection 7 (183 cm distance from center)

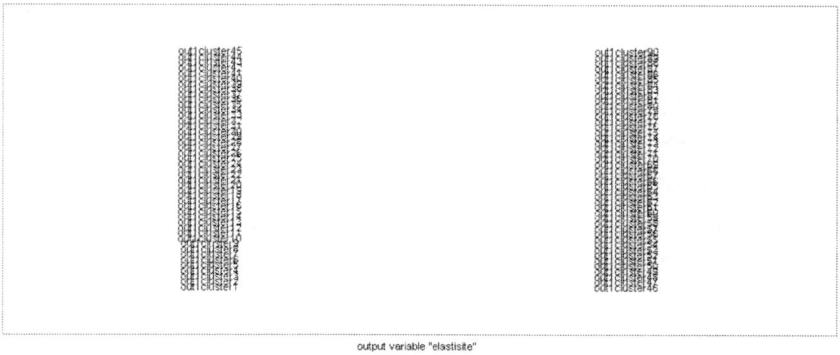

Fig. 29. The membership functions plots of layer moduli

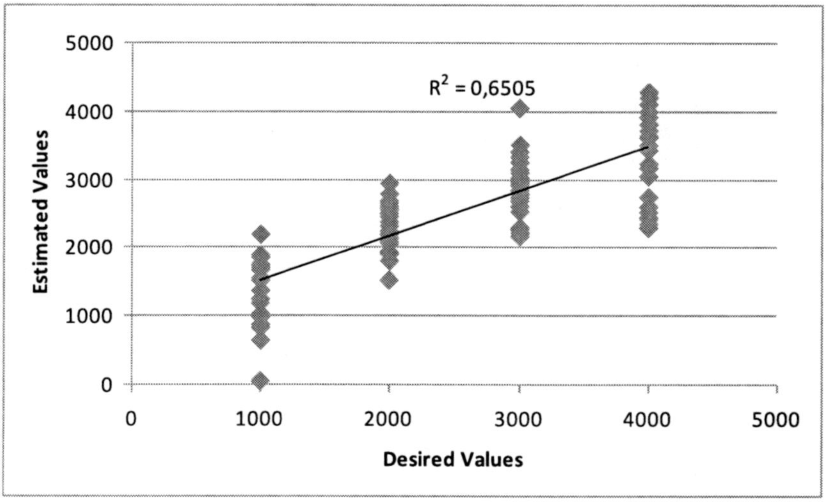

Fig. 30. Comparison of the estimated values (MLR) and desired values for training set for layer moduli

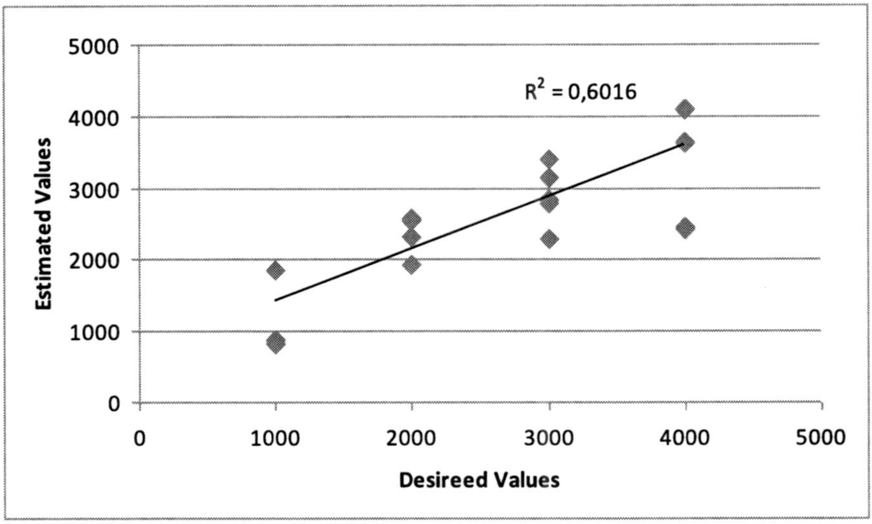

Fig. 31. Comparison of the estimated values (MLR) and desired values for testing set for layer moduli

Also, multiple linear regression model (MLR) is developed to estimate layer moduli for the same training and testing data set. The results are given in Figures 30 and 31.

Comparison of between obtained and ANFIS model for layer moduli are given in Figures 32 and 33 for training and testing sets, respectively. As can be seen

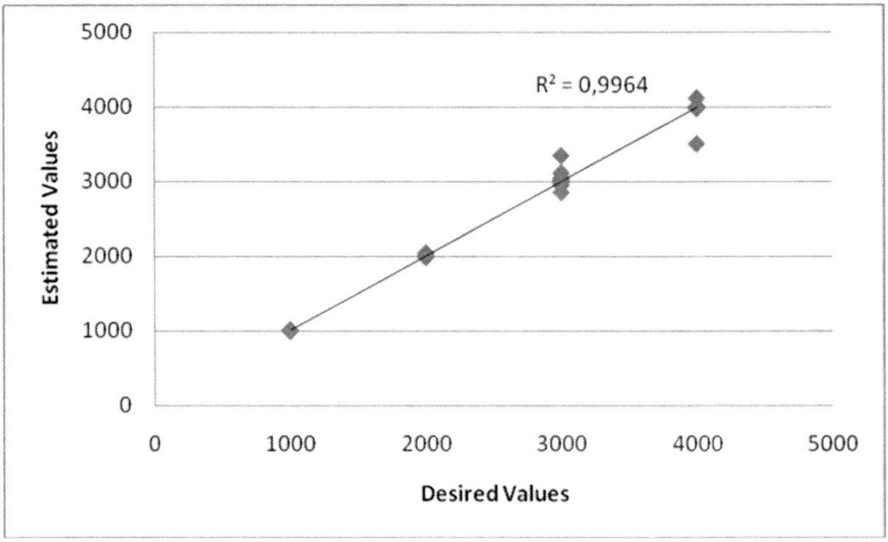

Fig. 32. Comparison of the estimated values (ANFIS) and desired values for training set for layer moduli

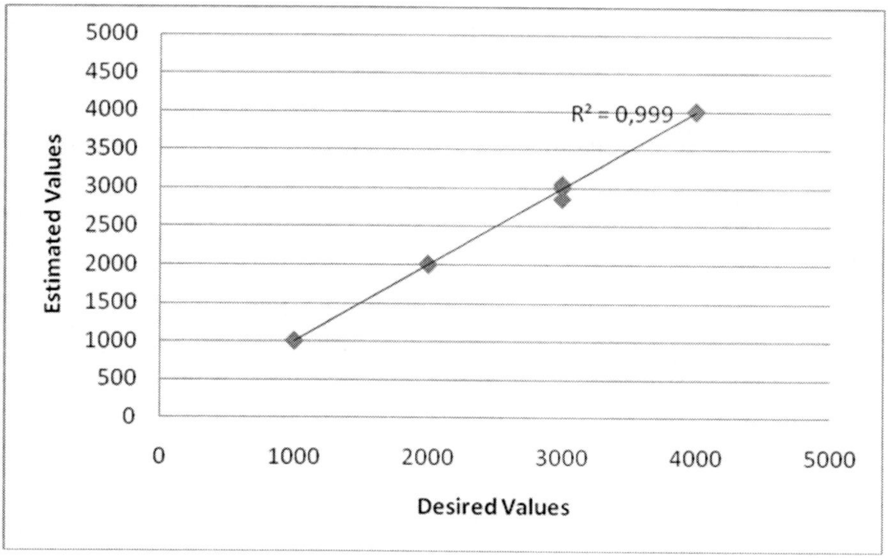

Fig. 33. Comparison of the estimated values (ANFIS) and desired values for testing set for layer moduli

from the figures, ANFIS model gives the satisfactory pavement surface layer moduli prediction.

As seen from the figures above, ANFIS gives excellent results than MLR values for layer moduli estimation.

6 Conclusions

In the present study, two ANFIS models have been presented for determining flexible pavement surface layer thickness and moduli. The models were established by using ANFIS approach. ANFIS approach provides a fundamentally different way to backcalculate pavement layer thickness and moduli from FWD deflection basins. Unlike conventional approaches that backcalculate layer thickness and moduli by attempting to march theoretical and experimental deflection basins, ANFIS model simply maps deflection basins into their corresponding layer thickness and moduli. Results show that wearing course thickness and moduli of flexible pavement regression values of the models are extremely acceptable.

Some models used for this type of problems are based on some simplifying assumptions that cannot reflect the reality. Solutions of the problems which do not have a formula or function about the solution can be easily and realistically performed using these approaches presented here.

Multiple linear regression model (MLR) is also performed both layer thickness and moduli estimation in the study so as to compare with ANFIS results. Same data set is utilized for the two different modeling techniques. Comparing the models, performance of the MLR models fall into lower limit in view of the literature while ANFIS models give excellent estimation performance.

Based on the obtained results, it may be concluded that the proposed ANFIS model can assist and guide to meet highway pavement engineer requests. The model can be effectively used for the backcalculation of the layer thickness and moduli of the flexible pavements.

References

[1] Huang, Y.H.: Pavement Analysis and Design. Prentice-Hall, New Jersey (1993)
[2] Arraigada, S., Partl, M.N., Angelone, S.M., Martinez, F.: Evaluation of Accelerometers to Determine Pavement Deflections Under Traffic Loads, Materials and Structures. Article in Press (2008), doi:10.1617/s11527-008-9423-5
[3] Reddy, B.B., Veeraragavan, A.: Structural Performance of In-service Flexible Pavements. Journal of Transportation Engineering, ASCE 123(2), 156–167 (1997)
[4] Madanat, S., Prozzi, J.A., Han, M.: Effect of Performance Model Accuracy on Optimal Pavement Design. Computer-Aided Civil and Infrastructure Engineering 17, 22–30 (2002)
[5] Hassan, J.P., Mousa, R.M., Gadallah, A.A.: Comparative Analysis of Using AASHTO and WESDEF Approaches in Backcalculation of Pavement Layer Moduli. Journal of Transportation Engineering, ASCE 129(3), 322–329 (2003)
[6] Chang, J., Lin, J., Chung, W., Chen, D.: Evaluating the Structural Stregth of Flexible Pavements in Taiwan Using the Falling Weight Deflectometer. The Int. Journal of Pavement Engineering 3(3), 131–141 (2002)
[7] Ceylan, H., Guclu, A., Tutumluer, E., Thompson, M.R.: Use of Artificial Neural Networks for Backcalculation of Pavement Layer Moduli. In: 2004 FWD Users Group Meeting. University Inn, West Lafayette (2004)
[8] Ceylan, H.: Analysis and design of concrete pavement systems using artificial neural networks, Ph.D. Dissertation, University of Illinois at Urbana-Champaign (2002)
[9] Meier, R.W., Rix, G.J.: Backcalculation of Flexible Pavement Moduli Using Artificial Neural Networks, TRR 1448, 75-82 (1994)
[10] Meier, R.W., Rix, G.J.: Backcalculation of Flexible Pavement Moduli from Dynamic Deflection Basins Using Artificial Neural Networks, TRR 1473, 72-81 (1995)
[11] Meier, R.W., Alexander, D.R., Freeman, R.B.: Using Artificial Neural Networks as a Forward Approach to Backcalculation, TRR 1570, 126-133 (1999)
[12] Saltan, M., Tığdemir, M., Karaşahin, M.: Artificial Neural Network Application for Flexible Pavement Thickness Modeling. Turkish J. Eng. Env. Sci. 26, 243–248 (2002)
[13] Saltan, M., Terzi, S.: Backcalculation of pavement layer parameters using Artificial Neural Networks. Ind. J. Eng. & Mat. Sci. 11(1), 38–42 (2004)
[14] Saltan, M., Terzi, S.: Comparative analysis of using artificial neural networks (ANN) and gene expression programming (GEP) in backcalculation of pavement layer thickness. Ind. J. Eng. & Mat. Sci. 12(1), 42–50 (2005)
[15] Jang, J.: ANFIS: adaptive-network-based fuzzy inference system. IEEE Transactions on Systems Management and Cybernetics 23(3), 665–685 (1993)

[16] Kaur, D., Chou, E.: Applying Neuro-Fuzzy Techniques for Intelligent Highway Pavement Performance Prediction Model. In: 42th Midwest Symposium on Circuits and Systems, Las Cruces, NM, USA, vol. 2, pp. 922–924 (1999)
[17] Göktepe, A.B., Ağar, E., Lav, A.H.: Comparison of Multilayer Perceptron and Adaptive Neuro-Fuzzy System on Backcalculating the Mechanical Properties of Flexible Pavements, ARI The Bulletin of the Istanbul Technical University 54 (3), 65–77 (2005)
[18] McMullen, D., Snaith, M.S., Burrow, J.C.: Back Analysis Techniques for Pavement Condition Determination. In: The 1986 Int.Conf. on Bearing Capacity of Roads and Airfields, Plymouth, England, pp. 335–344 (1986)
[19] Inoue, T., Matsui, K.: Structural Analysis of asphalt Pavement by FWD and Backcalculation of Elastic Layered Model. In: 3rd Int. Conf. on Bearing Capacity of Roads and Airfields, Trondheim, Norway, pp. 425–434 (1990)
[20] Zhou, H., Hicks, R.g., Bell, C.a.: Development of A Backcalculation Program and Its Verification. In: 3rd Int.Conf. on Bearing Capacity of Roads and Airfields, Trondheim, Norway, pp. 391–400 (1990)
[21] Kang, Y.W.: Multi-frequency backcalculation of pavement-layer moduli. Journal of Transport Eng., ASCE 124(1), 73–81 (1998)
[22] Mahoney, J.P., Winters, B.C., Jackson, N.C., Pierce, L.M.: Some Observations About Backcalculation and Use of A Stiff Layer Condition, TRR 1384, 8-14 (1993)
[23] Uzan, J., Lytton, R.L., Germann, F.P.: General Procedure for Backcalculating Layer Moduli. In: Nondestructive testing of pavements and backcalculation of moduli (ASTM STP 1026, USA), pp. 217–228 (1998)
[24] Harichandran, R.S., Mahmood, T., Raab, A.R., Baladi, G.Y.: Modified Newton Algorithm for Backcalculation of Pavement Layer Properties, TRR 1384, 15–22 (1993)
[25] Rakesh, N., Jain, A.K., Reddy, M.A., Reddy, K.S.: Artificial Neural Networks-Genetic Algorithm Based Model for Backcalculation of Pavement Layer Moduli. Int. journal of Pavement engineering 7(3), 221–230 (2006)
[26] Li, G., Li, Y., Metcalf, J.B., Pang, S.: Elastic Modulus Prediction of Asphalt Concrete. Journal of Transportation Engineering, ASCE 11(3), 236–241 (1999)
[27] Attoh-okine, N.O., Roddis, W.M.K.: Uncertainties of Asphalt Layer Thickness Determination in flexible Pavements-Influence Diagram Approach. Civil Engineering and Environmental systems 15, 107–124 (1998)
[28] Wang, F., Lytton, R.L.: System Identification Method for Backcalculating Pavement Layer Properties, TRR 1384, 1–7 (1993)
[29] Alkasawneh, W.: Backcalculation of Pavement Moduli Using Genetic Algorithms, PhD Thesis, The Graduate Faculty of The University of Akron, USA (2007)
[30] Backcalculation, `http://training.ce.washington.edu/wsdot/` (Last Access, June 15, 2009)
[31] Jung, F.W.: Interpretation of Deflection Basin for Real-World Materials in Flexible Pavements, Technical Report, RR-242, Ministry of Transportation, Research and Development Branch, Canada (1990)
[32] Noureldin, A.S.: New Scenario for Backcalculation of Layer Moduli of Flexible Pavements, TRR 1384, pp. 23–28 (1993)
[33] Garg, N., Thompson, M.R.: Structural Response of LVR Flexible Pavements at Mn/Road Project. Journal of Transportation Engineering, ASCE 125(3), 238–244 (1998)

[34] Karadelis, J.N.: A Numerical Model for the Computation of Concrete Pavement Moduli: A Nondestructive and Assessment Method. NDT&E International 33, 77–84 (2000)
[35] Sharma, S., Das, A.: Backcalculation of Pavement Layer moduli from Falling Weight Deflectometer Data Using an artificial Neural Network. Canadian Journal of Civil Engineering 35(1), 57–66 (2007)
[36] Briggs, R.C., Johnson, R.F., Stubstad, R.N., Pierce, L.: A Comparison of the Rolling Weight Deflectometer with the Falling Weight Deflectometer. In: Nondestructive Testing of Pavements and Backcalculation of Moduli, ASTM STP 1375, vol. 3, pp. 444–456 (2000)
[37] Bay, J.A., Stokoe II, K.H.: Continuous Profiling of Flexible and Rigid Highway and Airport Pavements with the Rolling Dynamic Deflectometer. In: Nondestructive Testing of Pavements and Backcalculation of Moduli, ASTM STP 1375, vol. 3, pp. 429–443 (2000)
[38] Tsoukalas, L.H., Uhrig, R.E.: Fuzzy and Neural Approaches in Engineering, p. 587. A Wiley-Interscience Publications, John Wiley & Sons, Inc, New York (1997)
[39] Lin, C.T., Lee, C.S.G.: Neural Fuzzy Systems. Prentice Hall PTR 797, New Jersey (1995)
[40] Akbulut, S., Hasiloglu, A.S., Pamukcu, S.: Data generation for shear modulus and damping ratio in reinforced sands using adaptive neuro-fuzzy inference system. Soil Dynamics and Earthquake Engineering 24, 805–814 (2004)
[41] Saltan, M.: Analytical Evaluation of Flexible Pavements, PhD Thesis, S. Demirel University, Turkey (1999)

Case Studies of Asphalt Pavement Analysis/Design with Application of the Genetic Algorithm

Bor-Wen Tsai[1], John T. Harvey[2], and Carl L. Monismith[3]

[1] University of Califronia Pavement Research Center, Institute of Transportation Studies, University of California, Berkeley, CA 94804
bwtsai@berkeley.edu

[2] University of Califronia Pavement Research Center, Department of Civil and Environmental Engineering, University of California, Davis, CA 95616
jtharvey@ucdavis.edu

[3] University of Califronia Pavement Research Center, Institute of Transportation Studies, University of California, Berkeley, CA 94804
clm@maxwell.berkeley.edu

Abstract. The primary purpose of this study is to demonstrate the applicability of the genetic algorithm (GA) to solve nonlinear optimization problems encountered in asphalt pavement design. The fundamentals of the GA are briefly discussed, and four case studies are presented. The first case study is an example showing the backcalculation of layer moduli with deflection data from a falling weight deflectometer and a layered-elastic program. The second case study demonstrates how to construct the master curve, either from a mix flexural frequency sweep test or from a binder rheometer test, and how to apply that master curve in pavement design. The third case shows how to apply the GA to characterize the binder discrete relaxation spectrum with a generalized Maxwell solid model. The last case study illustrates how to apply the GA to define the mix fatigue damage process of a flexural controlled-deformation beam fatigue test and the permanent shear strain accumulation process of a controlled-load repetitive simple shear test with constant height using a three-stage Weibull approach, and how to apply the three-stage Weibull approach in predicting pavement performance. The results indicate that the GA is promising and successful in resolving the nonlinear optimization problem although the GA presents some difficulty in terms of computing efficiency in the case study of backcalulation of layer moduli.

1 Introduction

The Genetic Algorithm (GA) has long been used as an optimization tool in resolving numerical problems, especially the problem of nonlinear optimization. However, little has been done to apply GAs to asphalt pavement design until recently. The backcalculation of pavement layer moduli, design of pavement structures, and scheduling of pavement maintenance operations are the major applications of

K. Gopalakrishnan et al. (Eds.): Intel. & Soft Comp. in Infra. Sys. Eng., SCI 259, pp. 205–238.
springerlink.com

GAs. This paper presents the use of GAs for the backcalculation of pavement moduli focused in the sensitivity analysis of GAs, the characterization of mix master curve, the binder relaxation spectrum, and the material performance models of asphalt concrete using three-stage Weibull approach.

Kameyama et al. developed a method for backcalculating pavement layer moduli from surface deflections with the GA [1], with the deflections calculated by layered-elastic theory as the input condition. FWA et al. [2] stressed that although the strength of the GA-based methods for backcalculation lies in its superior global search ability, its computation time needs to be reduced before the method can be considered for routine backcalculation analysis in practice. The GA has replaced traditional calculus methods used to search for best-fit stiffness profiles of in situ pavement systems based on non-destructive test methods [3].

GAs have been used by Liu and Wang [4] for design of flexible pavement structure and by Hadi and Arfiadi [5] for design of rigid structures.

Shekharan used GAs to develop solutions to pavement deterioration models [6], and Attoh-Okine presented the application of the GA in predicting roughness progression in flexible pavements [7]. GAs have been merged with artificial neural networks and fuzzy logic to efficiently identify the distresses to be treated [8]. The advantages of such approaches in the design of decision support systems have been described by Loia et al. [9].

The GA is a technique inspired by the Darwinian theory of survival of the fittest [10]. It mimics the natural process of evolution to develop an optimum solution. GAs operate on a set of randomly generated solutions. For each of the solutions, the values of fitness function, which indicate the proximity of the solution to the optimum solution, are evaluated. The solutions with good fitness values are combined in an attempt to produce a better solution set. Replacing solutions with poor fitness with new solutions completes one iteration, or generation of the algorithm. This process is repeated until a sufficiently good fit is obtained.

The general procedure to conduct a GA analysis is [11]:

1. *Define the problem.* The parameter definition and fitness function associated with the problem should be clearly identified before conducting a GA-based analysis. The rest of the GA procedure is to find an optimum set of parameters— that is, a good gene—that minimize the fitness function. Intuitively, the residual sum of squares (*RSS*) is a good choice for the fitness function, given that the objective is to have the measurements and predictions as close as possible to model fitting.
2. *Generate N (even number) genetic starting strings.* From the problem definition, a total of p parameters $(t1, t2, t3, \ldots, tp)$ are selected to construct a gene string (or gene). A gene $\boldsymbol{\Lambda}_i$ consists of values $\left\{S_{t1}^{(i)}, S_{t2}^{(i)}, S_{t3}^{(i)}, \ldots, S_{tp}^{(i)}\right\}$ of p parameters, which are generated by using a uniform distribution over a specified range of each parameter. A gene pool is defined as a set of genes, that is, $\{\boldsymbol{\Lambda}_1, \boldsymbol{\Lambda}_2, \ldots, \boldsymbol{\Lambda}_N\}$:

Gene #1: $\mathbf{\Lambda}_1 = \left\{S_{t1}^{(1)}, S_{t2}^{(1)}, S_{t3}^{(1)}, \ldots, S_{tp}^{(1)}\right\}$

Gene #2: $\mathbf{\Lambda}_2 = \left\{S_{t1}^{(2)}, S_{t2}^{(2)}, S_{t3}^{(2)}, \ldots, S_{tp}^{(2)}\right\}$

…………

Gene #*N*: $\mathbf{\Lambda}_N = \left\{S_{t1}^{(N)}, S_{t2}^{(N)}, S_{t3}^{(N)}, \ldots, S_{tp}^{(N)}\right\}$

3. *Compute the fitness of each gene*: $\Pi(\mathbf{\Lambda}_1), \Pi(\mathbf{\Lambda}_2), \ldots, \Pi(\mathbf{\Lambda}_N)$. Fitness is the term used to measure the goodness of fit of the problem-specified fitness function $\Pi(\mathbf{\Lambda}_i)$. Most of time, it is appropriate to define the fitness function as the *RSS* of the response variables.
4. *Rank the genes according to fitness:* $\mathbf{\Lambda}_1^*, \mathbf{\Lambda}_2^*, \ldots, \mathbf{\Lambda}_N^*$.
5. *Mate nearest ranked pairs (produce offspring).* The offspring $\mathbf{\Theta}_i$ is generated as a linear combination of its parents' genes as indicated as follows. The reason to have even number of genes (Step 2) is clear.

$$\mathbf{\Theta}_1 \equiv \Phi^{(1)}\mathbf{\Lambda}_1^* + \left(1 - \Phi^{(1)}\right)\mathbf{\Lambda}_2^*$$

$$\mathbf{\Theta}_2 \equiv \Phi^{(2)}\mathbf{\Lambda}_1^* + \left(1 - \Phi^{(2)}\right)\mathbf{\Lambda}_2^*$$

…………

$$\mathbf{\Theta}_{N-1} \equiv \Phi^{(N\text{-}1)}\mathbf{\Lambda}_{N-1}^* + \left(1 - \Phi^{(N-1)}\right)\mathbf{\Lambda}_N^*$$

$$\mathbf{\Theta}_N \equiv \Phi^{(N)}\mathbf{\Lambda}_{N-1}^* + \left(1 - \Phi^{(N)}\right)\mathbf{\Lambda}_N^*$$

where, $\Phi^{(i)} = A\phi^{(i)}$ and $0 \le \phi^{(i)} \le 1$ (a random number from uniform distribution). Here, *A* is set at $\sqrt{2}$. If $A \le 1$ then there is no mutation.
6. *Check design range (or constraints)*: $S_{t1}^- \le S_{t1}^{(i)} \le S_{t1}^+$; … ; $S_{tp}^- \le S_{tp}^{(i)} \le S_{tp}^+$, *where* S_{ti}^- *is the left constraint and* S_{ti}^+ *the right constraint.* If the variable is out of range, set equal to (or preferably a little less than) constraint. It was found that the precision of the solution was affected by the range specified for each parameter. A broad parameter range is suggested for a novice with limited knowledge of the parameter. An experienced user is able to narrow down the parameter range to increase the probability of finding the optimum solution. If an inappropriate range is specified, it could lead to a local (or restrained) optimum.
7. *Discard the bottom M (< N) genes.*
8. *Repeat with top N – M gene pool plus M new genes.*

Several issues can be addressed regarding the GA:

1. The size of gene pool (or gene size), the number of generations, and the number of discarded genes for each generation run are going to affect the computational time and precision of the solution.

2. Because of the superior global search ability of GA-based method for a nonlinear problem, to obtain a more accurate solution, GA can be run first and the Newton algorithm, which is a gradient-based iteration method for solving a root-finding problem [12], can be used to refine GA solutions. However, conducting the GA alone with more generations or with a larger gene pool will also give a satisfactory solution.
3. The GA is a simple and powerful algorithm for solving the nonlinear optimization problem with constraints regardless of its disadvantages: (*a*) computational inefficiency with respect to time (not always the case) and (*b*) input of appropriate parameter range.

The generalized GA-based procedure is worth keeping in mind. In the following sections, four case studies of GA application are demonstrated. These cases begin with the problem definition and follow with numeric examples to explain how the GA works.

2 Case 1: Backcalculation of Layer Mouduli Using FWD Data

The FWD has been widely used to characterize pavement response. A variety of approaches have been used to backcalculate elastic moduli that minimize the difference between the measured peak surface deflections from the FWD and deflections calculated using a static model of the FWD on a pavement characterized by layer elastic theory—for example, MODULUS [documented by J. Uzan in Appendix E: MODULUS User's Guide of NCHRP Report 327 [13]—or the Odemark-Boussinesq approach—for example, ELMOD [14, 15] or CalBack [16]. An alternative is to use the GA approach to find an optimum set of layer moduli that minimizes the *RSS* of the difference between measured and calculated peak surface deflections. As indicated in Figure 1a, an FWD was used on a three-layer pavement structure to characterize the layer moduli. The surface deflection was instrumented at various distances. According to the GA procedure outlined previously, the follwing term are defined:

- Parameters—Three layer moduli (modulus of asphalt concrete E_{ac}, modulus of aggregate base E_{ab}, and modulus of subgrade E_{sg}).
- Fitness Function— $RSS = \sum_i \left(D_i - \hat{D}_i\right)^2 = \min$, where D_i is the FWD deflection at i^{th} position and $\hat{D}_i$ is the corresponding deflection calculated from the layered-elastic program (ELSYM5).

Application of the GA procedures for this problem includes the following:

1. Specify ranges of parameters.
2. Generate genes.
3. Calculate pavement response and evaluate the fitness function.
4. Rank and mate genes.
5. Discard bad genes and replace with new genes.
6. Repet Steps 1 through 5 until the specified number of generations is completed.

The ranges of parameters (E_{ac}, E_{ab}, and E_{sg}) were determined according to the user's own experience and judgment. For each gene in a gene pool, ELSYM5 was used to calculate the pavement response—in this case the surface deflection $\hat{D}_i$—and then the fitness function was evaluated. According to the evaluated fitness, the parent genes were ranked and then mated to generate better offspring. The mated genes with larger fitness were discarded and replaced with the new genes generated randomly by using the uniform distribution constrained by their specified ranges as in Step 2. The whole process was iterated until the specified number of generations was completed.

This example FWD deflection basin was measured at 15.6°C pavement surface temperature. Figure 1a schematically illustrates the FWD configuration, problem definition, and the pavement structure. The properties and parameter ranges of this three-layer pavement system are as follows:

Layer	Thickness (mm)	Poisson's ratio	Parameter range (MPa)
Asphalt concrete	100	0.35	2,068 to 3,447
Aggregate base	400	0.40	345 to 1,379
Subgrade		0.45	69 to 345

Figure 1b presents the backcalculation result when a heavy FWD load (54.6 kN) is applied to the pavement. As indicated in Figure 1b, the calculated response is quite satisfactory compared with the measured deflections. In this case, the nonlinearity of the stress-dependent subgrade was not considered. However, the nonlinearity can be resolved with a finite element program as the stress-strain engine and following the same GA procedure as presented here.

An analysis of variance (ANOVA) was conducted to evaluate the effects of gene size, number of generations, and the FWD loading levels on the parameter variation. Each factor has three levels. The experimental design was:

- Load factor [H(54.6 kN), M (39.1 kN), and L (22.3 kN)].
- Genes factor(500, 1,000, and 2,000); and
- Generations factor (50, 100, and 200).

For each combination of factor levels, 10 replicates were conducted for a total of 270 GA runs. The number of discarded genes was half the total number of genes. Figure 2 presents the design plots of mean of layer moduli and the coefficient of variation (CV) of layer moduli.

Several findings of ANOVA can be addressed as follows:

1. The mean of E_{ac} is sensitive to the gene size and number of generations. On the contrary, the means of E_{ab} and E_{sg} are relatively insensitive. The mean of E_{ac} is positively proportional to the load level. However, the means of E_{ab} and E_{sg} are negatively proportional to the load level.

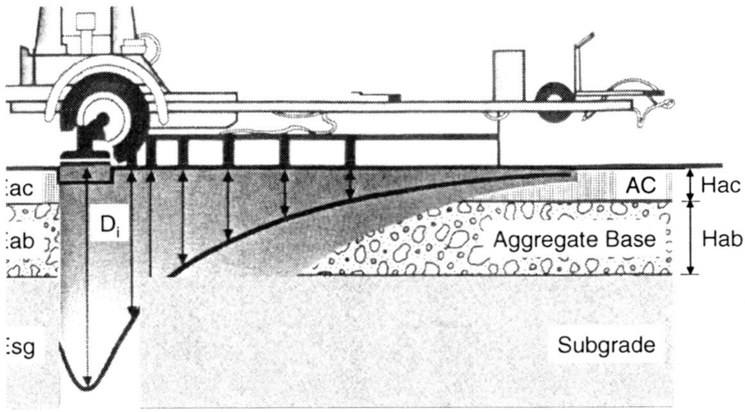

(a)

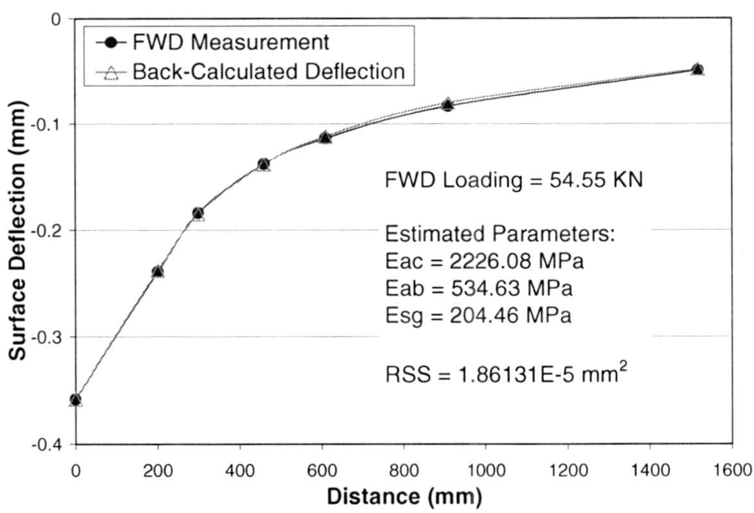

(b)

Fig. 1. (a) Schematic of FWD and pavement structure (AC: asphalt concrete) and (b) surface deflection fitting results after 50 generations with 500 genes.

2. In general, all three parameters (E_{ac}, E_{ab}, and E_{sg}) share a common tendency: the bigger the genes and the higher the number of generations, the lower the CV of the moduli (i.e., the smaller the modulus variation in all three parameters).
3. For the asphalt concrete layer, the lower the loading level, the lower the CV of moduli. Conversely, for the aggregate base and subgrade layers, the low loading level causes a higher CV than the high loading level.

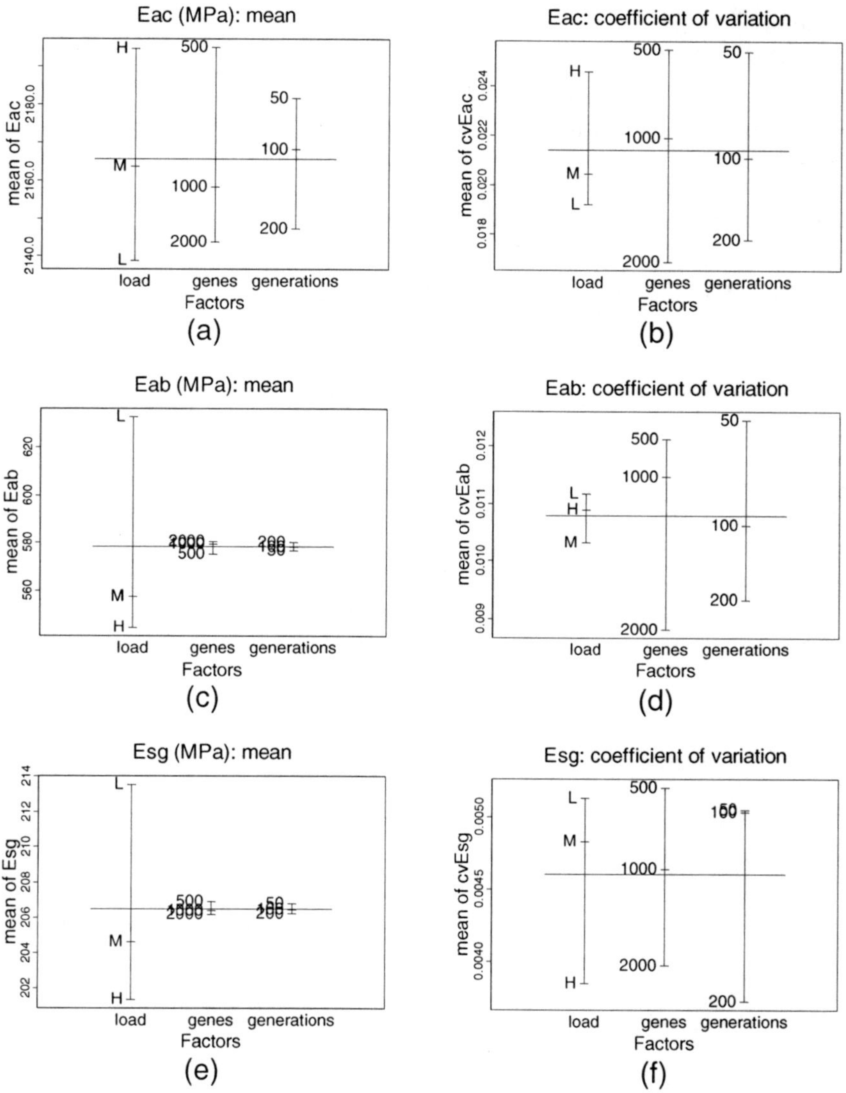

Fig. 2. Design plots of means and CVs of three parameters (E_{ac}, E_{ab}, and E_{sg}) showing effects of load level, gene size, and number of generations.

4. On average, the magnitude of the CV of the asphalt concrete layer is about two times that of the aggregate base layer and about four times that of the subgrade layer.
5. One can certainly conclude from the ANOVA that the bigger the size of genes and generations, the smaller the modulus variation. However, a trade-off exists between the computational time and the modulus variation. The computational time is a considerable disadvantage for the backcalculation of layer moduli

because the layered-elastic program needs to be called up whenever the fitness of a gene is calculated.

In spite of the GA's superior global search ability, computation efficiency is still its Achille's heel when compared to other backcalulation programs, such as ELMOD and CalBack, or the method using the Kalman filter [17]. However, the accuracy of moduli obtained through these various approaches still needs to be verified with in situ measured stress-strain data.

3 Case2: Master Curve and Its Application

The complex modulus master curve obtained from flexural frequency sweep tests, which are considered nondestructive tests, is a useful tool to characterize the effects of loading frequency and temperature on the initial stiffness of an asphalt mix. A function that describes flexural stiffness as a function of temperature and time of loading can be used for pavement design.

Flexural frequency sweep tests are conducted mostly from 15 to 0.01 Hz at three or four temperature levels. Under the assumption that asphalt mix is a time-temperature-rheologically simple material, the curves can be shifted horizontally (or vertically if necessary) relative to one of the test temperatures to obtain the full spectrum of complex moduli. The ratio of relaxation times at different temperatures is termed as a_T. The temperature shift factor $\ln a_T$ is expressed in the form $\ln a_T = -C_1(T - T_{ref})/(C_2 + T - T_{ref})$, where C_1 and C_2 are two constants [18]. The questions then are what should be the horizontal (or vertical if necessary) distances and how should the goodness of fit be measured?

The most intuitive measure to set up the fitness function is *RSS*, while comparing the observed dependent variable data (y_i) with the corresponding predicted values ($\hat{y}_i$). With limited knowledge about the shifted master curve—no exact mathematical function is known—the generalized additive model with spline fitting becomes the best candidate to calculate the predicted $\hat{y}_i$ for fitness function [19]. In this case, the parameters and fitness function are defined as follows:

- Parameters—Horizontal shifts S_1, S_2, and S_3 (Figure 3a).
- Fitness function—$RSS = \sum (y_i - \hat{y}_i)^2 = \min$, where y_i is the measured complex modulus and $\hat{y}_i$ is the fitted complex modulus.

As an example, Figure 3a presents the flexural frequency sweep test results for a typical California mix with asphalt from a Valley source. The flexural frequency sweep test results presented here are a summary from two flexural fatigue test setups, namely conventional (without support) for temperatures of 5, 20, and 30°C and modified (with support) for temperatures of 20, 30, and 40°C [20]. "With support" means an aluminum support is placed beneath the specimen to prevent the creep effect from occurring because of specimen self-weight while it is subjected

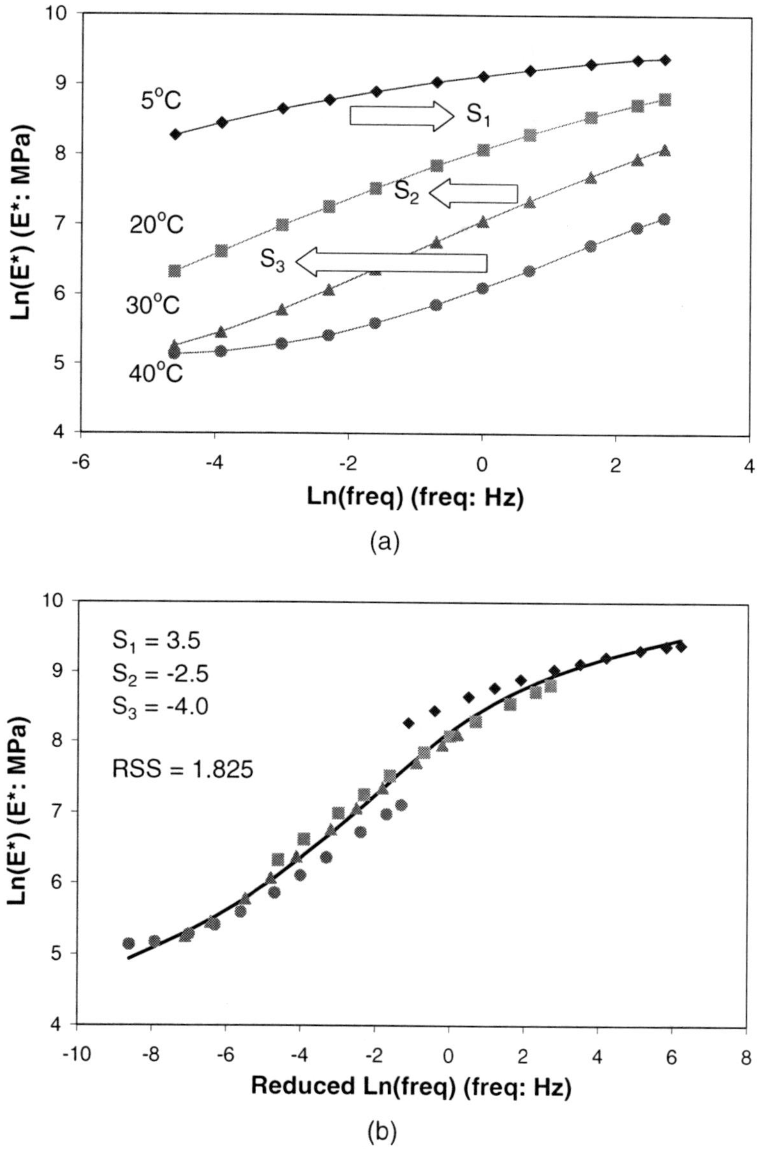

Fig. 3. (a) Parameter definition and flexural frequency sweep test results at 5°C, 20°C, 30°C, and 40°C and (b) intermediate fitting result (freq: frequency).

to high temperatures. The loading frequencies used are 15, 10, 5, 2, 1, 0.5, 0.2, 0.1, 0.05, 0.02, and 0.01 Hz. The upper limit of 15 Hz is a constraint imposed by the capacity of the test machine. The general principles for conducting frequency sweep tests are "quick to slow in loading frequency and hot to cold in temperature (if one beam is used)." The temperatures were set at 40, 30, 20, and 5°C.

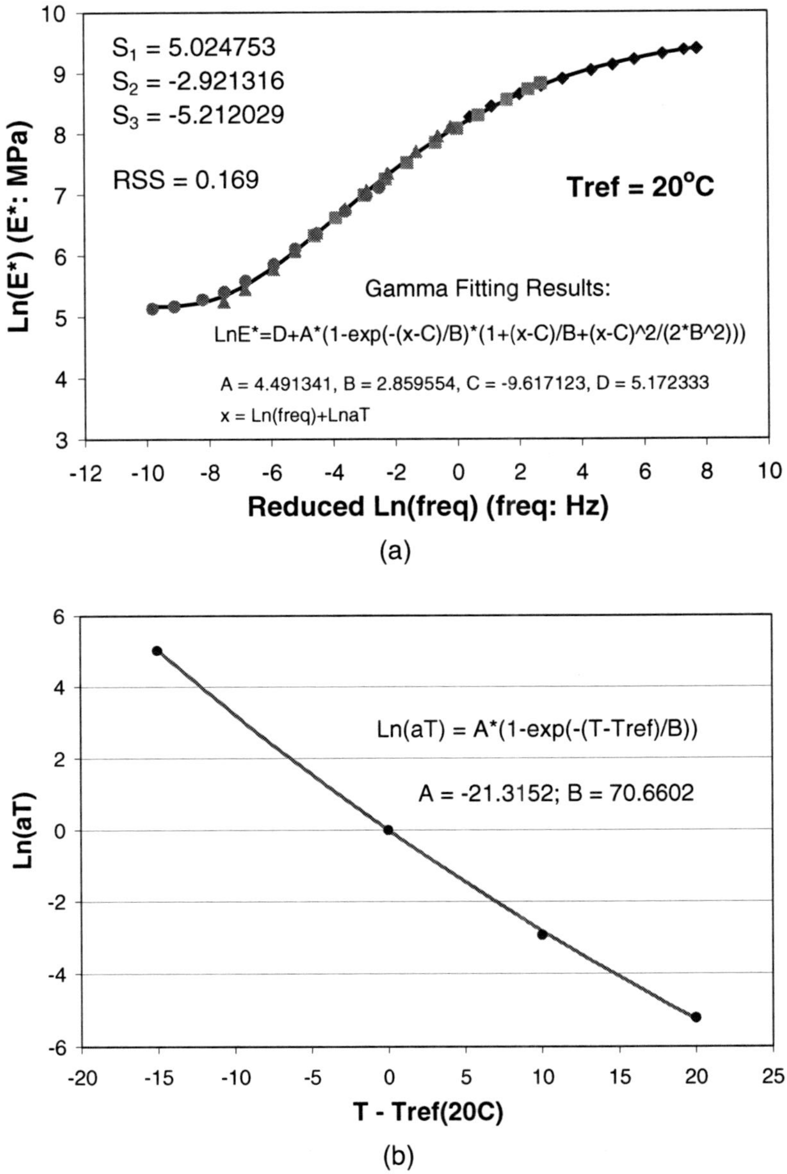

Fig. 4. (a) Fitting results after 50 generations with 100 genes and gamma fitting afterward and (b) temperature-shifting relationship with a gamma-type fitting (freq: frequency).

Taking the E^* master curve as an example, the parameters to be fit are the horizontal shifts for different temperatures—that is, $\Lambda = \{S_1, S_2, S_3\}$. In this case, three parameters were defined (as indicated in Figure 3a), S_1 shifting to the right and S_2 and S_3 shifting to the left according to the reference temperature 20°C. The shifting is signed: positive to the right and negative to the left. Figure 3b illustrates an intermediate fitting result with a set of parameters {3.5, -2.5, -4.0}, which results in a *RSS* of 1.825. After 50 generations (100 genes at each generation), an optimum parameter set {5.0248, -2.9213, -5.2120} with *RSS* 0.169 produced a quite satisfactory shifting result, as indicated in Figure 4a. No function was specified for the complex modulus master curve. Instead, a generalized additive model with spline fitting was used to calculate *RSS*. However, for the purpose of further analysis, gamma fitting was suggested for the master curve as well as the time-temperature shifting relationship.

The main purpose in conducting the gamma nonlinear fitting of the data of frequency sweep tests is to find a suitable mathematical function that can represent the relationship of the complex modulus and reduced loading frequency at a reference temperature. The gamma distribution function with shape parameter n (a positive number) and scale parameter β is expressed in the following form [21]:

$$y = F(x) = \begin{cases} 1 - \exp\left(\dfrac{-x}{\beta}\right) \cdot \displaystyle\sum_{m=0}^{n-1} \frac{x^m}{\beta^m m!} & , x \geq 0 \\ 0 & , x < 0 \end{cases}$$

where $F(x)$ is the gamma distribution function and m is an index number.

The characteristics of distribution function are

- The values of y range from 0 to 1,
- The x values start from zero, and
- The y value is asymptotic to 1 as x increases.

These aspects of a distribution function are employed in constructing the nonlinear fitting.

The modified gamma fitting of E^* master curve has the following formulation:

$$\ln E^* = D + A \cdot \left(1 - \exp\left(-\frac{x - C}{B}\right) \cdot \sum_{m=0}^{n-1} \frac{(x - C)^m}{B^m m!}\right)$$

where x = reduced loading frequency (Hz) in natural logarithm and A, B, C, and D are mathematically determined coefficients.

The modified gamma fitting of temperature-shifting relationship has the same formulation as used in the E^* master curve but with $n = 1$, $D = 0$ and can be expressed as follows:

$$Ln(a_T) = A \cdot \left(1 - \exp\left(-\frac{T - T_{ref}}{B}\right)\right)$$

where $Ln(a_T)$ is the temperature shift factor, A and B are the mathematically determined coefficients, and T_{ref} is the reference temperature (20°C).

Figure 4 demonstrates the appropriate nonlinear fit by using the modified gamma fitting for the master curve and the time-temperature shifting relationship. The advantage of using the gamma function is to provide a specified continuous mathematic function that is easy to incorporate into programming and calculation. In a discrete sense, using the spline function when estimating the GA parameters can do the same job.

The major purpose of finding the master curve and the time-temperature shift factor is to correct the loading frequency and temperature effects when conducting a pavement performance prediction.

As an example, Figure 5 indicates the way to correct the initial stiffness with various loading frequencies and temperatures. The general steps to calculate interpolatively the initial stiffness at the loading frequency ω and temperature T are as follows:

1. Find the temperature difference with respect to the reference temperature.
2. Map the temperature difference through the time-temperature relationship function to obtain the shift factor (Figure 5a).
3. Locate the loading frequency ω in Figure 5b; add up the signed shift factor; and then map through the master curve function to obtain the corrected stiffness.

For example, suppose that the temperature at the bottom of asphalt concrete layer is 25°C, which is 5°C greater than the reference temperature 20°C; hence, from Figure 5a the time-temperature relationship we have $\ln a_T = -1.452$. By adding this value with the in situ loading frequency, for example, 10 Hz in this case, we then have the corrected initial stiffness at 25°C and 10 Hz as illustrated in Figure 5b. Notice that for a specified loading frequency, the correction moves to the left at higher temperature (relative to the reference temperature) and moves to the right at lower temperature.

With the GA, it is relatively easy to incorporate the parameters for vertical shifts and conduct the same Gamma nonlinear fitting procedure to the shifted master curve. For this GA application, the computational time is not an issue because no serious calculation is involved as in the modulus-backcalulation case.

4 Case 3: Discrete Relaxation Spectrum of Asphalt Binder

Characterization or modeling of asphalt binder properties often requires the relaxation time spectrum, $G(t)$, instead of using the dynamic moduli, $G'(\omega)$and $G''(\omega)$, which are easy to measure from the dynamic shear rheometer test in the frequency domain. In addition, the relaxation model in the time domain is easier to interpret than in the frequency domain. The relaxation spectrum not only enables conversion of time to frequency (or vice versa) but also yields information about molecular structure. The GA provides a relatively easy and powerful way to solve the parameters of a discrete relaxation spectrum g_i, λ_i of a generalized Maxwell solid model by fitting dynamic mechanical data. Instead of reinventing the wheel theoretically in the field of asphalt rheology, the main point of this GA application

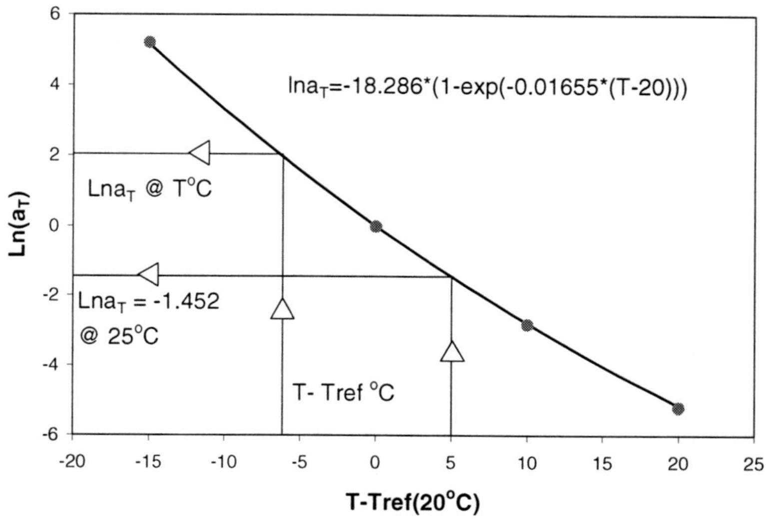

(a)

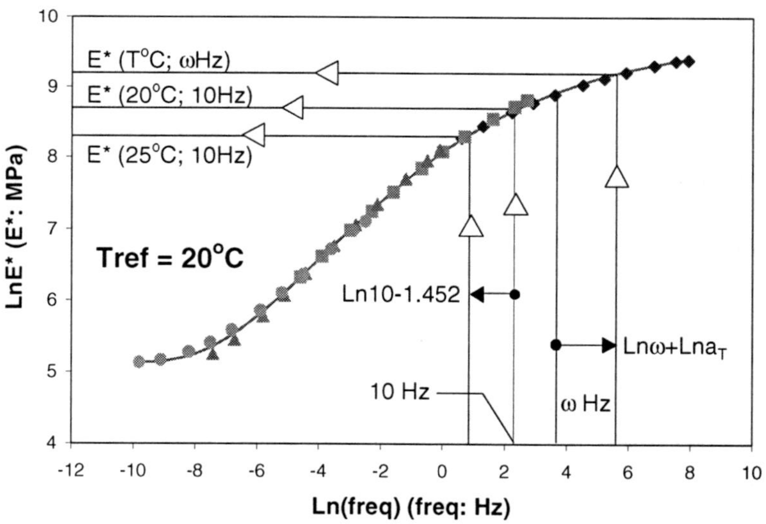

(b)

Fig. 5. (a) Temperature shift factors at T°C and 25°C and (b) initial stiffness correction at T°C and 25°C at loading frequency of 10 Hz.

is to demonstrate the relatively intuitive way to find the parameters of relaxation spectrum.

The relaxation modulus, $G(t)$, of a generalized Maxwell solid model can be written as a discrete set of exponential decays as follows [19]:

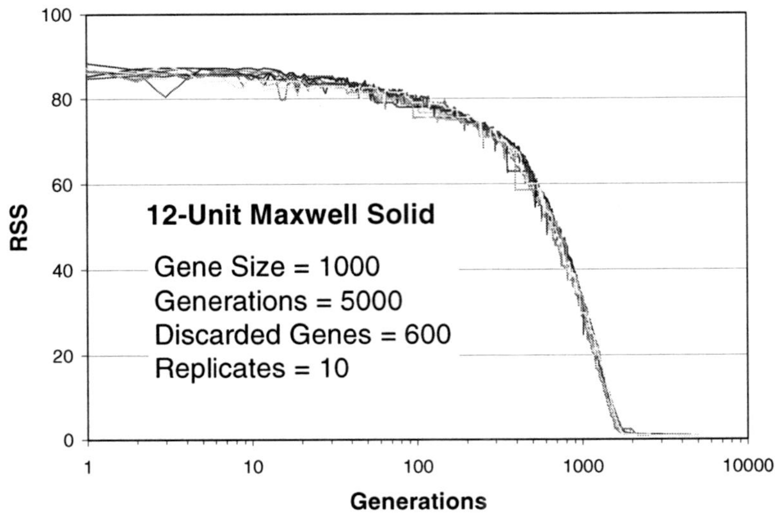

Fig. 6. RSS convergence trends of 10 GA replicates against generations, with a 12-unit generalized Maxwell solid model.

$$G(t) = G_e + \sum_{i=1}^{N} g_i \exp(-t/\lambda_i)$$

The formulation consists of the equilibrium modulus G_e which is finite and greater than 0 for solids and N relaxation modes defined by their relaxation strengths g_i and their relaxation times λ_i. The dynamic moduli, storage modulus G' and loss modulus G'' can be expressed as follows:

$$G'(\omega) = G_e + \sum_{i=1}^{N} g_i \frac{(\omega\lambda_i)^2}{1+(\omega\lambda_i)^2} \tag{1}$$

$$G''(\omega) = \sum_{i=1}^{N} g_i \frac{\omega\lambda_i}{1+(\omega\lambda_i)^2} \tag{2}$$

Hence, the problem definition for GA application is to find the parameters of a discrete relaxation spectrum by simply fitting Equations 1 and 2 to G' and G'' data. Because the ranges of parameters might be over several decades, it is suggested that the ratio residual $G'(\omega_j)/\hat{G}'_j - 1$ be used instead of the normal residual $G'(\omega_j) - \hat{G}'_j$. As before, it is necessary to minimize *RSS* to obtain the optimum model. It should be remembered that the G' and G'' curves need to be satisfied simultaneously by the same set of parameters; so, *RSS* is the sum of squares of ratio residuals of G' and G''. For GA application, the following are defined:

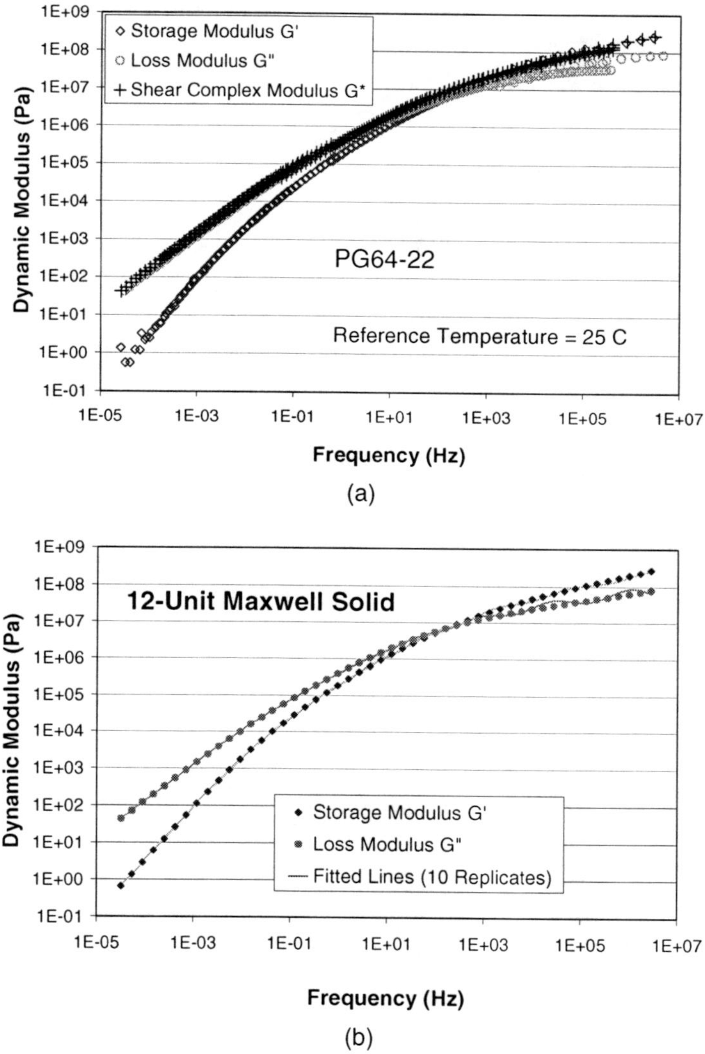

Fig. 7. (a) Plots of master curves of shear complex modulus G^*, storage modulus G', and loss modulus G'' with PG64-22 asphalt binder and (b) G' and G'' fitting results using a 12-unit generalized Maxwell solid model.

- Parameters—The model contains $2N+1$ parameters including N pairs of (g_i, λ_i) and G_e
- Fitness function— $RSS = \sum_{j=1}^{m}\left(\left[\frac{G'(\omega_j)}{\hat{G}'_j}-1\right]^2+\left[\frac{G''(\omega_j)}{\hat{G}''_j}-1\right]^2\right) = \min$, where $\hat{G}'_j$ and $\hat{G}''_j$ are the measured data at m frequencies

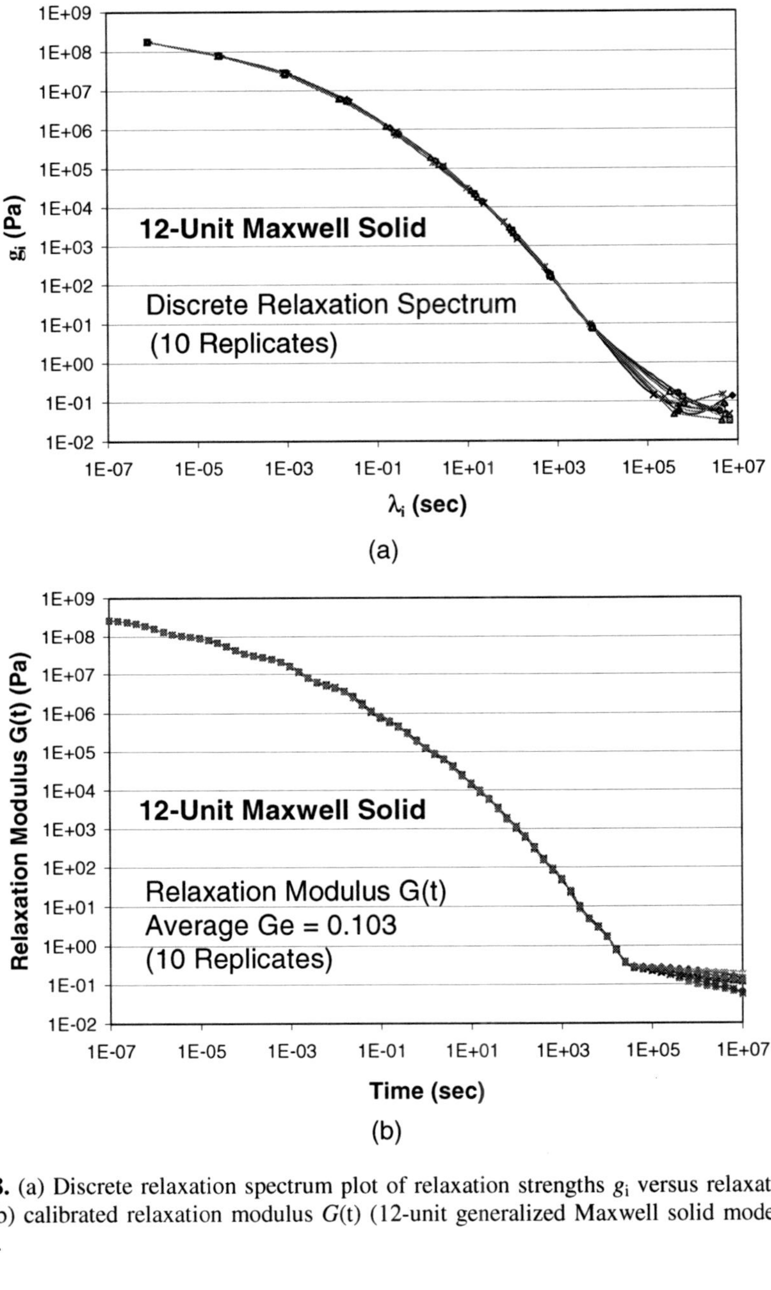

Fig. 8. (a) Discrete relaxation spectrum plot of relaxation strengths g_i versus relaxation λ_i and (b) calibrated relaxation modulus G(t) (12-unit generalized Maxwell solid model was used).

In this case, a total of 10 GA runs was conducted with a gene size of 1,000, 5,000 generations, and 600 discarded genes. A 12-unit generalized Maxwell solid model was adopted in this demonstration case (i.e., a total of 25 parameters). The asphalt binder studied here was a PG64-22. The convergence trends of *RSS* of these 10 GA runs present similar patterns as indicated in Figure 6. The *RSS* stabilizes around 2,000 generations. It indicates that the result cannot be better off if more than one run is simulated. This statement is also verified by the fitting results of G' and G'' data presented in Figure 7b. Because the fitting results of 10 replicates are so close, the fitted lines overlap each other in Figure 7b.

Figure 7a presents the shifted master curves (also using GA) for complex modulus, storage modulus, and loss modulus. Baumgaertel and Winter [22] suggested that,

- the initial number of relaxation models be chosen empirically between 1 and 2 decades and
- the negative g_i values be deleted, which is not necessary in GA analysis because one can specify the g_i range.

At the beginning, nothing was known about the magnitudes of g_i and λ_i; thus, a broad range (10^{-9} to 10^{9}) was specified for each g_i and the range of λ_i was specified over 3 decades and with 2 decades of overlap. The total range of λ_i was in the range 10^{-7} to 10^{7}. The equilibrium modulus G_e was set in the range 0 to 10^{8}. The detail parameter range is listed in Table 1. Table 1 also lists the statistical summary of the 10 GA runs.

As indicated in Figure 7b, other than the phenomenon that the loss modulus G'' has a wavy fitting at the high frequencies, the fitting results for G' and G'' data are quite satisfactory and consistent for all 10 replicates. This waviness phenomenon was also reported by Stastna et al. in their inspection of the Maxwell model [23]. The average *RSS* is about 0.898, the standard deviation is about 0.0504, and

Table 1. Statistical summary and parameter input range of GA application in discrete relaxation spectrum

i	g_i (Pa)					λ_i (sec)				
	μ	σ	CV	Parameter Range begin	Parameter Range end	μ	σ	CV	Parameter Range begin	Parameter Range end
1	1.768E+08	1.111E+06	0.006	1E-09	1E+09	8.226E-07	1.145E-08	0.014	1E-07	1E-04
2	7.730E+07	1.071E+05	0.001	1E-09	1E+09	3.179E-05	1.371E-06	0.043	1E-06	1E-03
3	2.735E+07	9.542E+05	0.035	1E-09	1E+09	9.668E-04	6.406E-05	0.066	1E-05	1E-02
4	5.586E+06	3.650E+05	0.065	1E-09	1E+09	2.057E-02	3.397E-03	0.165	1E-04	1E-01
5	8.668E+05	1.593E+05	0.184	1E-09	1E+09	2.609E-01	4.708E-02	0.180	1E-03	1E+00
6	1.214E+05	4.107E+04	0.338	1E-09	1E+09	3.181E+00	2.433E+00	0.765	1E-02	1E+01
7	3.029E+04	3.861E+04	1.275	1E-09	1E+09	1.599E+01	6.111E+00	0.382	1E-01	1E+02
8	2.384E+03	7.710E+02	0.323	1E-09	1E+09	1.019E+02	1.830E+01	0.180	1E+00	1E+03
9	1.941E+02	3.892E+01	0.200	1E-09	1E+09	6.547E+02	6.622E+01	0.101	1E+01	1E+04
10	8.390E+00	9.254E-01	0.110	1E-09	1E+09	5.626E+03	3.308E+02	0.059	1E+02	1E+05
11	1.149E-01	4.721E-02	0.411	1E-09	1E+09	4.072E+05	1.601E+05	0.393	1E+03	1E+06
12	6.965E-02	4.099E-02	0.588	1E-09	1E+09	5.449E+06	1.134E+06	0.208	1E+04	1E+07
Ge	1.029E-01	4.868E-02	0.473	0E+00	1E+08					
RSS	8.979E-01	5.041E-02	0.056							

the CV is 0.056. Figure 8a plots g_i versus λ_i to construct a discrete relaxation spectrum. Note that noisy discrepancies occur at high relaxation times. The same disturbance is also observed in the plot of relaxation modulus $G(t)$ shown in Figure 8b; in addition, an obvious discontinuity happens around 4,000 seconds.

The fitting results indicate that using the GA approach in solving the discrete relaxation spectrum is quite promising. The GA approach can not only avoid the potential ill-condition problems while solving a system of equations, but also encourage novices in their attempts to solve the question easily and intuitively without any advance knowledge. The fitting results also suggest that, practically, no benefit can be achieved if more than one run is simulated. In addition, computational time is not an issue for a GA approach for this type of problem.

5 Case 4: Three-Stage Weibull Approach and Its Application

Fatigue cracking and rutting have long been recognized as the two major distress types occurring in asphalt concrete pavement. To characterize asphalt-aggregate mixes relative to these two distresses, the Strategic Highway Research Program (SHRP) project has recommended the flexural controlled-stress/strain (or controlled-load/deformation) beam fatigue test for fatigue response [24] and the repetitive simple shear test with constant height (RSST-CH) for measuring permanent deformation chacteristics [25]. The fatigue damage process (or stiffness deterioration process) and the permanent deformation accumulation process are the two most important material properties which can be used in mechanistic-empirical models to predict the pavement performance. The three-stage Weibull approach introduced here can express these two distress processes with one common formulation.

5.1 Three-Stage Weibull Approach

A three-stage Weibull fatigue curve is a three-stage stiffness deterioration curve representing flexural fatigue damage formulated by the Weibull distribution function. The stiffness ratio (*SR*) at repetition *n*, which is defined as the ratio of stiffness at repetition *n* relative to the initial stiffness (determined roughly after 50 repetitions), is utilized as an index for characterizing the stiffness deterioration process. The use of stiffness ratio as an index has several advantages including:

1. Stiffness is easy to measure both in the laboratory and in the field, and
2. Stiffness is often used as an input for layered-elastic programs for pavement analysis, thus making it useful for programming fatigue performance simulations.

The stiffness deterioration curve obtained from the flexural controlled-deformation beam fatigue test, especially in the crack initiation phase, can be adequately expressed as a two-parameter one-stage Weibull distribution function with the following form:

$$SR = \exp\left(-\alpha n^{\beta}\right) \text{ or } \ln\left(-\ln SR\right) = \ln\alpha + \beta\ln n$$

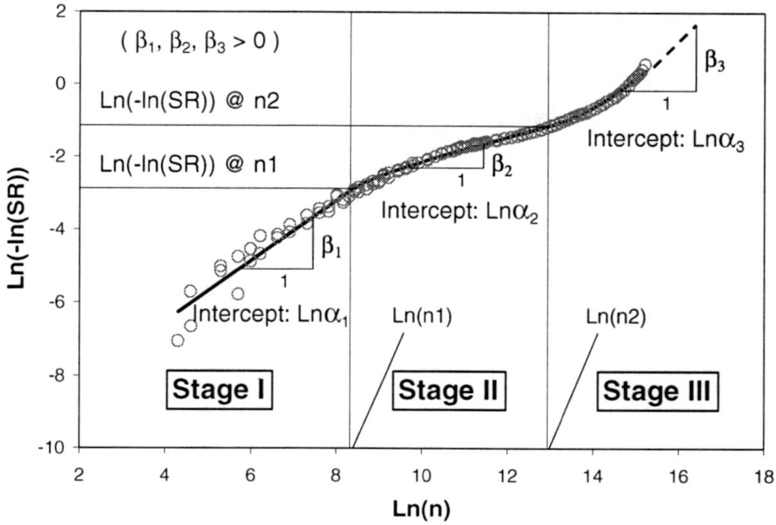

Fig. 9. Definition of a Weibull fatigue curve in three stages: Stage I—heating and reaching of temperature equilibrium under initial repetitions, Stage II—crack initiation developing, and Stage III—crack propagation.

where SR is the stiffness ratio, n is number of the loading repetitions, and α, β are the experiment-determined coefficients.

However, the one-stage Weibull equation does not appear to represent the damage process when:

1. The fatigue test has a prolonged initial phase;
2. The fatigue tests are conducted beyond a certain stiffness ratio threshold at which the fatigue cracks start to propagate or at the fatigue damage slows down as seen in certain mixes with modified binders.

Therefore, to describe the stiffness deterioration process in all three stages, an alternative is necessary. One of the approaches that serves this purpose is the application of the two-stage Weibull distribution function suggested by Jiang [26], with extension to a three-stage Weibull distribution function [27]. This extended distribution function has the following form:

$$\begin{cases} SR_1 = \exp\left(-\alpha_1 n^{\beta_1}\right), & 0 \le n < n_1 \\ SR_2 = \exp\left(-\alpha_2 \cdot (n-\gamma_1)^{\beta_2}\right), & n_1 \le n < n_2 \\ SR_3 = \exp\left(-\alpha_3 \cdot (n-\gamma_2)\beta_3\right), & n_2 \le n < \infty \end{cases}$$

That is,

$$\begin{cases} \ln\left(-\ln SR_1\right) = \ln\alpha_1 + \beta_1 \ln n, & 0 \le n < n_1 \\ \ln\left(-\ln SR_2\right) = \ln\alpha_2 + \beta_2 \ln\left(n-\gamma_1\right), & n_1 \le n < n_2 \\ \ln\left(-\ln SR_3\right) = \ln\alpha_3 + \beta_3 \ln\left(n-\gamma_2\right), & n_2 \le n < \infty \end{cases}$$

The equation consists of six fundamental parameters, α_1, α_2, α_3, β_1, β_2, and β_3, and four derivative parameters, n_1, n_2, γ_1, and γ_2 (as illustrated in Figure 9).

The equation needs to comply with the continuity conditions at n_1 and n_2:

1. $SR_1 = SR_2$ and $\frac{\partial SR_1}{\partial n} = \frac{\partial SR_2}{\partial n}$, when $n = n_1$.
2. $SR_2 = SR_3$ and $\frac{\partial SR_2}{\partial n} = \frac{\partial SR_3}{\partial n}$, when $n = n_2$.

With a series of mathematical manipulations, for the given α_1, α_2, α_3, β_1, β_2, and β_3, the four parameters n_1, n_2, γ_1, and γ_2 can be calculated sequentially as follows:

$$n_1 = \left[\frac{\alpha_2}{\alpha_1}\cdot\left(\frac{\beta_2}{\beta_1}\right)^{\beta_2}\right]^{\frac{1}{\beta_1-\beta_2}}$$

$$\Rightarrow \gamma_1 = \left(1-\frac{\beta_2}{\beta_1}\right)\cdot n_1$$

$$\Rightarrow n_2 = \gamma_1 + \left[\frac{\alpha_3}{\alpha_2}\cdot\left(\frac{\beta_3}{\beta_2}\right)^{\beta_3}\right]^{\frac{1}{\beta_2-\beta_3}}$$

$$\Rightarrow \gamma_2 = \left(1-\frac{\beta_3}{\beta_2}\right)\cdot n_2 + \frac{\beta_3}{\beta_2}\cdot\gamma_1$$

To fit the values of the parameters for a three-stage Weibull equation to a flexural controlled-deformation beam fatigue test using the GA, it is necessary to resolve the six fundamental parameters, α_1, α_2, α_3, β_1, β_2, and β_3. Following the GA procedure outlined previously, the following are defined:

- Parameters—Six Weibull fundamental parameters $\ln\alpha_1$, $\ln\alpha_2$, $\ln\alpha_3$, β_1, β_2, and β_3.
- Fitness Function—$RSS = \sum_i (y_i - \hat{y}_i)^2 = \min$, where y_i is the measured value of $\ln(-\ln SR)$ and $\hat{y}_i$ the corresponding fitted value.

The GA procedures for this problem include the following:

1. Specify the ranges of parameters, $\ln\alpha_1$, $\ln\alpha_2$, $\ln\alpha_3$, β_1, β_2, and β_3.
2. Define a gene, which is a set of these six parameters that are generated randomly by using the uniform distribution constrained by their specified ranges.
3. For each gene in a gene pool, calculate the four derivative parameters, n_1, γ_1, n_2 and γ_2, and then evaluate the fitness function.
4. Rank the genes according to fitness, mate the ranked genes, abandon the bad genes, and then replace the discarded genes with new genes.

5. Repeat Step 2 through Step 4 until the specified number of generations is reached.

A systematic method for determining the initial estimate for the three-stage Weibull fitting is used. Figure 10a plots $\ln(-\ln SR)$ versus $\ln n$ for a laboratory-mixed laboratory-compacted PG64-10 mix with 6.2 percent air-void content and 5.0 percent binder content tested at 10 Hz, 20°C, and 200 microstrain. The following steps are required to determine the three-stage Weibull parameters:

1. Find the asymptotic regression lines for both stages I and III and the regression line for stage II in the plot of $\ln(-\ln SR)$ versus $\ln n$. These intercepts and slopes of linear regression lines are the initial estimates of $\ln\alpha_1$, $\ln\alpha_2$, $\ln\alpha_3$, β_1, β_2, and β_3 (Figure 10b).
2. Calculate n_1, γ_1, n_2 and γ_2 sequentially.
3. Plot the fitting result with the real data and change the values of the parameters if modification is necessary (Figure 10c).
4. Fit with the GA using an appropriate program—for example, FORTRAN (Figure 10d).

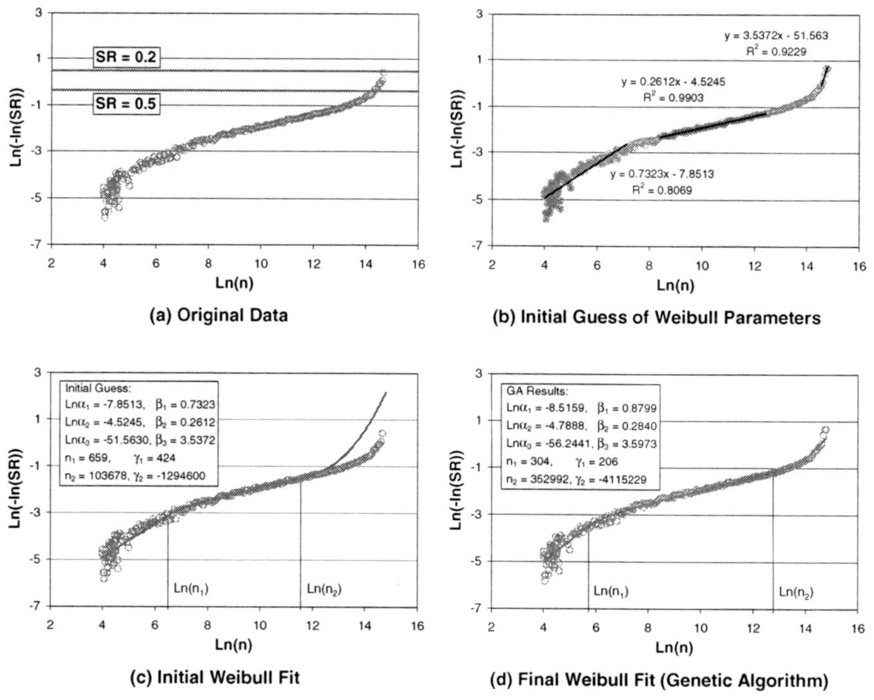

Fig. 10. Parameter estimation of a three-stage Weibull fatigue equation: (a) original data, (b) initial guess of Weibull parameters, (c) initial Weibull fit, and (d) final Weibull fit by using GA.

The parameter properties of the three-stage fatigue Weibull curves can be summarized as follows:

- β_1, β_2, and $\beta_3 > 0$
- $\ln\alpha_1$, $\ln\alpha_2$ and $\ln\alpha_3 < 0$
- $\begin{cases}\beta_2 < \beta_1 & \textit{if and only if} \quad \gamma_1 > 0 \\ \beta_2 > \beta_1 & \textit{if and only if} \quad \gamma_1 < 0\end{cases}$
- $\begin{cases}\beta_3 < \beta_2 & \textit{if and only if} \quad \gamma_2 > 0 \quad \text{(i.e., crack propagation has been suppressed)} \\ \beta_3 > \beta_2 & \textit{if and only if} \quad \gamma_2 < 0 \quad \text{(i.e., crack propagation occurs in stage III)}\end{cases}$

The definition of a three-stage Weibull shear curve (as presented in Figure 11) is similar to the definition of fatigue, except the permanent shear strain (*PSS*), rather than the stiffness ratio (*SR*), is used as the response variable to characterize the mix's rutting performance. The three-stage Weibull shear curve is generally a mirror image of a Weibull fatigue curve along the x-axis.

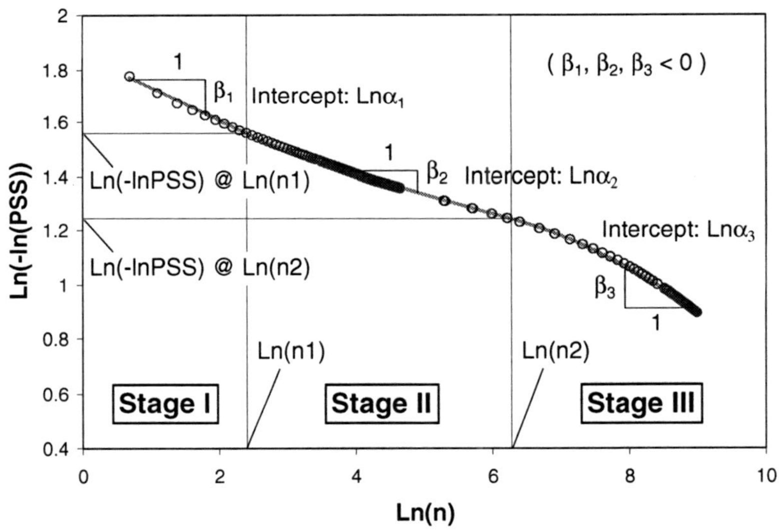

Fig. 11. Definition of a Weibull shear curve in three stages.

The parameter properties of the three-stage Weibull shear curve can be summarized in the following:

- β_1, β_2, and $\beta_3 < 0$
- $\ln\alpha_1$, $\ln\alpha_2$ and $\ln\alpha_3 > 0$
- $\begin{cases}|\beta_2| < |\beta_1| & \textit{if and only if} \quad \gamma_1 > 0 \\ |\beta_2| > |\beta_1| & \textit{if and only if} \quad \gamma_1 < 0\end{cases}$

- $\begin{cases} |\beta_3| < |\beta_2| & \textit{if and only if} \quad \gamma_2 > 0 \quad \text{(i.e., retarded shear damage at stage III)} \\ |\beta_3| > |\beta_2| & \textit{if and only if} \quad \gamma_2 < 0 \quad \text{(i.e., accelerated shear damage at stage III)} \end{cases}$

5.2 Application of Three-Stage Weibull Approach: Material Specification

One feasible application of using the Weibull parameters obtained from the GA is to establish the material performance specifications for fatigue cracking and rutting of asphalt concrete mixes.

The laboratory testing data used in the following discussion includes a total of 172 fatigue tests and a total of 177 RSST-CH tests subjected to various testing conditions and material properties [28]. The mix types include, conventional dense-graded asphalt concrete with AR4000 binder (AR4000-D), rubberized asphalt concrete with gap gradation (RAC-G), and dense-graded and gap-graded mixes with MB4 [meeting the Caltrans MB4 specification (2003)], MB15 (meeting the MB4 specification and containing 15 percent recycled tire rubber), and MAC15 (Southern California Green Book specification containing 15 percent recycled tire rubber) modified binders. All fatigue and RSST-CH tests were fit with a three-stage Weibull curve using the GA to obtain the optimized Weibull parameters. The three-stage Weibull parameters were then collected and statistically analyzed.

It was found that the use of parameters $Ln\alpha_2/\beta_2$ and $Ln\alpha_3/\beta_3$, the intercepts divided by slopes of a Weibull curve at stages II and III respectively, to characterize both fatigue and rutting performance seems to be very promising and rational especially from the point of view of mixes with polymer-modified or rubberized binder. The conventional two-point fatigue life modeling considers only initial (50th load repetitions) and end (50 percent stiffness reduction) points and neglects the fatigue damage process. For mixes using polymer-modified or rubberized binders, the traditional fatigue life definition of 50 percent loss of initial stiffness does not capture the improved crack propagation resistance of mixes; in addition, test duration to reach 50 percent stiffness reduction is very long and thus not feasible. Hence, to accurately characterize the fatigue performance of these mixes, the fatigue damage process becomes important.

Figures 12 and 13 illustrate the performance contour plots of fatigue [fatigue life (*Nf*) in natural logarithm] and shear [cycles to 5 percent permanent shear strain (*pct5*) in natural logarithm] test results as well as the overall mean of each mix. The plots indicate that the contour line values appear to be a linear combination of the $Ln\alpha_3/\beta_3$ and $Ln\alpha_2/\beta_2$ parameters. It is then suggested that the $Ln\alpha_3/\beta_3$ and $Ln\alpha_2/\beta_2$ parameters can be used for characterizing both fatigue and shear performance in a very consistent way, with increasingly negative values corresponding to improved performance for both fatigue tests and RSST-CH tests.

5.2.1 Rationales for Selecting Lnα3/β3 and Lnα2/β2 Parameters

The rationale for selecting $Ln\alpha_2/\beta_2$ and $Ln\alpha_3/\beta_3$ as the parameters to specify the fatigue/rutting performance of asphalt-aggregate mixes is based on the following

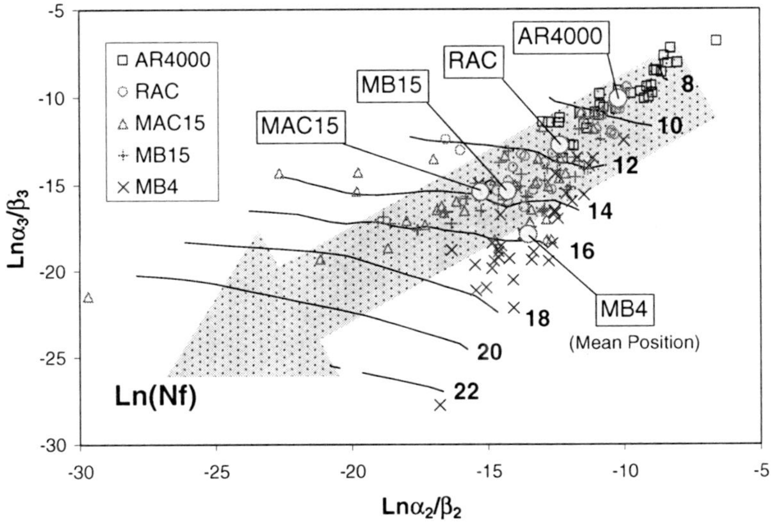

Fig. 12. Identificaiton of fatigue performance by using $Ln\alpha_3/\beta_3$ and $Ln\alpha_2/\beta_2$ parameters.

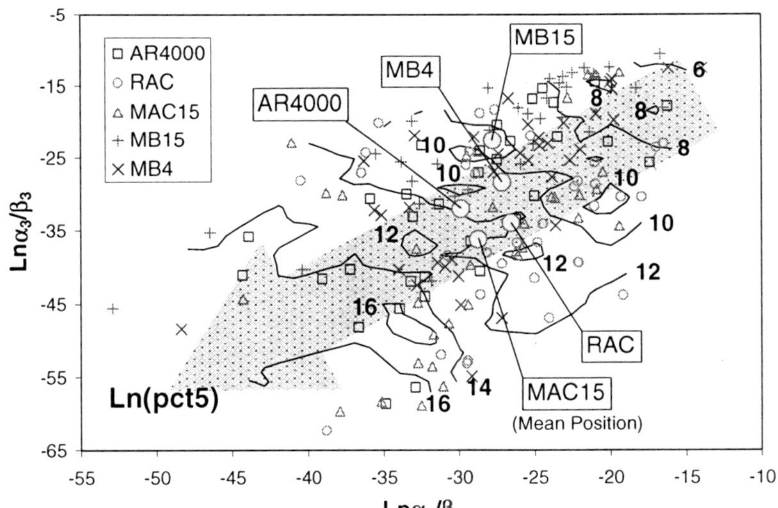

Fig. 13. Identification of shear performance by using $Ln\alpha_3/\beta_3$ and $Ln\alpha_2/\beta_2$ parameters.

hypothesis: "Stages II and III of a three-stage Weibull fatigue/shear curve are critical in determining the fatigue/rutting performance."

The following facts are provided to support the hypothesis:

1. Based on the observations from the summary boxplots of *SR* and accumulated *PSS* at separation points n_1 and n_2 as respectively plotted in Figures 14 and 15, it was found that (1) the deterioration of stiffness ratio at n_1 (i.e., stage I) is

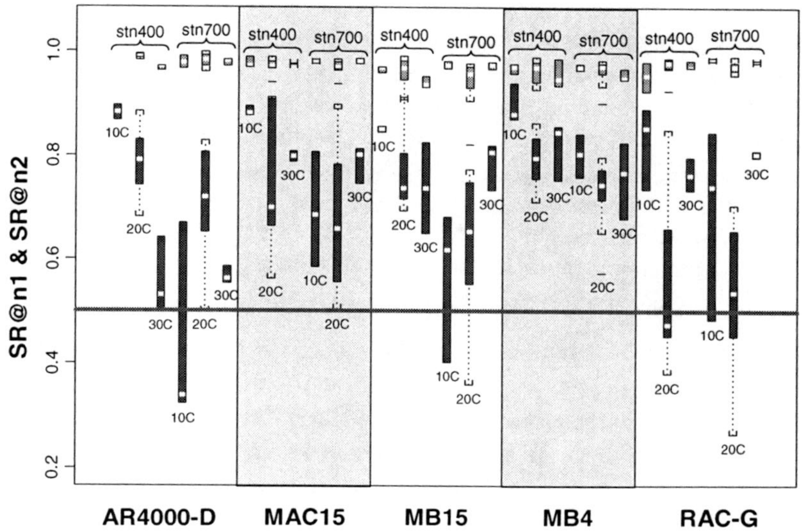

Fig. 14. Summary boxplots of stiffness ratios at stages I and II separations points [Note: (1) $SR @ n_1$ in grey and $SR @ n_2$ in dark and (2) stn400 and stn700 represent 400 and 700 microstrain].

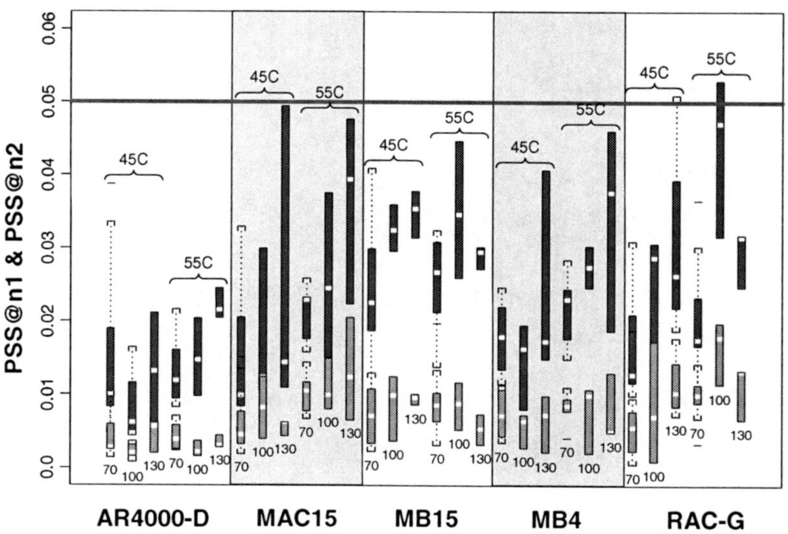

Fig. 15. Summary boxplots of permanent shear strain at stages I and II separation points [Note: (1) $PSS @ n_1$ in grey and $PSS @ n_2$ in dark and (2) the values of 70, 100, and 130 are test shear stress level in kPa].

relatively small when compared with the deterioration of stages II and II; likewise, the accumulated permanent shear strain at n_1 (i.e., stage I) is around 1 percent which is smaller than that accumulated at stages II and III; (2) with the exception of limited amount of tests, the stiffness ratio and accumulated permanent shear strain at n_2 do not cross the lines of $SR = 0.5$ and 5 percent permanent shear strain. Hence, the use of the fundamental Weibull parameters associated with stages II and III seems to be appropriate to characterize fatigue and shear performance.

2. Given that the intercepts and slopes of each stage ($\ln\alpha_1$, $\ln\alpha_2$, $\ln\alpha_3$, β_1, β_2, and β_3) can be used to determine the full Weibull curve, If stages II and III are dominant, then the parameters $\ln\alpha_2$, $\ln\alpha_3$, β_2, and β_3 should be considered as fundamental parameters.
3. The performance contour plots shown in Figures 12 and 13 indicate that the use of $Ln\alpha_3/\beta_3$ and $Ln\alpha_2/\beta_2$ parameters issues a very consistent pattern for fatigue and shear performance: the more negative the values of $Ln\alpha_3/\beta_3$ and $Ln\alpha_2/\beta_2$, the better the fatigue and shear performance. Figure 16, supported by Equations 3 through 6, schematically illustrates two shear (RSST) cases and two fatigue (FAT) cases at stage III verifying this statement. Sub-indices outside brackets indicate test results subjected to conditions 1 and 2; the horizontal lines represent the 5 percent permanent shear strain and $SR = 0.5$.

Case I (RSST): $(\beta_3)_1 = (\beta_3)_2 < 0$; $(\ln\alpha_3)_1 > (\ln\alpha_3)_2 > 0$ (Figure 16a)

$$\Rightarrow \frac{(\ln\alpha_3)_1}{(\beta_3)_1} < \frac{(\ln\alpha_3)_2}{(\beta_3)_2} < 0 \quad \Leftrightarrow \quad (\ln pct5)_1 > (\ln pct5)_2 \tag{3}$$

Case II (RSST): $(\ln\alpha_3)_1 = (\ln\alpha_3)_2 > 0$; $0 > (\beta_3)_1 > (\beta_3)_2$ (Figure 16b)

$$\Rightarrow \frac{(\ln\alpha_3)_1}{(\beta_3)_1} < \frac{(\ln\alpha_3)_2}{(\beta_3)_2} < 0 \quad \Leftrightarrow \quad (\ln pct5)_1 > (\ln pct5)_2 \tag{4}$$

Hence, the more negative the ratio of the $Ln\alpha_3/\beta_3$, the larger the $\ln pct5$ value for the RSST-CH test results.

Case III (FAT): $(\beta_3)_1 = (\beta_3)_2 > 0$; $0 > (\ln\alpha_3)_1 > (\ln\alpha_3)_2$ (Figure 16c)

$$\Rightarrow \frac{(\ln\alpha_3)_2}{(\beta_3)_2} < \frac{(\ln\alpha_3)_1}{(\beta_3)_1} < 0 \quad \Leftrightarrow \quad (\ln Nf)_2 > (\ln Nf)_1 \tag{5}$$

Case IV (FAT): $(\ln\alpha_3)_1 = (\ln\alpha_3)_2 < 0$; $(\beta_3)_1 > (\beta_3)_2 > 0$ (Figure 16d)

$$\Rightarrow \frac{(\ln\alpha_3)_2}{(\beta_3)_2} < \frac{(\ln\alpha_3)_1}{(\beta_3)_1} < 0 \quad \Leftrightarrow \quad (\ln Nf)_2 > (\ln Nf)_1 \tag{6}$$

Hence, the more negative the ratio of the $Ln\alpha_3/\beta_3$, the larger the $\ln Nf$ value for the fatigue test results.

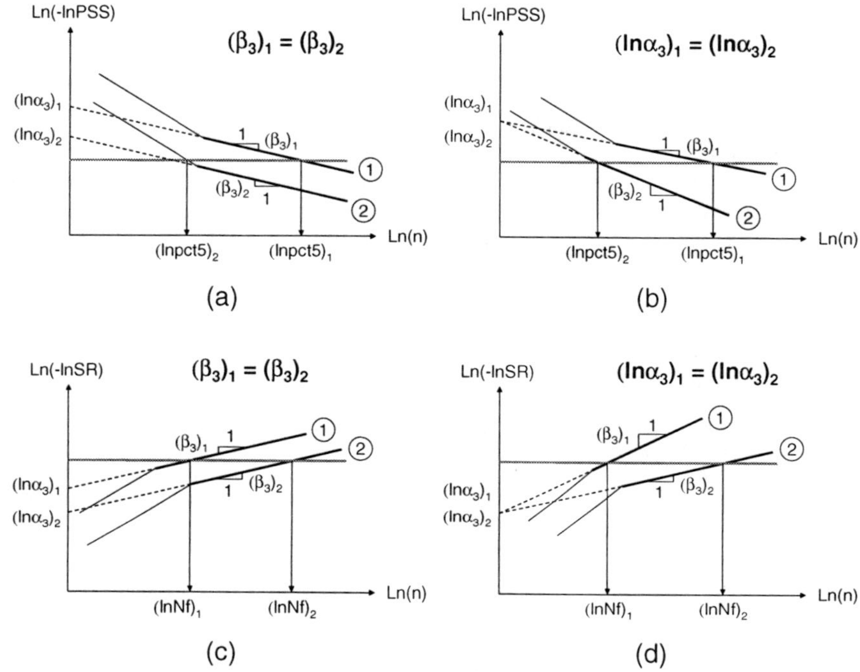

Fig. 16. Schematic interpretation between $Ln\alpha_3/\beta_3$ and $Ln(pct5)/Ln(Nf)$: (a) Case I (*RSST*) —$(\beta_3)_1 = (\beta_3)_2 < 0$ and $(\ln\alpha_3)_1 > (\ln\alpha_3)_2 > 0$, (b) Case II (*RSST*) —$(\ln\alpha_3)_1 = (\ln\alpha_3)_2 > 0$ and $0 > (\beta_3)_1 > (\beta_3)_2$, (c) Case III (FAT) —$(\beta_3)_1 = (\beta_3)_2 > 0$ and $0 > (\ln\alpha_3)_1 > (\ln\alpha_3)_2$, and (d) Case IV (FAT) —$(\ln\alpha_3)_1 = (\ln\alpha_3)_2 < 0$ and $(\beta_3)_1 > (\beta_3)_2 > 0$.

5.3 Application of Three-Stage Weibull Approach: Pavement Performance Prediction

The use of mechanistic-empirical methods in predicting in situ pavement performance has recently mushroomed in the pavement design community. Proper determination of material performance models of asphalt concrete using laboratory test results is a critical element to ensure the success of the mechanistic-empirical method. The three-stage Weibull approach introduced here can express not only the fatigue damage process but also the permanent shear strain accumulation process in one common formulation.

Tsai, et al. have demonstrated the applicability and effectiveness of the integrated two-stage Weibull approach in the fatigue pavement performance of asphalt concrete mixes [29, 30]. An integrated two-stage (stages II and III) Weibull model was established to take into account both crack initiation and crack propagation of laboratory flexural controlled-deformation beam fatigue tests at various testing conditions and material properties. Correction factors which calibrate the laboratory testing results and in situ pavement performance were obtained by comparing the simulated stiffness deterioration curve with the in situ stiffness deterioration utilizing accelerated pavement testing facilities—a Heavy Vehicle Simulator (HVS). Deflection data from multi-depth deflectometers (MDD) were used to back-calculate the stiffness deterioration of asphalt concrete based on the Odemark-Boussinesq method. Figure 17 presents the simulation results with the optimum correction factors and indicates the integrated two-stage Weibull approach is quite promising in fatigue pavement performance prediction.

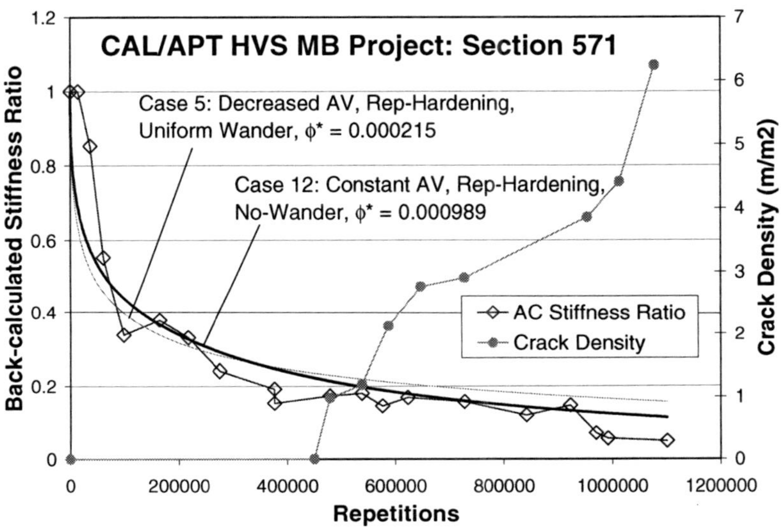

Fig. 17. The optimum predicted stiffness deterioration curve for cases 5 and 12 [27].

By applying the same conceptual method as that used with the integrated Weibull approach, Tsai et al. [31] have proposed a "recursive" modification to the linear sum of cycle ratios cumulative damage hypthesis (sometimes called Miner's law) to calibrate the fatigue surface cracking with accelerated pavement testing data obtained from the HVS. The main concept in the "recursive" Miner's law is to update the fatigue-resistant capacity at θth traffic application by multiplying a correction function, $\phi\exp\left(-\alpha\phi^{-\beta}\theta^{\beta}\right)$, where α and β are two distress-oriented structure deterioration parameters, θ is the cumulative traffic applications, and ϕ is the correction factor. Hence, for the in situ fatigue performance, the original Miner's law, $\sum\frac{n_i}{N_i}=1$, can be modified to have a recursive formulation as follows:

$$\sum_{i=1} \frac{n_i}{(Nf_{Field})_i} = \sum_{i=1} \frac{n_i}{\phi \cdot (Nf_{Lab})_i \cdot \exp\left(-\alpha\phi^{-\beta}\left(\sum_{j=1}^{i} n_j\right)^{\beta}\right)} = 1$$

As can be seen in Figure 18, the predicted evolution curves fit reasonably well with the observed surface cracking evolution curves of the HVS testing sections. Of great importance is that Tsai et al. have concluded that the correction factor should be a function of distress-oriented structure deterioration parameters associated with the whole pavement structure rather than the deterioration parameters of the asphalt concrete mix.

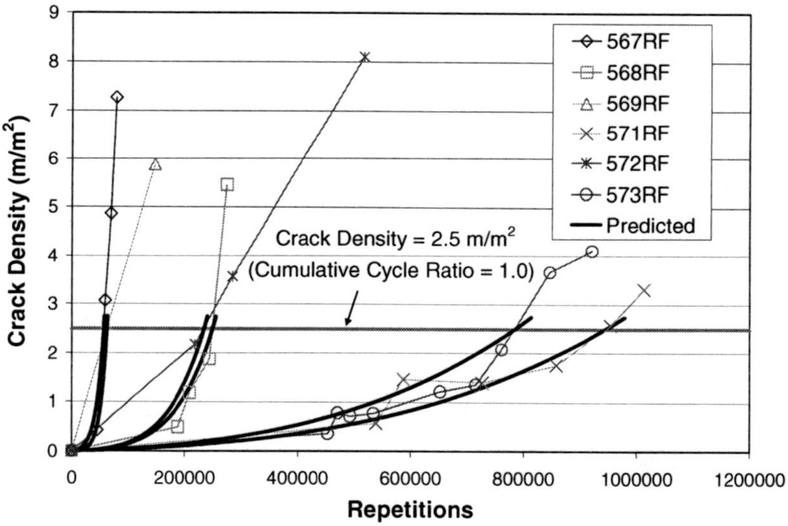

Fig. 18. Predicted fatigue surface cracking evolutions on HVS testing sections [28].

Coleri et al. have confirmed that the integrated Weibull approach is a successful and reliable method for predicting the in situ rutting performance of flexible pavements [32]. As illustrated in Figure 19, four HVS test sections [Sections 582RF (AR4000), 583RF (MB4), 584RF (MB4), and 585RF (MAC15)] with different overlays were analyzed. The general testing conditions are as follows:

- Tire pressure: 720 kPa
- Tire load: 60 kN
- Center-to-center distance between wheels: 360 mm
- Temperature: 48±5°C
- Air-void content: 8±1%

The comparison of the fitted Weibull curves with the HVS test results for all sections is given in Figure 19, with two permanent deformation accumulation mechanisms: *PSS* hardening and repetition-hardening mechanisms. The results of

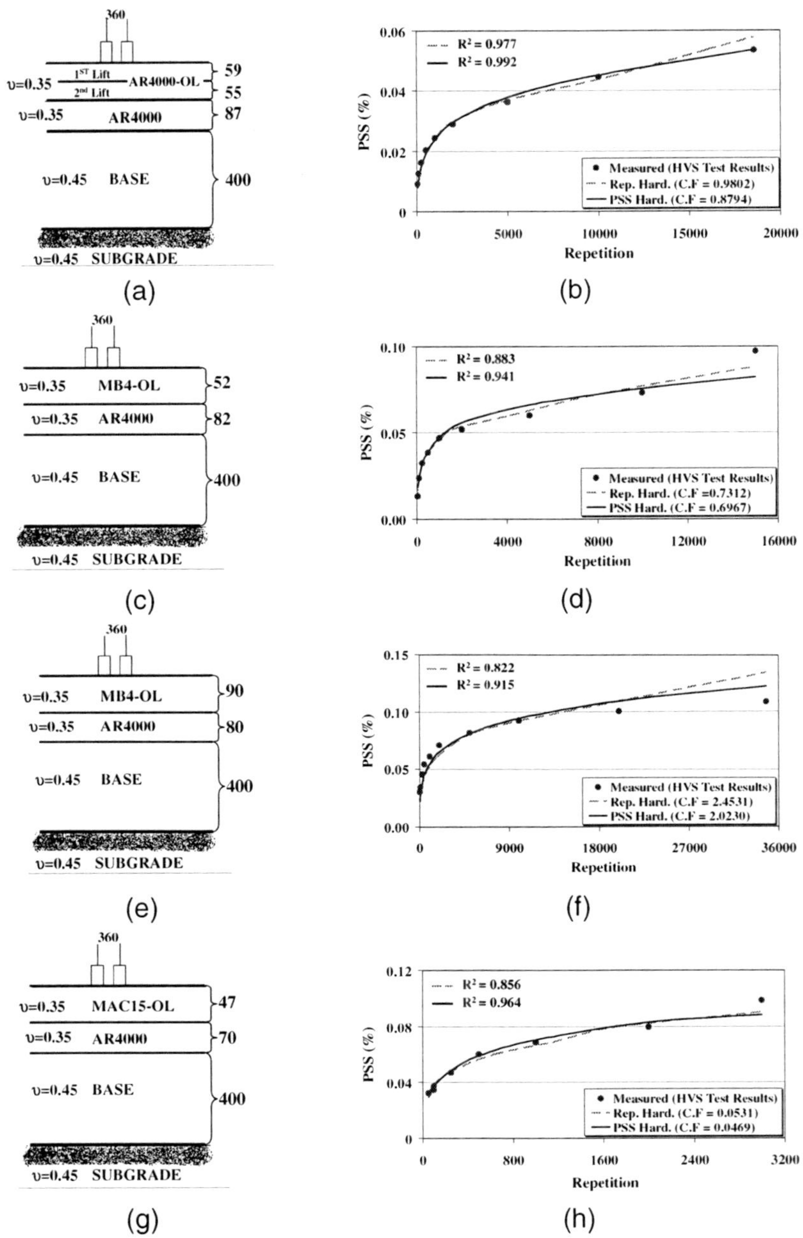

Fig. 19. Cross sections and HVS test results compared with calibrated Weibull model curves: (a, b) Section 582RF (AR4000), (c, d) Section 583RF (MB4), (e, f) Section 584RF (MB4), and (g, h) Section 585RF (MAC15) [29] [Note: Layer thickness is in mm].

the analysis prove that the integrated Weibull approach is quite satisfactory for in situ rutting performance prediction.

6 Findings and Discussions

From the preceding case studies, satisfactory results can be achieved by using the GA in solving the nonlinear optimization problems encountered in pavement design without finalizing the solutions by the Newton method.

Several findings are addressed below:

1. To conduct a GA-based application, the parameters and fitness functions should be clearly identified and defined.
2. Computational time is the primary problem of GAs. It is especially obvious when serious computation is involved, as in the case of backcalculation of layer moduli. This is because the number of times the layered-elastic program or finite element program is called up grows quickly as the gene size and number of generations increases. If, as in Case 2, 3, or 4, a one-time mathematical calculation is all that is required, the computational time is completely acceptable. It was found that the computational time of GAs dependends on how the number of genes, the number of parameters, the number of generations, and the specified range of each parameter are defined. Therefore, a trade-off usually exists between gene size and the number of generations.
3. One of the advantages of using the GA is that a user with expertise can specify the appropriate parameter range while a novice can input the parameter range with a conservative guess for the range. Either situation can lead to the same results. However, an incorrect specified parameter range could lead to an incorrect local optimum solution.
4. As a general rule, the bigger the gene size and number of generations, the smaller the variation in the predicted parameters, computational efficiency notwithstanding.
5. With an appropriate gene size and number of generations, no extra benefit is added if more GA replicates are conducted. To choose the appropriate gene size and number of generations to provide satisfactory results, it is necessary for users to go through a trial-and-error procedure.

Generally speaking, the GA is an intuitive and powerful tool to solve the nonlinear optimization problems, although in some cases, for example in Case 1, the computational time is barely tolerable. To apply the GA successfully, extreme caution must be used with regard to two issues. First, the inappropriate or incorrect parameter input range could lead to an incorrect local optimum solution, and thus a rather wide parameter range is required if no knowledge is available to develop a reasonable initial guess for the parameters. Second, the parameters and fitness function need to be defined clearly at the beginning. In addition, judgment is needed to determine input and to evaluate output.

Regardless of the computation time of certain GA applications, several tasks, especially in model fitting, could be worthy to focus in the future:

- Apply the GA approach developed here to extend the data base of mix master curves and binder relaxation spectra of various mixes and binders.
- Apply the GA approach in searching the nonlinear models of distress-oriented structure deterioration parameters as a function of traffic loading and environment covariates.
- Apply the GA approach in establishing the asphalt mix performance specifications for fatigue cracking and rutting, especially for mixes with modified or rubberized binders.

Acknowledgements

This research was sponsored by the California Department of Transportation (Caltrans) as part of the work of the University of California Pavement Research Center. The authors wish to thank Caltrans for support. The opinions and conclusions expressed in this paper are those of the authors and do not necessarily represent those of Caltrans.

References

[1] Kameyama, S., Himeno, K., Kasahara, A., Maruyama, T.: Back-calculation of Pavement Layer Moduli using Genetic Algorithms. In: Eighth International Conference on Asphalt Pavements (1997)
[2] Fwa, T.F., Tan, C.Y., Chan, W.T.: Backcalculation Analysis of Pavement-Layer Moduli Using Genetic Algorithm. Transportation Research Record: Journal of the Transportation Research Board, No. 1570, Transportation Research Board of the National Academies, Washington, D.C., 134–142 (2008)
[3] Hunaidi, O.: Evolution-Based Genetic Algorithms for Analysis of Non-destructive Surface Wave Tests on Pavements. NDT & e International 31(4), 273–280 (1998)
[4] Liu, M.-Y., Wang, S.-Y.: Genetic Optimization Method of Asphalt Pavement Based on Rutting and Cracking Control. Journal Wuhan University of Technology, Materials Science Edition 18(1), 72-75 (2003)
[5] Hadi, M.N.S., Arfiadi, Y.: Optimum Rigid Pavement Design by Genetic Algorithms. Computers & Structures 79(17), 1617–1624 (2001)
[6] Shekharan, A.R.: Solution of Pavement Deterioration Equations by Genetic Algorithms. Transportation Research Record 1699, 101–106 (2000)
[7] Attoh-Okine, N.O., Appea, A.K.: Predicting Roughness Progression Models in Flexible Pavements—An Evolutionary Algorithm Approach. In: Intelligent Engineering Systems through Artificial Neural Networks. ASME, Fairfield, NJ, USA, vol. 8, pp. 845–853 (1998)
[8] Sundin, S., Braban-Ledoux, C.: Artificial Intelligence-Based Decision Support Technologies in Pavement Management. Computer-Aided Civil & Infrastructure Engineering 16(2), 143–157 (2001)
[9] Loia, V., Sessa, S., Staiano, A., Tagliaferri, R.: Merging Fuzzy Logic, Neural Networks, and Genetic Computation in the Design of a Decision-Support System. International Journal of Intelligent Systems 15(7), 575–594 (2000)

[10] Sait, S.M., Youssef, H.: Iterative Computer Algorithms with Application in Engineering: Solving Combinatorial and Optimization Problems. IEEE Computer Society Press, Los Alamitos (1999)
[11] Zohdi, T.I.: Multiscale Modeling and Design of New Materials: Theoretical and Numerical Analyses, Class note for ME-290B, University of California, Berkeley (2003)
[12] Burden, R.L., Faires, J.D.: Numerical Analysis, 4th edn. PWS-KENT Publishing Company (1989)
[13] Lytton, R.L., Germann, F.P., Chou, Y.J., Stoffels, S.M.: NCHRP Report 327: Determining Asphaltic Concrete Pavement Structural Properties by Nondestructive Testing. TRB, National Research Council, Washington, D.C (1990)
[14] ELMOD Version 5.1.54. Dynatest International, Glostrup, Denmark
[15] Ullidtz, P.: Modelling Flexible Pavement Response and Performance. Polyteknisk Forlag, Oslo, Norway (1998)
[16] Lu, Q., Ullidtz, P., Basheer, I., Ghuzlan, K., Signore, J.M.: CalBack: Enhancing Caltrans Mechanistic-Empirical Pavement Design Process with New Backcalculation Software. Journal of Transportation Engineering, ASCE 135(7) (2009)
[17] Choi, J.W., Wu, R., Pestana, J.M.: Application of Constrained Extended Kalman Filter Moduli Backcalculation. In: CD-ROM for the 86th Annual Meeting of the Transportation Research Board, Washington, D.C (2007)
[18] Ferry, J.D.: Viscoelastic Properties of Polymeric Material. Wiley, New York (1980)
[19] Hastie, T.J., Tibshirani, R.J.: Generalized Additive Models, 1st edn. Chapman and Hall, Boca Raton (1990)
[20] Tsai, B.-W.: High Temperature Fatigue and Fatigue Damage Process of Aggregate-Asphalt Mixes. Ph.D. Dissertation, University of California, Berkeley (2001)
[21] Stone, C.J.: A Course in Probability and Statistics. Duxbury Press, Pacific Grove (1996)
[22] Baumgaertel, M., Winter, H.H.: Determination of Discrete Relaxation and Retardation Time Spectra from Dynamic Mechanical Data. Rheologica Acta 28, 511–519 (1989)
[23] Stastna, J., Zanzotto, L., Berti, J.: How Good Are Some Rheological Models of Dynamic Material Functions of Asphalt? Journal of the Association of Asphalt Paving Technologists 66, 458–485 (1997)
[24] Tayebali, A.A., Deacon, J.A., Coplanz, J.S., Harvey, J.T., Fin, F.N., Monimith, C.L.: Fatigue Respone of Asphalt-Aggregae Mixes. Report SHRP-A-404. SHRP. National Research Council, Washington, D.C (1994)
[25] Sousa, J.B., Deacon, J.A., Weissman, S., Harvey, J.T., Monismith, C.L., Leahy, R.B., Paulsen, G., Coplantz, J.S.: Permanent Deformation Response of Asphalt-Aggregate Mixes. Report SHRP-A-415. SHRP, National Research Council, Washington, D.C (1994)
[26] Jiang, R., Murthy, D.N.: Reliability Modeling Involving Two Weibull Distributions. Reliability Engineering and System Safety 47, 187–198 (1995)
[27] Tsai, B.-W., Harvey, J.T., Monismith, C.L.: Using the Three-Stage Weibull Equation and Tree-Based Model to Characterize the Mix Fatigue Damage Process. In: Transportation Research Record: Journal of the Transportation Research Board, No. 1929, Transportation Research Board of the National Academies, Washington, D.C, pp. 227–237 (2005)

[28] Jones, D., Tsai, B.-W., Ullidtz, P., Wu, R., Harvey, J.T., Monismith, C.L.: Reflective Cracking Study: Second-Level Analysis Report. California Department of Transportation, and Pavement Research Center, University of California, UC Davis and UC Berkeley (2007)

[29] Tsai, B.-W., Harvey, J.T., Monismith, C.L.: Two-Stage Weibll Approach for Asphalt Concrete Fatigue Performance Preidction. Journal of the Association of Asphalt Paving Technologists 73, 623–655 (2004)

[30] Tsai, B.-W., Bejarano, M.O., Harvey, J.T., Monismith, C.L.: Prediction and Calibration of Pavement Fatigue Performance Using Two-Stage Weibull Approach. Journal of the Association of Asphalt Paving Technologists 74, 697–732 (2005)

[31] Tsai, B.-W., Harvey, J.T., Monismith, C.L., Bejarano, M.O.: Calibration of Fatigue Surface Cracking Usng Simplified Recursive Miner's Law. Journal of the Association of Asphalt Paving Technologists 76, 693–735 (2007)

[32] Coler, E., Tsai, B.-W., Monismith, C.L.: Pavement Rutting Performance Prediction by Integrated Weibull Approach. In: Transportation Research Record: Journal of the Transportation Research Board, No. 2087, Transportation Research Board of the National Academies, Washington, D.C., pp. 120–130 (2008)

Extended Kalman Filter and Its Application in Pavement Engineering

Rongzong Wu[1], Jae Woong Choi[2], and John T. Harvey[3]

[1] University of California Pavement Research Center, Department of Civil and Environmental Engineering, University of California at Davis, One Shields Ave, Davis, CA 95616, U.S.A.
rzwu@ucdavis.edu
[2] MMI Engineering, 475 14th Street, Suite 400, Oakland, CA, 94612-1940, U.S.A.
jaewoong27@hotmail.com
[3] University of California Pavement Research Center, Department of Civil and Environmental Engineering, University of California at Davis, One Shields Ave, Davis, CA 95616, U.S.A.
jtharvey@ucdavis.edu

Abstract. Kalman filter is a signal processing technique that estimates the state of a dynamic system from a series of noisy measurements. It is used in a wide range of engineering applications from radar to computer vision. This chapter demonstrates the application of a model identification procedure based on extended Kalman filter (EKF) and weighted global iteration (WGI) technique in pavement engineering. In particular, EKF-WGI is used to perform layer moduli back-calculation from falling weight deflectometer (FWD) data and to identify model parameters for Generalized Maxwell Model for hot mix asphalt using frequency sweep test data. In both cases, EKF-WGI is shown to provide consistent results that are independent of the seed values for both linear and nonlinear problems. It is believed that EKF-WGI provides an efficient, consistent and robust tool for optimization that has many potential applications.

Keywords: Kalman filter, weighted global iteration, falling weight deflectometer, layer moduli back-calculation, model parameter identification, generalized Maxwel model.

1 Introduction

The Kalman filter is one of the most well-known mathematical tools that can be used for estimation from noisy measurements. It is named after Rudolph E. Kalman, who in 1960 published his famous paper describing a recursive solution to the discrete-data linear filtering problem [1]. The Kalman filter is essentially a set of mathematical equations that implement a predictor-corrector type estimator that is optimal in the sense that it minimizes the estimated error covariance - when some presumed conditions are met. Since the time of its introduction, the Kalman filter has been the subject of extensive research and application, particularly in the area of autonomous or assisted navigation. This is likely due in large part to advances in digital computing that made the use of the filter practical, but also to the relative

K. Gopalakrishnan et al. (Eds.): Intel. & Soft Comp. in Infra. Sys. Eng., SCI 259, pp. 239–253.
springerlink.com

simplicity and robust nature of the filter itself. Rarely do the conditions necessary for optimality actually exist, and yet the filter apparently works well for many applications in spite of this situation [2]. More details on the Kalman filter can be found in [3-9]. In addition, for many years University of North Carolina (UNC) has maintained a web site dedicated to the Kalman filter [10]. Among other things, readers are strongly encouraged to check out the introduction paper on Kalman Filter listed at the website. The paper provides an excellent starting point for understanding how Kalman Filter works.

One of the early applications of Kalman Filter was the trajectory estimation and control problem for the Apollo project. The Kalman Filter was successfully conducted as a part of onboard guidance during Apollo program [11]. Since then, it has been widely used as a core part of navigation system in aircraft, ships, and spacecraft [11-13]. The Kalman Filter is also used for predicting the likely future courses of dynamic systems that people are not likely to control, such as the tracking object in machine vision [14], parameter estimation of freeway traffic model [15], or predicting the prices of traded commodities [16]. In addition to that, it has been widely used as a main part of control in manufacturing process, automation, and robotics [17, 18]. The focus of this chapter is the application of Kalman Filter in pavement engineering.

Pavement engineers are commonly faced with the parameter identification problems. Specifically, when a model is developed to describe a certain type of phenomenon, the model parameters need to be determined based on some measured data. Typically, these models are non-linear. A weighted global iteration procedure based on extended Kalman filter (EKF-WGI) has been developed for this purpose by engineers in other fields of civil engineering.

EKF-WGI was first used for estimating parameters of a running load and beam system and a hysteretic restoring system of non-degrading type [19]. Since then, several researchers adopted this method to earthquake engineering for estimating material properties and also applied to structure health monitoring [20-22]. Especially, Loh and Tsaur [20] applied the EKF-WGI algorithm for identifying parameters in an equivalent linear system, a bilinear hysteretic restoring system and a system with stiffness degradation effect. Koh et al. [23] extensively used the proposed method for estimating dynamic properties of 3-D structure system by using the ‘ improved condensation’ method.

Although EKF-WGI has been extensively applied in other fields, its application in pavement engineering has been relatively rare. The purpose of this chapter is to evaluate the performance of EKF-WGI when used to solve some common model identification problems in pavement engineering. It is hoped that this will spark more interest in the procedure and lead to many more applications in pavement engineering.

In the following sections, EKF-WGI is first introduced and then two examples are included to show its application: layer stiffness back-calculation and regression of Generalized Maxwell Model parameters for hot mix asphalt (HMA).

2 Model Identification Procedure Based on the Extended Kalman Filter

2.1 The Kalman Filter Method

The Kalman Filter is a mathematical tool for finding the best estimate for the state of a system in a discrete time process that is governed by a linear stochastic differential equation (i.e., [1]). The filter is essentially described by the state and measurement equations given by:

(State equation) $$\boldsymbol{x}_k = \boldsymbol{F}_{k-1}\boldsymbol{x}_{k-1} + \boldsymbol{w}_{k-1} \tag{2.1}$$
(Measurement equation) $$\boldsymbol{z}_k = \boldsymbol{H}_k\boldsymbol{x}_k + \boldsymbol{v}_k \tag{2.2}$$

where subscripts k-1 and k indicate time steps, $\boldsymbol{x}$ is the state vector describing the current state of the system, $\boldsymbol{z}$ is the measurement vector that includes observable quantities that depend on the current state vector, matrix $\boldsymbol{F}$ is the dynamic coefficient matrix relating the state vectors between subsequent time steps and matrix $\boldsymbol{H}$ is the measurement sensitivity matrix relating the measurement vector to the state vector (e.g., [9], [24]). The vectors $\boldsymbol{w}$ and $\boldsymbol{v}$ represent the inherent system and measurement noise respectively, and they are considered independent processes with normal (i.e., Gaussian) distributions of zero mean and covariance matrices $\boldsymbol{Q}$ and $\boldsymbol{R}$, respectively:

(Noise Information) $$\boldsymbol{w}_k = N(\boldsymbol{0}, \boldsymbol{Q}_k); \boldsymbol{v}_k = N(\boldsymbol{0}, \boldsymbol{R}_k) \tag{2.3}$$

where $N(a, b)$ indicates a random quantity following normal distribution with mean a and covariance matrices b. In the Kalman Filter, there are two estimates for the state vector $\boldsymbol{x}$: the mean $\bar{\boldsymbol{x}}$ and the error covariance matrix $\boldsymbol{P}$ defined as:

$$\bar{\boldsymbol{x}} = E(\boldsymbol{x}) \tag{2.4}$$
$$\boldsymbol{P} = E\left\{(\boldsymbol{x} - \bar{\boldsymbol{x}})(\boldsymbol{x} - \bar{\boldsymbol{x}})^T\right\} \tag{2.5}$$

where $E(\cdot)$ denotes the "expected" value. The Kalman filter operates by propagating estimates for the state vector $\boldsymbol{x}$, from some given initial values at time step 0 up to the current time step k using time update (predictor) and measurement update (corrector). In other words, a sequence of estimates for $\bar{\boldsymbol{x}}$ and $\boldsymbol{P}$ values are obtained by repeating time updates and measurement updates.

Time update for step k provides the a-priori estimates based on the measurement information up to time step $k-1$, while measurement update provides a-posteriori estimates based on the measurement information at time step k. A two-subscript system is introduced to distinguish between these two sets of estimates. Specifically, the a-priori time estimates for step k are denoted as $\bar{\boldsymbol{x}}_{k|k-1}$ and $\boldsymbol{P}_{k|k-1}$, while the a-posteriori measurement estimates are denoted as $\bar{\boldsymbol{x}}_{k|k}$ and $\boldsymbol{P}_{k|k}$. Time updates are calculated by the following equations:

$$\bar{\boldsymbol{x}}_{k|k-1} = \boldsymbol{F}_{k-1}\bar{\boldsymbol{x}}_{k-1|k-1} + \boldsymbol{w}_{k-1} \tag{2.6}$$
$$\boldsymbol{P}_{k|k-1} = \boldsymbol{F}_{k-1}\boldsymbol{P}_{k-1|k-1}\boldsymbol{F}^T_{k-1} + \boldsymbol{Q}_{k-1} \tag{2.7}$$

With the additional measurement at step k, the time updates are corrected using the following equations:

$$\bar{\boldsymbol{x}}_{k|k} = \bar{\boldsymbol{x}}_{k|k-1} + \boldsymbol{K}_k(\boldsymbol{z}_k - \boldsymbol{H}_k\bar{\boldsymbol{x}}_{k|k-1}) \tag{2.8}$$

$$\boldsymbol{P}_{k|k} = (\boldsymbol{I} - \boldsymbol{K}_k\boldsymbol{H}_k)\boldsymbol{P}_{k|k-1} \tag{2.9}$$

where $\boldsymbol{I}$ is the identity matrix. The equation above shows that the a-priori estimate $\bar{\boldsymbol{x}}$ is corrected by $\boldsymbol{K}_k(\boldsymbol{z}_k - \boldsymbol{H}_k\bar{\boldsymbol{x}}_{k|k-1})$, which is a combination of the Kalman Gain, $\boldsymbol{K}_k$ and the difference in expected and actual measurements: $\boldsymbol{z}_k - \boldsymbol{H}_k\bar{\boldsymbol{x}}_{k|k-1}$. The Kalman Gain is calculated by:

$$\boldsymbol{K}_k = \boldsymbol{P}_{k|k-1}\boldsymbol{H}_k^T(\boldsymbol{R}_k + \boldsymbol{H}_k\boldsymbol{P}_{k|k-1}\boldsymbol{H}_k^T)^{-1} \tag{2.10}$$

Derivation of these equations can be found in the introduction paper listed on the UNC website [10].

2.2 The Extended Kalman Filter Methodology

In the extended Kalman filter (EKF), the state transition and observation models need not be linear functions of the state but may instead be (differentiable) functions:

$$\text{(State equation)} \quad \boldsymbol{x}_k = \boldsymbol{f}_{k-1}(\boldsymbol{x}_{k-1}) + \boldsymbol{w}_{k-1} \tag{2.11}$$

$$\text{(Measurement equation)} \quad \boldsymbol{z}_k = \boldsymbol{h}_k(\boldsymbol{x}_k) + \boldsymbol{v}_k \tag{2.12}$$

where $\boldsymbol{f}$ is a function relating state vectors between subsequent time steps and $\boldsymbol{h}$ is a function relating the measurement vector to the state vector. Matrices $\boldsymbol{F}$ and $\boldsymbol{H}$ are in turn replaced by the Jacobians evaluated with current predicted states. This process essentially linearizes functions $\boldsymbol{f}$ and $\boldsymbol{h}$ around the current estimate:

$$\boldsymbol{F}_{k-1} = \left.\frac{\partial \boldsymbol{f}_{k-1}}{\partial \boldsymbol{x}}\right|_{\boldsymbol{x}=\boldsymbol{x}_{k-1|k-1}} \tag{2.13}$$

$$\boldsymbol{H}_k = \left.\frac{\partial \boldsymbol{h}_k}{\partial \boldsymbol{x}}\right|_{\boldsymbol{x}=\boldsymbol{x}_{k|k-1}} \tag{2.14}$$

2.3 Model Parameter Identification Using EKF

When the EKF is used for model parameter identification, state vector is essentially a collection of all of the parameters needed to be identified. Since model parameters do not change between different measurements, the state equation can be simplified:

$$\text{(State equation)} \quad \boldsymbol{x}_k = \boldsymbol{x}_{k-1} + \boldsymbol{w} \tag{2.15}$$

which implies $\boldsymbol{F} \equiv \boldsymbol{I}$ and $\boldsymbol{w}$ is constant for all time steps. The model function $\boldsymbol{h}$ that links model parameters to observed quantities typically remains nonlinear.

2.4 EKF with Weighted Global Iteration Procedure for Model Parameter Identification

The EKF method can sometimes converge very slowly when the system equations (2.11) and (2.12) are highly nonlinear. To improve numerical stability, the iterative use of EKF was introduced for model identification applications. This method uses all the measurement data available and applies the EKF iteratively until the model parameters converge. The same set of measurement information is used repeatedly for measurement update process. Between iteration steps, the estimates, mean and covariance at the end of current time step are used for initial guess for next iteration step. Therefore, a better estimation can be obtained at the end of each iteration step [25]. This approach was successfully used to identify the hydrodynamic coefficient matrices associated with non-linear drag and linear inertia forces of an offshore structure subjected to wave forces [26].

The iterative use of EKF can be further improved by applying a global weight to the a-posteriori estimates of $\boldsymbol{P}$ for each iteration [19]:

$$\boldsymbol{P}_{k|k} = \omega \cdot \boldsymbol{P}_{k|k} \tag{2.16}$$

This simple modification proposed by [19] has been shown to increase the rate of convergence and effectively prevents oscillation of the solution. The modified procedure is called EKF with weighted global iteration (EKF-WGI) procedure.

The proposed method has been used for estimating parameters of a running load and beam system and a hysteretic restoring system of non-degrading type [19]. Since it was introduced, several researchers adopted this method to earthquake engineering for estimating material properties and also applied to structure health monitoring [20-22, 27-33]. Especially, Loh and Tsaur [20] applied the EKF-WGI algorithm for identifying parameters in an equivalent linear system, a bilinear hysteretic restoring system and a system with stiffness degradation effect. Koh et al. [23] extensively used the proposed method for estimating dynamic properties of 3-D structure system by using the "improved condensation" method.

2.5 Applying Constraints

The standard EKF method estimates the mean of the state vector over the infinite domain $[-\infty, +\infty]^n$. When the method is used to estimate (or identify) physical quantities, such as material parameters for a constitutive law, it is necessary to constrain the state vector by performing appropriate variable transformations. For example, material parameters are usually bounded (e.g., positive stiffness) or exist within a prescribed range (e.g., Poisson's ratio). A positive constraint can be enforced by applying an exponential transformation:

$$x = \exp(\alpha) \tag{2.17}$$

and a range constraint on interval [a, b] can be enforced by applying an inverse tangent transformation:

$$x = a + (b - a) \cdot \left[\frac{1}{2} + \frac{1}{\pi}\arctan(\alpha)\right] \tag{2.18}$$

3 Layer Moduli Back-Calculation with EKF-WGI

The EKF-WGI procedure for model parameter identification is now applied to perform layer stiffness back-calculation using Falling Weight Deflectometer (FWD) data. FWD is one of the popular devices for measuring surface deflection. It applies an impulse load on the pavement surface and measures surface deflections from horizontally arrayed geophones. Usually, seven geophones are used for capturing surface deflections. Layer moduli can be back-calculated by finding a set of moduli that allows calculated deflections to match with the measured ones, with known layer thicknesses.

Layer moduli back-calculation is a model parameter identification problem. Layer moduli form the state vector $\boldsymbol{x}$, surface deflections form the observation vector $\boldsymbol{z}$. The model function $\boldsymbol{h}$ corresponds to the displacement engine that links pavement structure to surface deflections. In this work, the expected surface deflection was calculated using multi-layer elastic theory, which assumes that all layers are linear elastic. Despite the complex nonlinear behavior of materials in real situations, multi-layer elastic theory captures pavement response reasonably well and is widely used in pavement engineering for the benefit of computational simplicity. In this study, an open source program called LEOP for layer elastic theory is used [34] to calculate pavement response. Accordingly, the function $\boldsymbol{h}$ represents LEOP. It certainly can be any of the many displacement calculation programs. The Jacobian of $\boldsymbol{h}$ over $\boldsymbol{x}$, i.e., matrix $\boldsymbol{H}$, is determined using numerical differentiation with the central difference scheme.

While adapting EKF-WGI for back-calculation of layer moduli, the following set of parameters were used:

$\boldsymbol{Q} = q \cdot \boldsymbol{I}$ with $q = 2 \times 10^{-6}$ mm, essentially a very small number;

$\boldsymbol{R} = \delta \cdot \boldsymbol{I}$ with $\delta = 0.01$ mm, to represent the accuracy in deflection measurements;

$\boldsymbol{P}_{0|0} = p \cdot \boldsymbol{I}$ with $p = 0.8$ (mm^2); and

$\omega = 500$

Note that δ has physical meanings and should be set appropriately based on the capacity of the specific FWD, while other parameters are numerical parameters and can be set somewhat arbitrarily as long as the procedure works. Convergence criterion is defined through the maximum error between the measured and estimated surface deflections:

$$\theta_k = \max \left| \boldsymbol{z} - \boldsymbol{h}(\boldsymbol{x}_{k|k-1}) \right| \tag{3.1}$$

Convergence is achieved when $\theta_k < TOL \cdot \delta$ is satisfied. TOL was set to 0.5 during this study.

Table 1. Performance of the back-calculation procedure based on EKF-WGI*

Case No.	Layer Moduli (MPa)	Seed Moduli (MPa)	Converged Moduli (MPa)	Number of Iterations	Max. Rel. Error in Moduli (%)
1a	5000, 250, 60	100 for all	4997, 250, 60	6	0.12
1b	5000, 250, 60	1000 for all	4997, 250, 60	11	0.12
1c	5000, 250, 60	10000 for all	5003, 251, 60	7	0.47
1d	5000, 250, 60	1000, 200, 100	5014, 253, 60	4	1.24
2a	5000, 30000, 60	100 for all	4974, 30114, 60	7	0.52
2b	5000, 30000, 60	1000 for all	4990, 30009, 60	6	0.20
2c	5000, 30000, 60	10000 for all	4997, 30012, 60	11	0.11
2d	5000, 30000, 60	1000, 10000, 100	4993, 30027, 60	5	0.14

*: Layer thicknesses: 150 mm, 300 mm and semi-infinite

The performance of the EKF-WGI procedure is evaluated using simulated FWD deflection data using the LEOP program. Specifically, a 40 kN circular load with radius of 150 mm is applied and surface deflections are "measured" at radial distances of 0, 203, 304, 457, 610, 914, and 1524 mm from the center of the loading plate. The back-calculation program is written in Matlab [35] because of its strong support for matrix operations.

To illustrate the effectiveness of EKF-WGI, a three-layer pavement system is used. The layer thicknesses and moduli are listed in Table 1. Two types of pavement structures are evaluated: a typical flexible pavement (cases 1a to 1d) and a typical composite pavement (cases 2a to 2d). Four different combinations of seed moduli were used to evaluate their impact on the resulting back-calculated moduli.

As shown in Table 1, the EKF-WGI procedure successfully back-calculated accurate layer moduli for both structure types and all the selections of seed values. Using different seed moduli only affects the number of iterations required to reach convergence, while the resulting moduli are not affected. For both structures, using seed values that better match the actual moduli led to faster convergence.

Table 2. Iteration history for Case 2b

Time Step (k)	Layer Moduli (MPa)	Maximum Error in Surface Deflection θ_k (mm)	Note
0	1000, 1000, 1000	1.157E-1	
1	998, 988, 963	1.149E-1	
2	2677, 2180, 21	3.933E-1	
3	4018, 6744, 41	1.140E-1	
4	5414, 15770, 57	2.141E-2	
5	4753, 29247, 60	1.154E-3	Converged
6	4990, 30009, 60	1.082E-6	Final Update

An example of the iteration process is shown in Table 2 for case 2b (defined in Table 1). The table shows a fast and stable iteration history without oscillation even though the seed moduli are quite different from the true moduli.

Although the examples presented here demonstrated a stable and objective back-calculation procedure, the ultimate test for EKF-WGI should be done using actual FWD data. This will however be left for further publications. Another question remains to be answered is the performance of EKF-WGI for models with strong nonlinearity. This will be investigated in the next section.

4 Nonlinear Model Parameter Identification with EKF-WGI

To evaluate its behavior in nonlinear systems, EKF-WGI is now used to identify General Maxwell Model (GMM) parameters for hot mix asphalt (HMA) using stiffness data measured from beam bending frequency sweep tests [36]. An n-branch GMM composed of one elastic in parallel with n-1 Maxwell element can be fully defined by the following 2n-1 parameters:

Relaxation times for the n-1 Maxwell elements: λ_i for i=1 to n-1;
Stiffness of the springs for the n-1 Maxwell elements: E_i for i=1 to n-1; and
The spring stiffness for the elastic branch: E_∞

Under sinusoidal loading with angular frequency of w, the storage and loss modulus for a GMM unit can be calculated as:

- (Storage modulus) $E' = E_\infty + \Sigma_{i=1}^{n-1} E'_i$ (4.1)

- (Loss modulus) $E'' = \Sigma_{i=1}^{n-1} E''_i$ (4.2)

where:

$$E'_i = E_i w \lambda_i \cdot \frac{w\lambda_i}{1 + w^2\lambda_i^2} \tag{4.3}$$

and:

$$E''_i = E_i w \lambda_i \cdot \frac{1}{1 + w^2\lambda_i^2} \tag{4.4}$$

The amplitude of complex modulus, E^*, can then be calculated as the root of the sum of squares of E' and E'':

$$E^* = \sqrt{E'^2 + E''^2} \tag{4.5}$$

and the phase angle is:

$$\beta = \arctan \frac{E''}{E'} \tag{4.6}$$

Frequency sweep tests are typically conducted under different temperatures. To account for the effect of temperature on HMA stiffness, w needs to be converted to reduced angular frequency following the Mechanistic Empirical Pavement Design Guide (MEPDG) developed under NCHRP 1-37A project [37]. Considering the fact that loading time $t = 1/w$ and Equation 2.2.4 in MEPDG:

$$\log(w_{red}) = \log(w) + a_T[\log(\eta) - \log(\eta_{ref})] \tag{4.7}$$

where subscript *red* indicates reduced quantity and *ref* indicates quantities for reference temperature, a_T is the temperature shift factor, and η is the viscosity. Viscosity for a given temperature can be calculated using Equation 2.2.5 in MEPDG:

$$\log(\eta) = 10^{A+VTS\cdot\log(T_R)} \tag{4.8}$$

where *A* and *VTS* are constants and T_R is Rankine temperature. In this study, *A*=10.5254 and *VTS*= - 3.5047 assuming a 40-50 penetration grade for RTFOT binder. The reference temperature is chosen to be 20°C.

Frequency sweep data for a dense graded HMA with PG 64-10 binder are used here to evaluate the performance of EKF-WGI in identifying all of the unknown parameters. The frequency sweep test results are listed below in

Table 3 and Table 4 for amplitude of the complex moduli and phase angles respectively. Apparently, the tests were performed under 10, 20 and 30°C with twp replicates for each temperature.

Based on past experiences, an eight-branch GMM is used here to characterize the HMA stiffness. The relaxation times are predetermined as:

$$\lambda_i = 5.0 \times 10^{i-7} \text{ for } i\text{=1 to 7} \tag{4.9}$$

The remaining unknowns include the following *n*+1 quantities: E_i (*i*=1 to *n*-1), E_∞ and a_T. These unknowns form the state vector for this parameter identification problem. Exponential transformation was used to enforce positive definite constraints for all of the unknowns.

There are several possible choices for the measurement quantities for this problem: (a). E^* alone; (b). E^* and phase angle β; and (c). E' and E''. In this study, the measurement vector include both the storage and lost moduli for all of the loading temperatures and angular frequencies, i.e., option (c) is chosen. This provides a more complete set of model response and at the mean time maintains better uniformity within the measurement vector.

After some trial runs, it was found that the following set of parameters allows EKF-WGI to reach a solution that matches the calculated and measured measurement quantities:

$\boldsymbol{Q} = q \cdot \boldsymbol{I}$ with $q = 2 \times 10^{-6}$ MPa, essentially a very small number;

$\boldsymbol{R} = \delta \cdot \boldsymbol{I}$ with $\delta = 10$ MPa, to represent the accuracy in modulus measurements;

$\boldsymbol{P}_{0|0} = p \cdot \boldsymbol{I}$ with $p = 0.8$ MPa^2; and

$\omega = 2$

Table 3. Measured amplitude of the complex moduli for a dense graded PG 64-10 HMA tested under different temperatures. (Unit: MPa) (Note: header indicates specimen name and testing temperature)

Frequency (Hz)	641013A2 10°C	64101A2 10°C	64103C1 20°C	64106B2 20°C	64105A2 30°C	641013A1 30°C
0.01	3,089	2,839	908	924	232	204
0.02	3,772	3,519	1,234	1,289	291	282
0.05	4,674	4,442	1,737	1,806	461	420
0.1	5,501	5,257	2,232	2,307	626	595
0.2	6,397	6,159	2,857	2,886	869	826
0.5	7,546	7,270	3,784	3,751	1,312	1,243
1	8,473	8,219	4,620	4,477	1,722	1,663
2	9,331	9,171	5,496	5,215	2,292	2,195
5	10,506	10,251	6,758	6,296	3,171	3,014
10	11,463	11,183	7,672	7,122	3,931	3,724
15	12,330	12,092	8,458	7,753	4,433	4,189

Table 4. Measured phase angles for the a dense graded PG 64-10 HMA (Unit: Degree). (Note: header indicates specimen name and testing temperature)

Frequency (Hz)	641013A2 10°C	64101A2 10°C	64103C1 20°C	64106B2 20°C	64105A2 30°C	641013A1 30°C
0.01	28	30	43	38	47	55
0.02	27	28	41	36	50	48
0.05	24	25	38	34	48	47
0.1	19	20	34	30	44	46
0.2	19	20	33	29	43	43
0.5	17	17	29	26	41	41
1	15	15	26	23	37	37
2	13	14	23	21	35	34
5	12	12	20	18	31	31
10	12	12	19	17	28	28
15	11	11	19	15	27	27

Table 5. Effect of seed values on identified model parameters using eight-branch GMM

Case No.	1	2	3	4
Seed E_i (MPa)	1,000	10,000	1,000	1,000
Seed a_T	2.0	2.0	1.0	5.0
E_1	3,558	5,991	37,928	16
E_2	4,305	4,012	636	21,996
E_3	3,248	3,256	3,291	0
E_4	3,063	3,061	3,065	3,974
E_5	3,155	3,155	3,154	3,019
E_6	2,236	2,236	2,236	2,310
E_7	1,329	1,329	1,329	1,312
E_∞	329	329	329	332
a_T	1.092	1.092	1.092	1.067
RMS	0.268	0.268	0.268	0.270

The most critical option above is the global weight ω. It was found that ω has to be small in order for the EKF-WGI procedure to run properly. Although δ should match the measurement accuracy for loss and storage moduli, it was found that as long as it is non-zero changing its value only affects E_1 and E_2, i.e., the spring stiffnesses for the two Maxwell Elements with shortest relaxation times in the eight-branch GMM unit.

The convergence test here is defined based on the RMS (root mean square) of relative error in measurement quantities. In particular:

$$RMS = \sqrt{\frac{1}{2m}\Sigma_{i=1}^{m}\left[\left(\frac{E'_{calc}}{E'_{meas}}-1\right)^2+\left(\frac{E''_{calc}}{E''_{meas}}-1\right)^2\right]} \quad (4.10)$$

where m is the total number of stiffness measurements and equals 66 for this specific set of data. Convergence is achieved when $RMS < 1.0 \times 10^{-6}$ or RMS is not decreasing anymore over the last four iterations.

A sensitivity study was carried out to evaluate the effect of seed values on the identified model parameters. The results for eight-branch GMM unit are shown in Table 5. The table indicates that:

Changing seed E_i from 1,000 to 10,000 only causes changes in identified E_1 and E_2;

Changing seed a_T from 2.0 to 1.0 only causes changes in identified E_1 and E_2;

Changing seed a_T from 2.0 to 5.0 only causes changes in identified E_1, E_2, E_3 and a_T.

Table 6. Effect of seed values on identified model parameters using six-branch GMM

Case No.	1	2	3	4
Seed E_i (MPa)	1,000	10,000	1,000	1,000
Seed a_1	2.0	2.0	1.0	5.0
E_1	3,950	3,950	3,951	3,950
E_2	2,908	2,908	2,908	2,909
E_3	3,196	3,196	3,196	3,196
E_4	2,229	2,229	2,229	2,229
E_5	1,330	1,330	1,330	1,330
E_∞	328	328	328	328
a_1	1.084	1.084	1.084	1.084
RMS	0.266	0.266	0.266	0.266
No. of Iterations	19	23	18	23

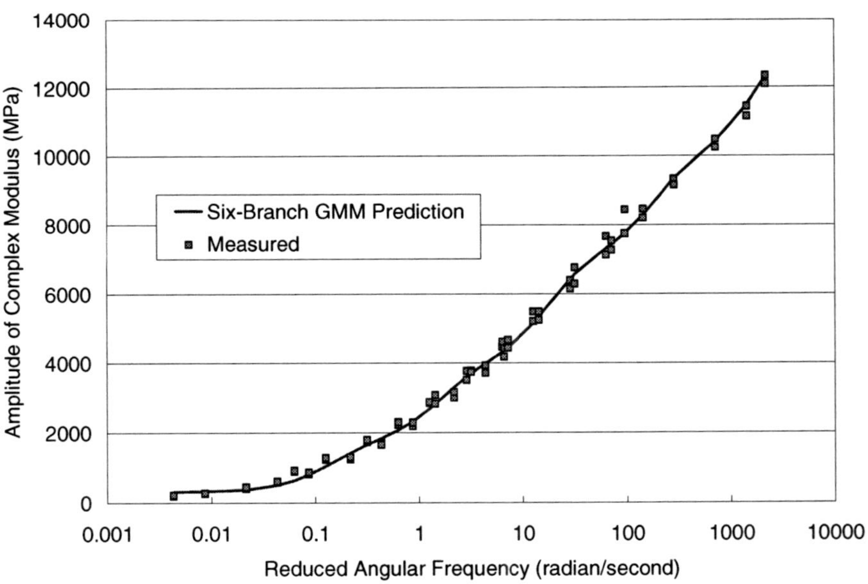

Fig. 1. Comparison of measured and calculated amplitude of complex modulus

For all of the cases evaluated, the final RMS are not significantly affected by changes in E_1 and E_2;

Based on these results, it is believed that the first two branches in the eight-branch GMM are not needed because they do not have significant effect on the overall stiffness of the GMM unit. Accordingly, a six-branch GMM unit is used instead to characterize the HMA stiffness. The relaxation times are changed to:

$$\lambda_i = 5.0 \times 10^{i-5} \text{ for } i\text{=1 to 5} \tag{4.11}$$

The effects of seed values on identified model parameters for the six-branch GMM unit are shown in Table 6. As shown in the table, the identified model parameters practically are not dependent on the seed values at all.

The comparison between measured and calculated amplitude of complex moduli with six-branch GMM unit is shown in Fig. 1, which indicates excellent fit between them.

5 Summary and Conclusions

In this chapter, a model parameter identification procedure based on the extended Kalman filter and a weighted global iteration (EKF-WGI) technique is presented. The procedure was adapted to perform layer moduli back-calculation using falling weight deflectometer (FWD) data and model parameter identification for Generalized Maxwell Model (GMM) from frequency sweep test data on hot mix asphalt (HMA). In both applications, EKF-WGI was successively applied although some of the options controlling EKF-WGI procedure need to be individually adjusted for each application.

Based on the results obtained in this chapter, EKF-WGI can provide seed-value independent results for both linear problems (e.g., layer moduli back-calculation) and non-linear problems (e.g. GMM model parameter identification) as long as the model parameters are significant. For the GMM case in particular, the relaxation times of the Maxwell Elements needs to be appropriately selected. It is believed that EKF-WGI provides an objective and robust algorithm for general model parameter identifications and has many more potential applications besides the examples shown here.

References

[1] Kalman, R.E.: A New Approach to Linear Filtering and Prediction Problems. Transactions of the ASME–Journal of Basic Engineering 82 (Series D), 35–45 (1960)

[2] Welch, G., Bishop, G.: SIGGRAPH 2001 Course 8: An Introduction to the Kalman Filter, University of North Carolina at Chapel Hill, Department of Computer Science. Chapel Hill (2001) NC 27599-3175

[3] Maybeck, P.S.: Stochastic models, estimation, and control. 1 (1979)

[4] Sorenson, H.W.: Least-Squares estimation: from Gauss to Kalman. IEEE Spectrum 7, 63–68 (1970)

[5] Gelb, A.: Applied Optimal Estimation. MIT Press, Cambridge (1974)

[6] Lewis, F.L.: Optimal Estimation with an Introductory to Stochastic Control Theory. John Wiley & Sons, Inc., Chichester (1986)

[7] Jacobs, O.L.R.: Introduction to Control Theory, 2nd edn. Oxford University Press, Oxford (1993)

[8] Brown, R.G., Hwang, P.Y.C.: Introduction to Random Signals and Applied Kalman Filtering: with MATLAB Exercises and Solutions, 3rd edn. Wiley & Sons, Inc., Chichester (1996)

[9] Grewal, M.S., Andrews, P.A.: Kalman Filtering Theory and Practice Using MATLAB, 2nd edn. John Wiley & Sons, Inc, New York (2001)

[10] University of North Carolina: The Kalman Filter, http://www.cs.unc.edu/~welch/kalman/

[11] McGee, L.A., Schmidt, S.F.: Discovery of the Kalman Filter as a Practical Tool for Aerospace and Industry, Technical Memorandum 86847, National Aeronautics and Space Administration (1985)

[12] Asher, R.B., Maybeck, P.S., Mitchell, R.A.K.: Filtering for precision pointing and tracking with application for aircraft to satellite tracking. In: IEEE Conference Decision and Control, Houston, TX (1975)

[13] Halamandaris, H., Ozdes, D.: Kalman Fitering Applied to a Marine Navigation System. In: Proceedings of the 1969 National Marine Navigation Meeting, Second Symposium on Manned Deep Submergence Vehicles, San Diego, CA (1969)

[14] Alexander, H.L., Azarbayejani, A.J., Weigl, H.J.: Kalman-Filter-based machine vision for controlling free-flying unmanned remote vehicles. In: Proceedings of the American Control Conference, Chicago, IL (1992)

[15] Grewal, M.S., Payne, H.J.: Identification of parameters in a freeway traffic model. IEEE Transactions on Systems, Man, and Cybernetics SMC-6, 176–185 (1976)

[16] Potter, J.M.: Kalman Filtering Approach to Market Price Forecasting, Ph.D. Thesis, Iowa State University, Ames (1986)

[17] Prabhu, A.V., Edgar, T.F.: A new state estimation method for high-mix semiconductor manufacturing processes. Journal of Process Control 19(7), 1149–1161 (2009)

[18] Jeon, S., Masayoshi, T., Tetsuaki, K.: A new kinematic Kalman filter (KKF) for end-effector sensing of robotic manipulators. In: The ASME International Mechanical Engineering Congress and Exposition, Seattle, WA (2008)

[19] Hoshiya, M., Saito, E.: Structural Identification by Extended Kalman Filter. Journal of Engineering Mechanics 110(12), 1757–1772 (1984)

[20] Loh, C.-H., Tsaur, Y.-H.: Time Domain Estimation of Structural Parameters. Engineering Structures 10(2), 95–105 (1988)

[21] Bao, Z.-W., Shi, W.-Y.: Parameter identification and state estimation of structure models in shaking table test. In: The 5th International Conference on Structural Safety and Reliability, San Francisco (1989)

[22] Zhang, H.Z., et al.: Parameter identification of inelastic structures under dynamic loads. Earthquake Engineering & Structural Dynamics 31(5), 1113–1130 (2002)

[23] Koh, C.G., See, L.M., Balendra, T.: Determination of storey stiffness of three-dimensional frame buildings. Engineering Structures 17(3), 179–186 (1995)

[24] Anderson, B.D., Moore, J.D.: Optimal Filtering. Prentice Hall, Inc., New Jersey (1979)

[25] Carmichael, D.G.: State Estimation Problem in Experimental Structural Mechanics, National Bureau of Standards. Special Publication (1979)

[26] Yun, C.B., Shinozuka, M.: Identification of Nonlinear Structural Dynamic Systems. Mechanics based design of structures and machines 8(2), 187–203 (1980)

[27] Yun, C.-B.: Damage assessment of bridge structures by system identification, pp. 2179–2186 (1989)

[28] Koh, C.: Estimation of structural parameters in time domain. A substructure approach. Earthquake engineering structural dynamics 20(8), 787–801 (1991)

[29] Seibold, S.: Recursive identification of nonlinear systems by means of the Extended Kalman Filter. In: Proceedings of the International Modal Analysis Conference - IMAC, 1991, vol. 1, pp. 802–809 (1991)

[30] Oreta, A.W.C., Tanabe, T.-a.: Localized identification of structures by Kalman filter. Structural engineering/earthquake engineering 9(4), 217–226 (1993)

[31] Sutoh, A.: Dynamic parameter identification by the EK-WLI-FEM. Doboku Gakkai ronbunshū. 2. Journal of hydraulic, coastal and environmental engineering (477), 97–100 (1993)

[32] Lin, J.-S., Zhang, Y.: Nonlinear structural identification using extended kalman filter. Computers & Structures 52(4), 757–764 (1994)

[33] Oreta, A.W.C., Tanabe, T.-a.: Element Identification of Member Properties of Framed Structures. Journal of Structural Engineering 120(7), 1961–1976 (1994)

[34] Lea, J.: Open Source Pavement Engineering, `http://www.openpave.org/`

[35] The MathWorks, Matlab, `http://www.mathworks.com`

[36] Tayebali, A., Deacon, J.A., Monismith, C.L.: Development and evaluation of dynamic flexural beam fatigue test system. Transportation Research Record (1545), 89–97 (1996)

[37] ARA Inc.: Guide for Mechanistic-Empirical Design of New and Rehabilitated Pavement Structures, ERES Consultants Division, ARA Inc., National Cooperative Highway Research Program, Transportation Research Board, National Research Council (2004)

Hybrid Stochastic Global Optimization Scheme for Rapid Pavement Backcalculation

Kasthurirangan Gopalakrishnan

Iowa State University, Ames, IA 50011, USA
rangan@iastate.edu

Abstract. Over the years, several techniques have been proposed for back-calculation of pavement layer moduli which involves searching for the optimal combination of pavement layer stiffness solutions in an unsmooth, multimodal, complex search space. In recent years, researchers are actively deriving inspiration from nature, biology, physical systems, and social behavior of natural systems for developing computational techniques to solve complex optimization problems. Some well-known nature-inspired meta-heuristics, which are basically high-level strategies that guide the search process to efficiently explore the search space in order to find (near-) optimal solutions, include, but are not limited to: Genetic Algorithms (GA), Particle Swarm Optimization (PSO), Simulated Annealing (SA), Shuffled Complex Evolution (SCE), etc. Potential applications of such nature-inspired hybrid optimization approaches to pavement backcalculation are conceptually illustrated in this chapter which take advantage of the combined efficiency and accuracy achieved by integrating advanced pavement numerical modeling schemes, computational intelligence based surrogate mapping techniques, and stochastic nature-inspired meta-heuristics with global optimization strategies using a system-of-systems approach.

1 Introduction

The Falling Weight Deflectometer (FWD) is a Non-Destructive Test (NDT) device used by pavement engineers to evaluate the structural condition of roads and to determine the moduli or stiffness of pavement layers. In the field, the pavement deflection basins are obtained from the FWD measurements which require the use of backcalculation type structural analysis to determine pavement layer stiffnesses and as a result estimate pavement remaining life. The accuracy of pavement strength estimation from FWD measurements is particularly important since it impacts the design of pavement overlay thickness. The process of backcalculation involves comparison of FWD measured deflections with computed deflections (using a pavement response model) through an iterative optimization procedure to determine the representative pavement layer moduli which could have produced the measured FWD deflections. Over the years, several static, dynamic, and adaptive techniques have been proposed for backcalculation of pavement layer moduli and each has its own pros and cons. In recent years, nature-inspired heuristics are

K. Gopalakrishnan et al. (Eds.): Intel. & Soft Comp. in Infra. Sys. Eng., SCI 259, pp. 255–267.
springerlink.com

becoming popular for solving engineering optimization problems. In this chapter, potential applications of such nature-inspired hybrid optimization approaches to pavement backcalculation are conceptually illustrated.

2 Concept and Implementation

Backcalculation of pavement layer moduli from FWD deflections can be treated as a global optimization problem where the objective is to determine the unknown pavement layer moduli that minimize the difference between measured and computed deflections. Thus the objective (fitness) function or the cost function for the proposed hybrid optimization approach is the difference between measured FWD deflections (see Fig. 1) and computed pavement surface deflections. In this paper, the implementation of the hybrid optimization approach is discussed for a three-layered flexible pavement structure although it can be used for other pavement types with varying number of layers owing to its flexible and integrated modular systems approach. A typical three-layered flexible pavement structure consists of Hot-Mix Asphalt (HMA) surface layer, a granular base layer consisting of unbound aggregates, and the bottommost layer consisting of subgrade soils.

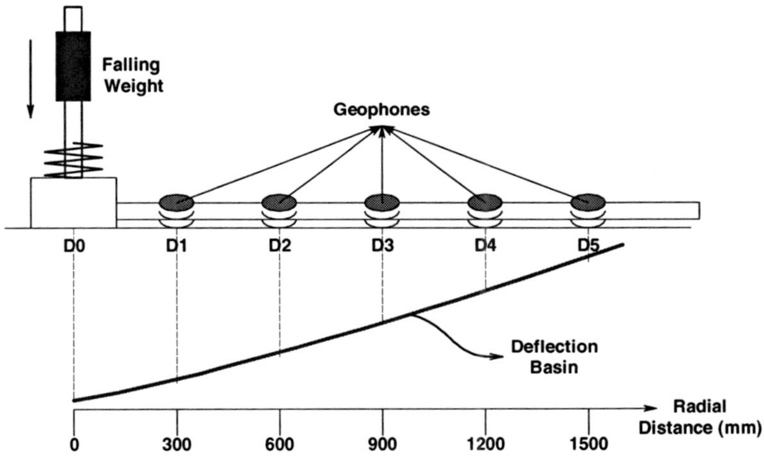

Fig. 1. Illustration of Falling Weight Deflectometer (FWD) test.

In the proposed hybrid optimization approach (see Fig. 2), a trained Neural Networks (NN) serves as a surrogate forward pavement response model that has learned the mapping between pavement layer elastic moduli and resulting pavement surface deflections for a variety of case scenarios generated using a 2-D axi-symmetric pavement finite element program [Raad and Figueroa 1980].

The NN module is an important component of the hybrid optimization scheme which significantly reduces the computational time required for forward calculation of deflections for each of the individuals in the generation. A brief discussion on the NN background is first presented followed by details related to the development of NN based surrogate forward models incorporated into hybrid optimization scheme.

NNs are parallel connectionist structures constructed to simulate the working network of neurons in human brain. They attempt to achieve superior performance via dense interconnection of non-linear computational elements operating in parallel and arranged in a pattern reminiscent of a biological neural network. The perceptrons or processing elements and interconnections are the two primary elements which make up a neural network. A single perceptron is mathematically represented as follows [Haykin 1999]:

$$y_k = \varphi(v_j) = \varphi\left(\sum_{i=1}^{n} x_i w_{ij} - b_j\right) \tag{1}$$

where x_i is input signal, w_{ij} is synaptic weight, b_j is bias value, v_j is activation potential, $\varphi()$ is activation function, y_k output signal, n is the number of neurons for previous layer, and k is the index of processing neuron.

Multilayer perceptrons (MLPs), frequently referred to as multi-layer feedforward neural networks, consist of an input layer, one or more hidden layer, and an output layer. For a given training data consisting of input-output vectors, values of synaptic weights in a MLP are iteratively updated by a learning algorithm to approximate the target behavior. This update process is usually performed by backpropagating the error signal layer by layer and adapting synaptic weights with respect to the magnitude of error signal [Goktepe et al. 2006]. Rumelhart et al [1986] presented the first backpropagation (BP) learning algorithm for use with MLP structures.

In the BP learning algorithm, the error energy used for monitoring the progress toward convergence is the generalized value of all errors that is calculated by the least-squares formulation and represented by a Mean Squared Error (MSE) as follows [Haykin 1999]:

$$MSE = \frac{1}{MP}\sum_{1}^{P}\sum_{k=1}^{M}(d_k - y_k)^2 \tag{2}$$

where M is the number of neurons in the output layer and P represents the total number of training patterns. Other performance measures such as the Root Mean Squared Error (RMSE), Average Absolute Error (AAE), etc. are also used.

An NN-based forward calculation procedure was developed to map the relation between input layer thicknesses and moduli and output surface deflections for passing on to the GA module during fitness evaluation. The goal is to simulate FWD loading using a numerical model for a wide variety of layer thicknesses and combinations of layer moduli encountered in the field resulting in a comprehensive synthetic solution set. A 2-D axi-symmetric FE program [Raad and Figueroa 1980] commonly used in the structural analysis of flexible pavements was employed to generate a comprehensive synthetic database of moduli-deflection solutions for wide ranges of layer thicknesses and layer moduli. Numerous research studies have validated that this FE model [Raad and Figueroa 1980] provides a realistic pavement structural response prediction for both highway and airfield pavements by incorporating stress-sensitive geomaterial models, the typical hardening behavior of nonlinear unbound aggregate bases and softening nature of subgrade soils under increasing stress states, and Mohr-Coulomb failure criteria to limit material strength [Gomez-Ramirez et al. 2002].

The synthetic database of FE solutions constituted the training and testing sets for developing NN-based models for rapid forward analysis of flexible pavements. A generic three-layer flexible pavement structure consisting of Hot-Mix Asphalt (HMA) surface layer, unbound aggregate base layer, and subgrade layer was modeled using the FE software [Raad and Figueroa 1980]. The top surface HMA layer was characterized as a linear elastic material with Young's Modulus, E_{HMA}, and Poisson ratio, ν. The K-θ model [Hicks and Monismith 1971] was used as the non-linear characterization model for the unbound aggregate layer:

$$E_R = K\theta^n \tag{3}$$

where E_R is resilient modulus (MPa), $\theta = \sigma_1 + \sigma_2 + \sigma_3 = \sigma_1 + 2\sigma_3$ = bulk stress, and K and n are multiple regression constants obtained from repeated load triaxial test data on granular materials. Based on the work reported by [Rada and Witczak 1981], K and n model parameters can be correlated to characterize the non-linear stress dependent behavior with only one model parameter. Thus, good quality granular materials, such as crushed stone, show higher K and lower n values, whereas the opposite applies for lower quality aggregates.

Fine-grained subgrade soils were modeled using the commonly used bi-linear resilient modulus model [Hoffman and Thompson 1982]:

$$\begin{aligned} E_R &= E_{Ri} + K_1.(\sigma_d - \sigma_{di}) \quad for\, \sigma_d < \sigma_{di} \\ E_R &= E_{Ri} + K_2.(\sigma_d - \sigma_{di}) \quad for\, \sigma_d > \sigma_{di} \end{aligned} \tag{4}$$

where E_{Ri} is the breakpoint resilient modulus, σ_d is the breakpoint deviator stress ($\sigma_d = \sigma_1 - \sigma_3$), σ_{di} is the breakpoint deviator stress, and K_1 and K_2 are statistically determined coefficients from laboratory tests. E_{Ri} can be used to classify fine-grained subgrade soils as being soft, medium or stiff.

Thus, HMA modulus, E_{HMA}, granular base K-θ model parameter K, and the subgrade break-point resilient moduli, E_{Ri}, were used as the layer stiffness inputs for all the FE runs. The 40-kN wheel load was applied as a uniform pressure of 552 kPa over a circular area of radius 150 mm simulating the FWD loading. A comprehensive FE synthetic database was generated by varying the HMA layer thickness (in the range of 75 to 700 mm), aggregate base layer thickness (in the range of 100 to 550 mm), E_{HMA} (in the range of 6.9 to 41.5 GPa), K (in the range of 21 to

82 MPa), and E_{Ri} (in the range of 7 to 105 MPa) for NN training and testing. Independent datasets were used for NN training and testing.

Details related to the development of optimal NN configuration for forward analysis can be found in [Gopalakrishnan and Thompson 2004]. In this study, the 5-40-40-6 architecture was chosen as the best architecture for the NN forward model based on its lowest training and testing MSEs for all six deflection output variables. The five inputs correspond to three layer stifnesses, HMA layer thickness and base layer thickness. Average Absolute Errors (AAEs) were calculated as sum of the individual absolute relative errors divided by the number of independent testing patterns (1,500 in this case).

$$Average\ Absolute\ Error\,(AAE), \% = \sum_{i=1}^{n} \left| \frac{y_{actual} - y_{predicted}}{y_{actual}} \right|_i * 100 \qquad (5)$$

Where i is the ith testing pattern among n testing patterns. The AAE values for ANN predicted output deflections were in the range of 0.2% – 0.4% with R^2 values above 0.999, indicating proper training and excellent prediction performance of the ANN surrogate forward calculation model.

Such NN forward surrogate models integrated into the inversion process offer a number of advantages over the traditional methods, due to their generalization capabilities, massive parallelism and potential to offer real-time solutions, thus making them perfect tools for rapidly analyzing the routinely collected FWD deflection data.

The Stochastic Global Optimization (SGO) algorithm (GA, PSO, SCE, etc.), in essence, finds the optimal values of the NN inputs (pavement layer moduli) iteratively such that the corresponding values of the network outputs (deflections) match the measured pavement surface deflections to minimize the differences between the measured and computer deflections. Although the error-minimization deflection-based objective function can be defined in a number of ways, a simple objective function representing sum of the squared differences between measured and computed deflections as shown in Equation 1 was selected for this study (where n = 6):

$$f = \sum_{i=1}^{n} (D_i - d_i)^2 \qquad (6)$$

The hybrid optimization framework was implemented in MATLAB. The input variables include six FWD measured surface deflections at 300-mm radial offsets starting from the center of the FWD loading plate, HMA surface and base layer thicknesses and the corresponding min-max ranges of pavement layer moduli. In the following sections, the individual SGO algorithms and their corresponding results are discussed for two nature-inspired optimization strategies.

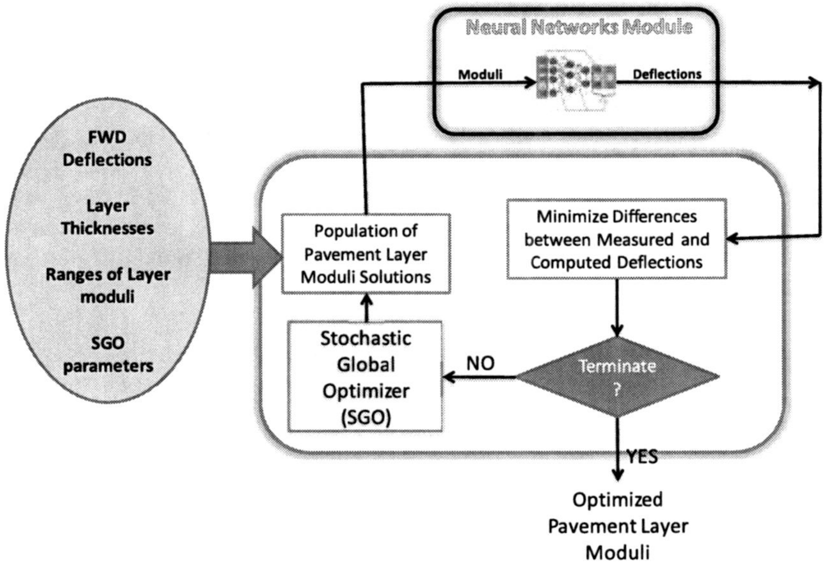

Fig. 2. Hybrid stochastic global optimization scheme for backcalculation

3 Stochastic Global Optimization (SGO)

3.1 Genetic Algorithms (GAs)

A significant number of previous research studies have employed GAs optimization strategy in backcalculating the mechanical properties of flexible pavement systems [Fwa et al. 1997, Reddy et al. 2002, Park et al. 2007, Alkasawneh 2007, Peckcan et al. 2008, Gopalakrishnan 2009, etc.]. Genetic algorithms are a part of evolutionary computing, a rapidly growing area of artificial intelligence. Categorized as global search heuristics, GAs use techniques inspired by evolutionary biology such as inheritance, mutation, selection, and crossover (also called recombination). Being robust search and optimization techniques, GAs are finding applications in a number of practical problems where calculus-based search methods are inefficient in searching for the optimal solution in a complex multimodal search space [Davis 1987, Goldberg 1989, Srinivas and Patnaik 1994, Holland 1975].

GAs begin with an initial random population of possible solutions to the problem and employs survival-of-the-fittest and exploitation of old knowledge in the gene pool to improve each generation's ability to solve the problem through a four-step process involving fitness evaluation, reproduction, recombination, and reproduction. After representing the optimization problem variables in suitable encoding mechanisms referred to as *chromosomes*, a set of trial solutions (often called as individuals) are generated and are forced to evolve towards an acceptable solution through the following four-step process [Frenzel 1993]:

- **Fitness Evaluation:** The mechanism for evaluating the fitness of individuals in the population in each generation is provided by the objective function, the function to be optimized. This fitness value is used in the next step in determining how many offsprings will be generated from any particular chromosome.
- **Reproduction:** In this step, a new population is created based upon the evaluation of the current one by taking advantage of the survival-of-the-fittest strategy, i.e., fitter solutions survive, while weaker ones perish. The two most popular methods, of calculating the number of offsprings that each chromosome will be allocated, are *ratioing* and *ranking*.
- **Recombination:** Reproduction simply produces multiple copies of existing chromosomes whereas recombination combines chromosomes from the population and produces new chromosomes while maintaining many of the features of the previous generation. The most common method for recombination is *crossover*. Subsections of two chromosomes randomly selected from the population are swapped about a randomly chosen crossover point governed by a specified crossover probability or rate. This is based on the assumption that the population as a whole contains the answer to the optimization problem and only by combining chromosomes will the best solution be found.
- **Mutation:** Mutation is only treated as a secondary genetic operator with the role of restoring lost genetic material. Under the assumption that the initial population didn't contain all the information necessary to solve the problem, new information is injected into the population by most randomly changing a fixed number of bits every generation based upon a specified mutation probability.

The GA module implemented in this study is capable of using either a floating point representation or a binary representation. First, the starting population is randomly generated. Each individual in the population, representing a set of pavement layer moduli, is passed on to the NN module for computing deflections which are then passed back to the GA module for fitness evaluation. Using the fitness function (Eq. 6), the GA module performs simulated evolution to determine the fitness of the solution strings.

In the current GA implementation [Gopalakrishnan 2009], it is possible to use a variety of crossover and mutation functions which include arithmetic crossover, heuristic crossover, simple crossover for crossover operator and boundary mutation, multi-non-uniform mutation, non-uniform mutation, and uniform mutation for mutation operator. Similarly, the implemented selection schemes include: roulette wheel, normalized geometric select, and tournament [Houck et al. 1995]. Based on literature review, the current study was conducted using the normalized geometric selection scheme with a probability of 0.08, arithmetic crossover and non-uniform mutation operators with variable probabilities. The size of the population and generation size were set to 80 and 100, respectively.

3.2 Particle Swarm Optimization (PSO)

PSO is a type of artificial intelligence method based on the collective behavior of decentralized, self-organized systems, has been proved to be an efficient method for many global optimization problems and in some cases it does not suffer the difficulties encountered by other evolutionary computation techniques. The PSO concept introduced by Kennedy and Eberhart [1995] draws its roots from artificial life (A-life), bird flocking, fish schooling, swarming theory, as well as genetic algorithms and evolutionary programming. Although it was originally introduced for optimization of nonlinear continuous functions, many advances in PSO development has enabled it to handle a wide class of complex engineering and science optimization problems.

Similar to GAs, a population of potential solutions to the problem under consideration is used to probe the search space in PSO. However, each individual of the population in PSO has an adaptable *velocity* (position change), according to which it moves in the search space. Moreover, each individual has a *memory*, remembering the best position of the search space it has ever visited [Eberhart and Shi 1998]. The movement of the individual is thus an aggregated acceleration towards its best previously visited position and towards the best individual of a topological neighborhood. Since the "acceleration" term was mainly used for particle systems in Particle Physics [Reeves 1983] and the term "swarm" for describing population, this algorithm was named as Particle Swarm Optimization. In essence, PSO employs a swarm of particles or possible solutions that fly through the feasible solution space to explore optimal solutions.

Two variants of the PSO algorithm were developed: one with a global neighborhood (Gbest model), and one with a local neighborhood. Each particle moves towards its best previous position and towards the best particle in the whole swarm, according to the global variant. In contrast, each particle moves towards its best previous position and towards the best particle in its restricted neighborhood, according to the local variant [Eberhart et al. 1996].

The global variant PSO algorithm, which is the most standard one, is described as follows. Suppose that the search space is D-dimensional, then the i-th particle of the swarm can be represented by a D-dimensional vector, $X_i = (x_{i1}, x_{i2}, \ldots, x_{iD})^T$. The *velocity* (position change) of this particle, can be represented by another D-dimensional vector $V_i = (v_{i1}, v_{i2}, \ldots, v_{iD})^T$. The best previously visited position of the i-th particle is denoted as $Pi = (p_{i1}, p_{i2}, \ldots, p_{iD})^T$. Defining g as the index of the best particle in the swarm (i.e., the g-th particle is the best), and let the superscripts denote the iteration number. Each particle updates its position based on its own best exploration, best swarm overall experience, and its previous velocity vector according to the following two equations [Eberhart et al. 1996] which define the initial version of the PSO algorithm:

$$v_{id}^{n+1} = v_{id}^{n} + c r_1^{n}(p_{id}^{n} - x_{id}^{n}) + c r_2^{n}(p_{gd}^{n} - x_{id}^{n}), \tag{7}$$

$$x_{id}^{n+1} = x_i^{d} + v_{id}^{n+1} \tag{8}$$

where $d = 1, 2,\ldots,D$; $i = 1, 2,\ldots,N$, and N is the size of the swarm; c is a positive constant, called *acceleration constant*; r_1, r_2 are random numbers, uniformly distributed in [0, 1]; and $n = 1, 2,\ldots,$ determines the iteration number. The performance of each particle is measured according to a pre-defined fitness function or objective function, which is related to the problem under consideration.

3.3 Shuffled Complex Evolution (SCE)

The SCE algorithm developed at the University of Arizona is reported to be an efficient global optimization method that can be used to handle non-linear problems with high-parameter dimensionality [Duan et al. 1992, Duan et al. 1993, Duan et al. 1994, Muttil and Liong 2004]. It consists of all the four principles for global optimization: the controlled random search, the implicit clustering, the complex shuffling, and the competitive evolution.

In SCE methodology, the search for the optimal solution begins with a randomly selected complex of points spanning the entire feasible space. The implicit clustering helps to concentrate the search in the most promising of the regions. The use of complex shuffling provides a freer and more extensive exploration of the search space in different directions, thereby reducing the chances of the search getting trapped in local optima. Three of these principles are coupled with the competitive complex evolution (CCE) algorithm, which is a statistical reproduction process employing the complex geometric shape to direct the search in the correct direction. The synthesis of these concepts makes the SCE algorithm not only effective and robust, but also flexible and efficient [Nunoo and Mrawira 2004].

The SCE control parameters should be determined in advance to achieve the required exploration process. These parameters include the number of points in a complex (m), the number of points in a sub complex (q), the number of complexes (p), the number of consecutive offspring generated by each sub complex (α), and the number of steps in-evolution taken by each complex (β). Duan et al [1994] provides guidelines for proper selection of these parameters.

The basic algorithm for SCE described by Duan et al [1993] can be outlined as follows:

1. An initial population of points is sampled randomly from the feasible solution space (Ω) in the real space (R^n).
2. The selected population is partitioned into one or more complexes, each containing a fixed number of points.
3. Each complex evolves according to a competitive complex evolution (CCE) algorithm.
4. The entire population is periodically shuffled and points are reassigned to complexes to share the information from the individual complexes.
5. Evolution and shuffling are repeated so that the entire population is close to convergence criteria, and are stopped if the convergence criteria are satisfied.

The CCE algorithm is a sub-route in SCE algorithm. CCE algorithm employs the downhill simplex method [Nelder and Mead 1965] in generating offsprings. The simplex method facilitates evolution of each complex independently in an improvement direction.

4 Preliminary Findings

Hypothetical data covering wide ranges of layer thicknesses and FWD deflections commonly encountered in the field were first used to evaluate the prediction accuracy of the developed hybrid stochastic global optimization schemes. A total of about 150 datasets were independently selected from the comprehensive synthetic FE solutions database to assess the prediction performance. The performance of hybrid SGO approach in backcalculating flexible pavement layer moduli is reported in Figs. 3 to 5. All three SGO methodologies were found to be efficient and robust for backcalculation application. As shown in the plots, in all three cases, the

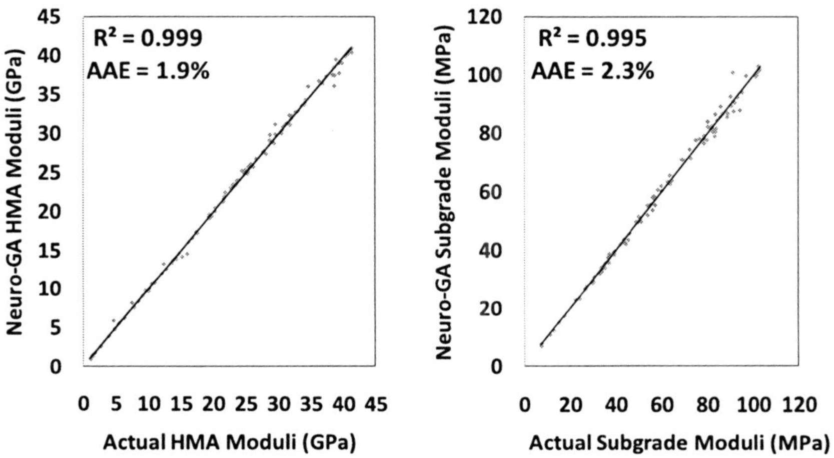

Fig. 3. Backcalculation of pavement moduli using hybrid Genetic Algorithms (GAs) approach [Gopalakrishnan 2009]

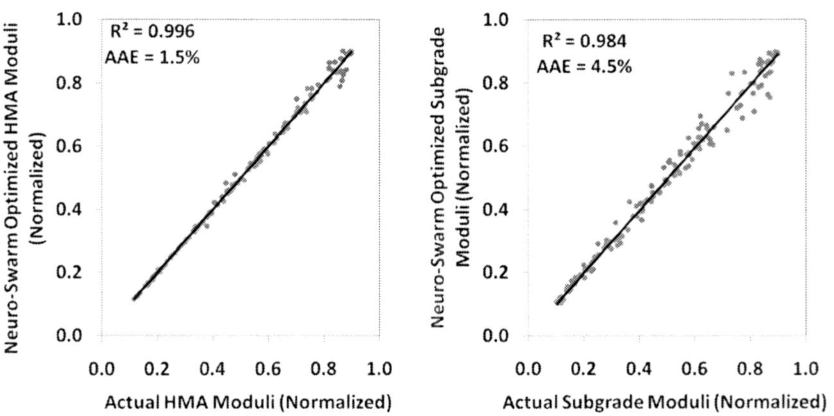

Fig. 4. Backcalculation of pavement moduli using hybrid Particle Swarm Optimization (PSO) approach

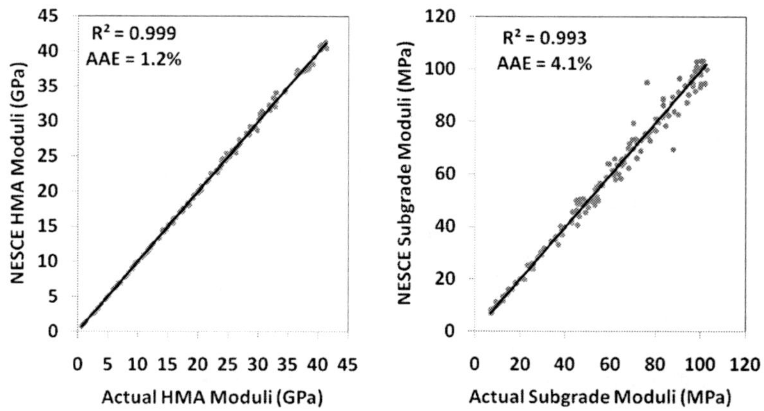

Fig. 5. Backcalculation of pavement moduli using hybrid Shuffled Complex Evolution (SCE) approach

150 SGO backcalculation predictions fell on the line of equality for the two pavement layer moduli thus indicating proper training and very good performance of the proposed hybrid backcalculation model.

5 Concluding Remarks

Backcalculation of pavement layer moduli is an ill-posed *inverse* engineering problem, where the unknown mechanical properties of the pavement structure are determined from the field-measured surface deflection profile generated by the system subjected to impulse (FWD) loading. Majority of the commercial backcalculation programs do not account for the non-linear response of pavement geomaterials under applied loading and employ an iterative optimization approach which is known to suffer from limitations such as dependency on the initial seed moduli and the possibility of local minimum solutions.

This chapter presented that the development of a new hybrid Neural Networks (NN)-Stochastic Global Optimization (SGO) approach for the backcalculation of pavement layer moduli. For the first time, efficient SGO techniques like PSO and SCE were applied to inversion of pavement non-destructive test deflection data and backcalculation of HMA layer moduli and stress-dependent, non-linear subgrade moduli. The developed hybrid scheme combines the robustness of SGO in global optimization with the computational efficiency of NNs. This allows the hybrid model to facilitate real-time non-destructive evaluation of pavement systems and makes it a very attractive alternative for handling the backcalculation problem. It was demonstrated using hypothetical data that the proposed hybrid backcalculation approach can successfully predict the flexible pavement surface layer moduli and non-linear subgrade moduli.

References

Alkasawneh, W.: Backcalculation of pavement moduli using genetic algorithms. Ph.D. dissertation, Dept. Civil Eng., Univ. of Akron, Akron, Ohio (2007)

Davis, L.: Genetic Algorithms and Simulated Annealing. Pitman, London (1987)

Duan, Q., Sorooshian, S., Gupta, V.K.: Effective and efficient global optimization for conceptual rainfall-runoff models. Water Resour. Res. 28(4), 1015–1031 (1992)

Duan, Q.A., Gupta, V.K., Sorooshian, S.: Shuffled complex evolution approach for effective and efficient global minimization. J. Optim. Theory Appl. 76(3), 501–521 (1993)

Duan, Q.A., Sorooshian, S., Gupta, V.K.: Optimal use of the SCE-UA global optimization method for calibrating watershed models. J. Hydrol. 158, 265–284 (1994)

Eberhart, R.C., Shi, Y.: Evolving artificial neural networks. In: Proc. 1998 Intl. Conf. on Neural Networks and Brain, Beijing, P.R.C., pp. 5–13 (1998)

Eberhart, R.C., Simpson, P.K., Dobbins, R.W.: Computational Intelligence PC Tools. Academic Press Professional, Boston (1996)

Frenzel, J.: Genetic Algorithms, A New Breed of Optimization. IEEE Potentials, 21–24 (October 1993)

Fwa, T.F., Tan, C.Y., Chan, W.T.: Backcalculation analysis of pavement-layer moduli using genetic algorithms. Transportation research record 1570, 134–142 (1997)

Goldberg, D.E.: Genetic Algorithms in Search, Optimization and Machine Learning. Addison Wesley, Reading (1989)

Goktepe, A.B., Agar, E., Lav, A.H.: Comparison of Multilayer Per-ceptron and Adaptive Neuro-Fuzzy System on Backcalculating the Mechanical Properties of Flexible Pavements, vol. 54(3), pp. 1–13. The Bulletin of the Istanbul Technical University, ARI (2006)

Gopalakrishnan, K., Thompson, M.R.: Backcalculation of airport flexible pavement nonlinear moduli using artificial neural networks. In: Proc. 17th International FLAIRS Conference, Miami Beach, Florida (2004)

Gopalakrishnan, K.: Backcalculation of Non-Linear Pavement Moduli Using Finite-Element Based Neuro-Genetic Hybrid Optimization. The Open Civil Engineering Journal 3, 83–92 (2009)

Gomez-Ramirez, F., Thompson, M.R., Bejarano, M.: ILLI-PAVE based flexible pavement design concepts for multiple wheel heavy gear load aircraft. In: Proc. Ninth International Conference on Asphalt Pavements, Copenhagen, Denmark (2002)

Haykin, S.: Neural networks: A comprehensive foundation. Prentice-Hall Inc., New Jersey (1999)

Hicks, R.G., Monismith, C.L.: Factors Influencing the Resilient Properties of Granular Materials. Transportation Research Record 345, 15–31 (1971)

Hoffman, M.S., Thompson, M.R.: Backcalculating nonlinear resilient moduli from deflection data. Transportation research record 852, 42–51 (1982)

Holland, J.H.: Adaption in Natural and Artificial Systems. Univ. Michigan Press, Ann Arbor (1975)

Houck, C., Joines, J., Kay, M.: A genetic algorithm for function optimization: A Matlab implementation. North Carolina State University, Raleigh, North Carolina, Research Rep. NCSU-IE TR 95-09 (1995)

Kennedy, J., Eberhart, R.C.: Particle swarm optimization. In: Proc. IEEE Int'l. Conf. on Neural Networks, Piscataway, NJ, vol. 4, pp. 1942–1948 (1995)

Muttil, N., Liong, S.Y.: Superior exploration–exploitation balance in shuffled complex evolution. J. Hydr. Engrg. 130(12), 1202–1205 (2004)

Nelder, J.A., Mead, R.: A simplex method for function minimization. Computer Journal 7, 308–313 (1965)

Nunoo, C., Mrawira, D.: Shuffled complex evolution algorithms in infrastructure works programming. ASCE Journal of Computing in Civil Engineering 18(3), 257–266 (2004)

Park, H.M., Park, S.W., Hwang, J.J.: Use of genetic algorithm and finite element method for backcalculating layer moduli in asphalt pavements. In: Proc. TRB 86th Annual Meeting Compendium of Papers, Washington, DC (2007)

Peckcan, O., Tutumluer, E., Thompson, M.R.: Nondestructive pavement evaluation using ILLI-PAVE based artificial neural network models. Illinois Center for Transportation, Champaign, IL, Research Rep. FHWA-ICT-08-022 (2008)

Raad, L., Figueroa, J.L.: Load response of transportation support systems. Transport. Eng. J. ASCE 106(TE1), 111–128 (1980)

Rada, G., Witczak, M.W.: Comprehensive evaluation of laboratory resilient moduli results for granular material. Transportation Research Record No. 810, pp. 23–33 (1981)

Reddy, M.A., Reddy, K.S., Pandey, B.B.: Backcalculation of pavement moduli using genetic algorithms. J. Highw. Res. Board. 66, 1–10 (2002)

Reeves, W.T.: Particle systems – a technique for modelling a class of fuzzy objects. ACM Transactions on Graphics 2(2), 91–108 (1983)

Rumelhart, D.E., Hinton, G.E., Williams, R.J.: Learning internal representation by error propogation. In: Rumelhart, D.E. (ed.) Parallel Distributed Processing, pp. 318–362. MIT Press, Cambridge (1986)

Srinivas, M., Patnaik, L.M.: Adaptive probabilities of crossover and mutation in genetic algorithms. IEEE Transactions on Systems, Man, and Cybernetics 24(4), 656–667 (1994)

Regression and Artificial Neural Network Modeling of Resilient Modulus of Subgrade Soils for Pavement Design Applications

Pranshoo Solanki[1], Musharraf Zaman[2], and Ali Ebrahimi[3]

[1] Doctoral Candidate, School of Civil Engineering and Environmental Science, The University of Oklahoma, 202 West Boyd Street, Room 334, Norman, OK, 73019
pranshoo@ou.edu

[2] David Ross Boyd Professor and Aaron Alexander Professor, Associate Dean for Research and Graduate Education, College of Engineering, The University of Oklahoma, 202 West Boyd Street, Room 107, Norman, OK, 73019
zaman@ou.edu

[3] President, Burgess Engineering and Testing, Inc., 2603 N. Shield Blvd., Moore, OK, 73160
ali@burgessengineer.com

Abstract. A combined laboratory and modeling study was undertaken to develop a database for common subgrade soils in Oklahoma and to develop relationships or models that could be used to estimate resilient modulus (M_R) from commonly used subgrade soil properties in Oklahoma. Sixty-three soil samples from 14 different sites throughout Oklahoma are collected and tested for the development of the database and models. Additionally, thirty-four soil samples from 3 different sites, located in Rogers and Woodward counties, are collected and tested to evaluate the developed models. The routine material parameters selected in the development of the models include moisture content (w), dry density (γ_d), plasticity index (PI), percent passing No. 200 sieve (P_{200}), and unconfined compressive strength (U_c). Bulk stress (θ) and deviatoric stress (σ_d) are used to identify the state of stress. A total of four, two regression models, namely, Polynomial and Factorial, and two feedforward-type artificial neural network (ANN) models, namely, Radial Basis Function Network (RBFN) and Multi-Layer Perceptrons Network (MLPN) are developed. A commercial software, STATISTICA 7.1, is used to develop these models. The strengths and weaknesses of the developed models are examined by comparing the predicted M_R values with the experimental values with respect to the R^2 values. An evaluation of the four models indicate that for the combined development and evaluation datasets, the MLPN model is a good model for evaluating M_R from the selected routinely determined properties. In order to illustrate the application of the developed model, the AASHTO flexible pavement design methodology is used to design asphalt concrete pavement sections.

1 Introduction

Empirical design methods for flexible pavement structures are primarily based on the equations that were developed largely from the AASHO Road Tests conducted

K. Gopalakrishnan et al. (Eds.): Intel. & Soft Comp. in Infra. Sys. Eng., SCI 259, pp. 269–304.
springerlink.com

in 1950's. These methods fail to reflect the dynamic nature of traffic loads. Therefore, the mechanistic design methods referred to as the "AASHTO Guide for Design of Pavement Structure" (AASHTO 1986) recommended the use of resilient modulus (M_R), a dynamic-strength parameter, to characterize the flexible pavement materials. The M_R accounts for the cyclic nature of vehicular traffic loading, and is defined as the ratio of deviatoric stress to recoverable elastic strain.

Several laboratory and field procedures are currently either used or evaluated for determining a design M_R value of subgrade soil. Direct laboratory methods used for evaluating M_R during the past two decades include resonant column, torsional shear, gyratory, and repeated load triaxial testing (AASHTO 1986, Kim and Stokoe 1992, George 1992, Kim et al.1997). Among these testing procedures, the M_R from repeated load triaxial test (RLTT) is used most frequently because of the repeatability of test results and its representation of field stress in controlled laboratory environments. RLTT is conducted in the laboratory on remolded or undisturbed samples according to different AASHTO test methods of which AASHTO T307 is used frequently (AASHTO 2004). The AASHTO T307 test method can be a time consuming and expensive test method, particularly for small projects.

In the new 2002 AASHTO guide, which is currently in the evaluation stage, a hierarchical approach is used to determine different design inputs including M_R (AASHTO 2004). It requires the evaluation of the engineering properties of subgrade soils in laboratory or field to pursue a Level-1 (most accurate) design. For a Level-2 (intermediate) design, however, the design inputs are user selected, possibly from agency database or from limited testing program or could be estimated through correlations (AASHTO 2004). A Level-3 design, which is the least accurate and generally not recommended, uses only the default values. For Level-2 designs, a regression model for M_R can be very useful as it provides the designer with significant flexibility in obtaining the design inputs for a project.

2 Objectives

The primary objective of this study is to develop a database and correlations or models for M_R of some commonly encountered subgrade soils in Oklahoma for Level-2 pavement design applications. Generally, there are two different modeling options, namely regression models and artificial neural networks (ANN). Both stress (deviatoric stress and bulk stress) and routine soil properties (unconfined compressive strength, dry density, moisture content, gradation, and Atterberg limits) are employed in developing these models. In this study, two regression models, namely, Polynomial and Factorial, were developed. In addition, two feed-forward-type ANN models, namely, Radial Basis Function Network (RBFN) and Multi-Layer Perceptrons Network (MLPN), were generated. A commercial software, STATISTICA 7.1, was used to develop these models. The strengths and the weaknesses of the developed models were examined using additional M_R test results that were not used in the development of these models. The models developed in this study are expected to be useful in the Level 2 designs of pavements in Oklahoma.

3 Review of Previous Studies

3.1 Regression Models

Several pertinent studies have previously been undertaken to develop empirical correlations to estimate M_R values in terms of other soil properties. One of the commonly used models to represent M_R is the power model (see e.g. Dunlap 1963, Seed et al. 1967, Thompson and Robnett 1976, Moossazadeh and Witczak1981, May and Witczak 1981, Uzan 1985, Farrar and Turner 1991, Yau and Quintus 2002, NCHRP 2003, Hopkins at el. 2004, Rahim and George 2004, and Khazanovich et al. 2006). Dunlap (1963) proposed the following correlation for M_R:

$$M_R = k_1(\sigma_3/P_a)^{k_2} \tag{1}$$

where, σ_3 is confining pressure, P_a is a reference pressure (e.g., atmospheric pressure) and k_1 and k_2 are regression coefficients.

A number of researchers (see e.g., Dingquing and Shelig 1994, Gomes and Gillet 1996, Paute and Hornych 1996, Rada and Witczak 1981, Raad et al. 1992 and Zaman et al. 1994) have utilized other soil property indices to estimate M_R. For example, Drumm et al. (1990) developed two regression models for M_R of fine-grained soils as a function of deviator stress and soil-index properties, namely, percentage passing No. 200 sieve (P_{200}), plasticity index (PI), dry density (γ_d), and unconfined compressive strength (U_c). A relatively small (twenty-two) number of these samples were used in developing these models.

In a similar study, Lee et al. (1997) investigated the M_R of cohesive soils, mainly clayey subgrade soils, with RLTT. Specimens were compacted using standard and modified proctor methods at near optimum moisture content (OMC) in a mold with a diameter of 38 mm (1.5 in) and a height of 100 mm (4.0 in). It was seen that the custom-compaction results were in close agreement with the maximum dry density (MDD) and the optimum moisture content (OMC) from the standard and modified Proctor tests. Regression analyses were conducted to obtain a relationship between M_R and the stress in unconfined compressive strength test causing 1% strain ($S_{U1.0\%}$) in laboratory compacted specimens. The relationship between M_R and $S_{U1.0\%}$ for a given soil was found to be unique regardless of moisture content and compaction effort. The results showed that the M_R and $S_{U1.0\%}$ vary with the moisture content in a similar manner. Furthermore, four different compactive efforts were used in that study, but a single relationship between M_R and $S_{U1.0\%}$ was obtained, as presented in equation (2):

$$M_R = 695.4\,(S_{U1.0\%}) - 5.93\,(S_{U1.0\%})^2 \tag{2}$$

where, M_R = resilient modulus at maximum axial stress of 41.4 kPa and confining pressure of 20.7 kPa; and $S_{U1.0\%}$ = stress causing 1% (strain kPa) in conventional U_C test.

Moreover, the relationship was similar for different cohesive soils, indicating that it may be applicable for different types of clayey soils. The limited data suggested that the same correlation might be used to estimate the M_R for both laboratory and field compacted conditions.

In a field study, Yau and Von Quintus (2002) proposed the following correlation using the M_R data obtained from the Long Term Pavement Performance (LTPP) test sections:

$$M_R = k_1 P_a (\theta/P_a)^{k2} [(\tau_{oct}/P_a)+1]^{k_3} \quad (3)$$

where, τ_{oct} is the octahedral shear stress, and k_1, k_2, and k_3, are the regression constants. Yau and Quintus (2002) expressed these regression constants as a function of moisture content, dry density, optimum dry density, liquid limit, percent silt, percent clay, and percent passing different sieve sizes. The soils were classified into three different groups (coarse grained sandy soils, fine grained silty soils, and fine grained clayey soils), and the regression constants were developed for each soil type.

Using data from six different pavement sections in Minnesota, Dai and Zollars (2002) suggested a similar model that used deviatoric stress (σ_d) as a stress variable instead of τ_{oct}. Also, they used only one sieve size (#200) instead of multiple sieve sizes used by Yau and Quintus (2002), and included PI and degree of saturation as model variables.

In Minnesota, Khazanovich et al. (2006) used M_R results for 23 samples from several locations and evaluated the regression constants for use in the mechanistic-empirical-based pavement designs. However, because the mineralogical and textural characteristics of soils in Oklahoma are different than those in Minnesota, those results may not be directly used for pavements in Oklahoma for a Level 2 design.

In a recent study, Malla and Joshi (2008) used long-term pavement performance information for 259 test specimens for developing model consisting bulk stress (θ) and τ_{oct} relating to soil properties such as moisture content (w), OMC, γ_d, MDD, liquid limit (LL), and PI. Predictions models were developed by conducting multiple linear regression analysis using computer software SAS.

3.2 Artificial Neural Network (ANN) Models

ANN has become an important modeling technique due to its success in many engineering applications including geotechnical engineering problems (see e.g., TRB 1999, Najjar et al. 2000, Shahin et al. 2004). One of the common artificial neural networks in use currently is the feedforward network. As evident from its name, a feedforward network only allows the data flow in the forward direction (Zurada 1992, Fausett 1994, Ripley 1996, StatSoft Inc. 2006). Based on the architecture, a number of feedforward networks are available such as multilayer perceptrons, radial basis function, probabilistic neural networks, generalized regression neural networks, and linear networks (TRB 1999, Shahin et al. 2004, StatSoft Inc. 2006, Sharma and Das 2008, Far et al. 2009).

ANN contains a number of simple, highly interconnected processing elements, known as "nodes" or "units." In a typical processing element, each input connection has a weighting value. With the weighting value, input data and bias value, a net input is described into the processing element. Then, a transfer function provides an output from the net input. Finally, a single output is produced and transmitted to other processing elements (Skapura 1996, Najjar et al. 2000, Shahin et al. 2001).

The weights between the processing elements are adjusted during the "training or learning" phase. In the training process, a number of epochs are performed in the network. After each epoch, the weights are adjusted and a sum of mean squared error between target and output values is calculated. The training process stops when the sum of mean squared error is minimized or falls within an acceptable range (Shahin et al. 2001, Shahin et al. 2004).

Different algorithms can be used to train a network. In general, the training algorithms can be divided into two types: supervised and unsupervised. The supervised algorithms adjust the weights and the thresholds using the input and target output values, while the unsupervised algorithms only use the input values. The supervised training algorithms include back propagation, conjugate gradient descent, Levenberg-Marquardt, Pseudo-inverse, etc. (Mehrotra et al. 1996, Shahin et al. 2004, StatSoft Inc. 2006).

A number of researches have utilized ANN technique in pavement applications. For example, Meier et al. (1996) augmented a computer program, WESDEF, with ANN models to backcalculate pavement layer moduli. The ANN models were trained to compute the layer moduli (M_R) from falling weight deflectometer (FWD) data from flexible pavements (Meier et al. 1996).

In a recent study, Sharma and Das (2008) used ANN models to backcalculate layer moduli with better accuracy compared with other software, namely, EVERCALC and ExPaS. In another recent pavement application, Far et al. (2009) utilized ANN for estimating the dynamic modulus of asphalt concrete. The results showed that the predicted and measured dynamic modulus values are in close agreement using ANN models.

4 Sources and Characteristics of Subgrade Soils

In the present study, a total of 97 bulk soils samples were collected from 16 different counties in Oklahoma. Of these, 63 samples from 14 different counties were used in the development of the regression models and are collectively referred to as the "development dataset." These sites were located in Adair, Alfalfa, Choctaw, Delaware, Greer, Jefferson, Kingfisher, Lincoln, Major, McClain, Noble, Okfuskee, Osage, and Rogers counties in Oklahoma. The remaining 34 soils from two different counties namely, Rogers and Woodward counties, were used for the evaluation of the regression models. Data for these soils are collectively referred to as the "evaluation dataset." A majority of soils in the development dataset was lean clay and lean clay with sands (Table 1). A majority of soils in the evaluation dataset, on the other hand, was lean clay, lean clay with sand and sandy lean clay. Bulk samples of different soil series, primarily from the B-horizon and the C-horizon, were collected from each site following the standard sampling method for pedological and geological soil survey (FHWA 2002, AASHTO 2004).

Table 1. Summary of USCS soil classification results for the development and evaluation dataset

Soil Classification		Number of Soils	
		Development Dataset	Evaluation Dataset
Unified Soil Classification System (USCS)			
Fat clay	CH	2	1
Fat clay with sand	CH	0	1
Sandy fat clay	CH	1	0
Lean clay	CL	23	8
Lean clay with sand	CL	22	8
Gravelly lean clay	CL	2	0
Sandy lean clay	CL	8	10
Sandy lean clay with gravel	CL	0	2
Silty clay with sand	CL-ML	1	0
Sandy silty clay	CL-ML	1	3
Clayey Sand	SC	1	1
Clayey Sand with gravel	SC	2	0
	Total :	**63**	**34**

5 Laboratory Testing and Results

The laboratory testing program included routine laboratory tests, namely grain size distribution (AASHTO T11 and AASHTO T27), Atterberg limits (AASHTO T89 and AASHTO T90) and standard proctor (ASTM D698), as well as resilient modulus (AASHTO T307) and unconfined compression (AASHTO T208). Using the proctor test results, two samples were prepared for each soil with different compaction conditions. One of these samples was compacted at the optimum moisture content (OMC) and 95% of the maximum dry density (MDD). For the other sample, the moisture content and dry density were set at 2% wet of OMC, representing the Oklahoma DOT in-construction stage requirements (Ebrahimi 2006). Specimens having a moisture variation of more than ±0.5 percent from the targeted moisture content and dry density less than 95% of MDD were discarded and new samples were compacted, evaluated and tested. Thus, a total of 126 M_R tests were conducted for 63 soils used in the development dataset. Likewise, 68 M_R tests were conducted for 34 soils in the evaluation dataset. A static compaction method (a modified version of the double plunger method) was used in sample preparation (AASHTO 2004). The unconfined compressive strength (U_C) test was conducted on the same sample, following the M_R testing. It is assumed that since the M_R strain is in the range of 1/ten thousand (mm/mm) the influence of M_R test on the UCS test would be negligible (Ebrahimi 2006).

Table 2. Basic statistical parameters for different soil properties

Dataset	No. of Soils	Mean	Me-dian	Min.	Max.	Std. Dev.	Skewness	Kurto-sis
Liquid Limit (LL)								
Development	63	34.	33.0	21	67	10.0	1.11	1.20
Evaluation	34	36.	35.0	24	52	8.4	0.18	-1.16
Plastic Limit (PL)								
Development	63	16.	15.0	9	27	3.5	1.11	1.48
Evaluation	34	15.	14.5	12	21	2.7	0.56	-1.00
Plasticity Index (PI)								
Development	63	18.	17.0	7	43	8.4	0.97	0.63
Evaluation	34	20.	21.5	6	36	8.9	-0.03	-1.22
Percent Passing No. 4 Sieve								
Development	63	98.	100.	73.	100	6.4	-3.47	10.79
Evaluation	34	97.	100.	79.	100	4.8	-2.77	7.89
Percent Passing No. 10 Sieve								
Development	63	96.	99.9	47.	100	9.3	-3.64	14.40
Evaluation	34	95.	99.3	74.	100	6.7	-1.87	3.05
Percent Passing No. 40 Sieve								
Development	63	93.	97.3	39.	99.	10.5	-3.27	12.38
Evaluation	34	88.	92.2	49.	100	11.4	-1.82	3.53
Percent Passing No. 200 Sieve								
Development	63	78.	83.0	36.	98.	15.0	-1.05	0.37
Evaluation	34	74.	74.8	37.	94.	13.1	-0.49	0.27
Group Index (GI)								
Development	63	13.	13.0	1	39	8.6	0.81	0.58
Evaluation	34	14.	13.0	1	31	9.6	0.29	-1.36
*Specimen Moisture Content (%)**								
Development	126	17.	17.2	10.	25.	3.0	0.39	-0.14
Evaluation	68	17.	17.5	11.	22.	2.9	-0.24	-0.94
*Specimen Dry Density (kg/m^3)**								
Development	126	16	1658	140	187	106.	-0.26	-0.44
Evaluation	68	16	1676	154	186	74.8	0.40	-0.60
*Unconfined Compressive Strength (kPa)**								
Development	126	20	193.	50.	443	70.8	0.68	0.50
Evaluation	68	16	159.	56.	357	64.4	0.78	0.81

*Specimens compacted at OMC and OMC+2% for each soil

A summary of the basic regression parameters for liquid limit, plastic limit, plasticity index, percentage passing #4 sieve, #10 sieve, #40 sieve, #200 sieve, group index, specimens moisture content, dry density and UC is listed in Table 2. Further regression details of different parameters (i.e. liquid limit, plastic limit, plasticity index, percentage passing #4, #10, #40, #200 sieve, specimens moisture content, dry density and UC) used in this study are given in Ebrahimi (2006).

Montgomery et al. (2006) recommended that datasets deviating from normal distribution would not affect the outcome of the analysis and the results would not be critically affected. Hence, in this study kurtosis and skewness were determined to select the input parameter for regression modeling. Kurtosis parameter is an indicator of heaviness of the tail. A perfectly normal distribution of data has a kurtosis of zero. A positive kurtosis is an indication of more observations on the tail end of the distribution curve, while a negative kurtosis is an indication of fewer observations on the tail end of the distribution curve. Skewness is a measure of distribution of the data. A skewness of zero indicates perfectly normal distribution of data. Negative value of skewness indicates the data skewed left and positive value indicates the data skewed right.

5.1 Atterberg Limits

The results from the liquid limit (LL) tests for the development dataset range from 21 to 67 with a mean of 34.5 and a standard deviation of 10.0. The range of LL for evaluation dataset is from, 24 to 52 with a mean of 36.5 and a standard deviation of 8.4. Table 2 presents the basic regression parameters for the two sets of data. Based on the skewness parameter, the LL values for the evaluation dataset (i.e., 0.18) are more normally distributed than the development dataset (i.e., 1.11). On the other hand, based on the kurtosis parameter, the LL data for development dataset and evaluation dataset are not distributed similarly. The development dataset has more data on the tail end with a kurtosis of 1.20. On the other hand, the evaluation dataset has less data on the tail end with a kurtosis of –1.16. These results indicate that the LL data are not perfectly normally distributed; however, the deviation is small and therefore, normally distributed theories may be applied in regression analysis.

Plastic Limit (PL) for the development dataset has a mean of 16.1 and a standard deviation of 3.5. The PL for the development dataset ranges from 9 to 27. The results for the evaluation dataset show a mean of 15.5 and a standard deviation of 2.7. The evaluation dataset ranges from 12 to 21 in PL values. The basic regression parameters for the two sets of data are presented in Table 2. The basic regression parameters and the figure show that the distributions for both datasets are close to normally distributed. The skewness of the development and the evaluation datasets are 1.11 and 0.56, respectively, which are close to zero. Moreover, the kurtosis of the development and the evaluation datasets are also close to zero. The kurtosis of the development and the evaluation datasets are 1.48 and –1.00, respectively.

The PI values for the development dataset ranged from 7 to 43, with a mean of 18.4 and a standard deviation of 8.4 (Table 2). Although the overall regression indicators for the evaluation dataset were comparable to those of the development dataset, the corresponding values for the Rogers County soils were significantly different from the Woodward County soils. Overall, the Rogers County soils were closer to the soils in the development dataset. From Table 2, the skewness and kurtosis for both datasets were negligible, so the distributions of the PI for both datasets may be considered normal. PI was used as an input parameter in the developed models.

5.2 Grain Size Distribution

Following specifications by the Oklahoma Department of Transportation (ODOT 2000), only selected sieves (#4, #10, #40, and #200) were used in the grain size distribution tests. Table 2 presents the grain size distribution results in terms of percent passing 4.75 mm (#4), 2.00 mm (#10), 0.425 mm (#40), and 0.075 mm (#200). The skewness (-1.87 to –3.64) and kurtosis (3.05 to 14.4) values of #10 sieve for both datasets were fairly high, indicating that these results are not normally distributed. Similar trends were observed for #4 and #40 sieves (Table 2). For #200 sieve, however, the overall skewness (-0.49 to –1.05) and kurtosis (0.27 to 0.37) values were much smaller, indicating that these data could be assumed normally distributed. Thus, from the grain size distribution tests, only percent passing #200 sieve (P_{200}) was used as an input parameter in the regression and ANN models.

5.3 Group Index (GI)

The GI values for the development dataset range from 1 to 39. The mean and standard deviation for the GI values are 13.2 and 8.6. The range of GI for the evaluation dataset is from 1 to 31, with a mean of 14.6 and a standard deviation of 9.6. The basic regression parameters for the two sets of data are presented in Table 2. The distribution of the GI values may be considered normally distributed since the deviation is small (Figure 4-19). The skewness of the development and the evaluation datasets are 0.81 and 0.29, respectively. The kurtosis values of the development and the evaluation datasets are 0.58 and -1.36, respectively. These values are close to the normal distribution values (i.e. 0).

5.4 Moisture Content (w)

After completion of the M_R test, the w of the tested specimen was evaluated, and used as an input parameter in the regression and ANN models. The moisture contents for the development dataset ranged from 10.5% to 25.3%, with a mean of 17.4% and a standard deviation of 3.0%. The w values for the evaluation dataset were in the range of 11.4% to 22.4%, with a mean of 17.3% and a standard deviation of 2.9%. The skewnesses of w data for the development and the evaluation datasets were 0.39 and –0.24, respectively. The corresponding kurtosis values were -0.14 and –0.94, respectively. These values indicate that the moisture content data were also approximately normally distributed.

Table 3. Basic statistical parameters for resilient modulus at each sequence for development and evaluation datasets

Sequence No.	Confining Pressure (kPa)	Axial Stress (kPa)	Resilient Modulus (MPa)			
			Mean	Minimum	Maximum	Standard Deviation
Development Dataset (126 Specimens)						
1	41.4	13.8	298.6	54.6	2042.3	347.6
2	41.4	27.6	86.8	34.5	229.0	31.7
3	41.4	41.4	72.3	24.6	163.8	28.2
4	41.4	55.2	63.4	24.3	155.7	28.0
5	41.4	68.9	57.9	20.7	152.6	26.8
6	27.6	13.8	269.4	41.9	2160.8	341.2
7	27.6	27.6	84.1	28.7	245.9	33.2
8	27.6	41.4	70.6	23.2	159.5	28.6
9	27.6	55.2	63.2	22.3	151.7	28.3
10	27.6	68.9	58.3	20.9	149.2	27.0
11	13.8	13.8	312.4	36.1	1892.2	393.3
12	13.8	27.6	93.3	23.7	979.9	87.9
13	13.8	41.4	76.9	19.8	727.9	65.9
14	13.8	55.2	63.9	19.6	159.8	29.6
15	13.8	68.9	59.4	18.2	190.9	29.3
Evaluation Dataset (68 Specimens)						
1	41.4	13.8	172.8	58.1	409.1	89.7
2	41.4	27.6	80.9	28.2	131.3	21.9
3	41.4	41.4	66.4	19.6	122.1	24.6
4	41.4	55.2	54.5	17.2	117.8	21.2
5	41.4	68.9	49.0	16.1	104.4	19.1
6	27.6	13.8	161.8	57.1	494.2	93.1
7	27.6	27.6	79.0	33.9	136.7	22.9
8	27.6	41.4	64.2	22.7	121.0	24.4
9	27.6	55.2	53.5	18.8	115.9	21.2
10	27.6	68.9	49.4	16.8	106.2	19.6
11	13.8	13.8	190.8	54.1	826.4	146.3
12	13.8	27.6	80.4	36.2	135.7	23.4
13	13.8	41.4	64.9	23.9	126.0	25.1
14	13.8	55.2	53.8	19.4	116.3	21.7
15	13.8	68.9	49.8	17.3	107.4	20.0

5.5 Dry Density (γ_d)

The γ_d values for the development dataset were in the range of 1404.2 kg/m^3 (87.7 pcf) to 1872.6 kg/m^3 (116.9 pcf), with a mean of 1658.0 kg/m^3 (103.5 pcf) and a standard deviation of 106.7 kg/m^3 (6.7 pcf). The corresponding range for the evaluation dataset was 1544.7 kg/m^3 (96.4 pcf) to 1862.9 kg/m^3 (116.3 pcf), with a mean of 1689.2 kg/m^3 (105.5 pcf) and a standard deviation of 74.8 kg/m^3 (4.7 pcf). For the both datasets, the γ_d values were normally distributed.

5.6 Unconfined Compressive Strength (U_C)

The U_c of soils in the development dataset varied between 50.9 kPa (7.38 psi) and 443.5 kPa (64.3 psi), with a mean of 204.7 kPa (29.7 psi) and a standard deviation of 70.8 kPa (10.3 psi). The U_c values of the evaluation dataset had a smaller range (56.5 kPa to 357.7 kPa or 8.2 psi to 51.9 psi) and lower mean (166.4 kPa or 24.1 psi) than that of the development dataset. The standard deviation (95.9 kPa or 13.9 psi) for the Woodward County soils, however, was relatively high. The skewness and kurtosis for the development dataset were 0.68 and 0.50, respectively. The corresponding skewness and kurtosis for the evaluation dataset were 0.78 and 0.81, respectively. Overall, the U_c values were normally distributed.

5.7 Resilient Modulus

The M_R test results for the development and the evaluation datasets are presented in Table 3. For the development dataset, a very high standard deviation, more than 340 MPa (49.3 ksi), is seen for the loading sequences 1, 6, and 11. A high standard deviation (more than 89 MPa or 12.9 ksi) is also observed for the evaluation dataset for the same loading sequences. For each of these loading sequences, the applied axial stress is the lowest (13.8 kPa or 2 psi), resulting in very small deformation of the sample that are difficult to measure due to electrical noise. As a result, the M_R values for these loading sequences were not used in developing the regression model.

6 Development of Models

6.1 Regression Models

In the present study, mainly two regression models were developed, namely, polynomial, and factorial. The M_R values were predicted using the evaluation dataset and then compared to the experimental M_R values. The R^2 and F values were utilized as the basis of comparing the developed models in regard to the goodness of fit and significance of the model, respectively (FHWA 2002, Tarefder et al. 2005). For each model, the experimental and the predicted M_R/P_a values were compared for the overall development dataset. Three specimens (MA-3B, NO-7A and OS-1B) were selected randomly for additional comparisons, so as to represent a wide range of scenarios (best, worst, and intermediate). For each specimen, both model predicted and experimental M_R values were plotted as a function of deviatoric stress.

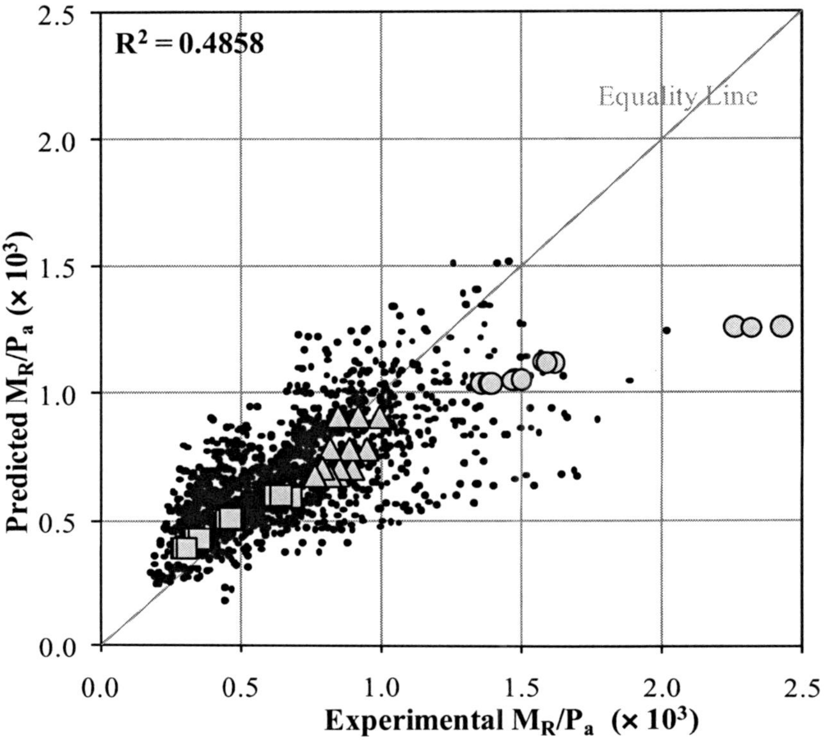

Fig. 1. Comparison of Experimental and Predicted M_R/P_a for Development Dataset: Polynomial Model

6.1.1 Polynomial Model

A polynomial model includes the basic components of a multiple linear regression model with the addition of higher order effects for the independent variables. Although a second order model may be adequate for many problems, a general polynomial model can have higher than second order terms (Carmichael and Stuart 1978, Ebrahimi 2006). In polynomial regression, higher order terms are added to the model to determine if they increase the associated R^2 significantly (Sokal and Rohlf 1995, Myers et al 2001, Montogomery et al. 2006). However, in most cases, polynomial models of orders greater than three are not practical (Sokal and Rohlf 1995).

Using the polynomial modeling option in STATISTICA 7.1, the resulting second order polynomial model is given by the following equation:

$$M_R/P_a = 15.8002 + 2.9994\, w - 7.4142\, w^2 - 18.3291\, (\gamma_d/\gamma_w) + 5.4596\, (\gamma_d/\gamma_w)^2 + 0.02191\, PI - 0.0003142\, PI^2 - 0.3705\, P_{200} - 0.009229\, P_{200}^{\;2} + 0.2628\, (U_c/P_a) - 0.01050(U_c/P_a)^2 - 2.0332(\sigma_d/P_a) + 1.62950(\sigma_d/P_a)^2 - 0.01181\, (\theta/P_a) + 0.004735(\theta/P_a)^2 \qquad (4)$$

The R^2 and F values for this model were found to be 0.4858 and 101.02, respectively. To examine if a higher order model was desired, a third order polynomial regression model was developed for the same development dataset. The R^2 and F values for the third order polynomial model changed to 0.4101 and 254.75, respectively. Specifically, the R^2 value for the third order polynomial regression model was worse than the corresponding values for both the multiple regression and the second order polynomial regression models. Also, the F value increased from the second order to the third order polynomial regression model indicating that the second order polynomial model was a better model (Sokal and Rohlf 1995, Fernandez-Juricic et al. 2003).

Figure 1 presents a comparison of experimental and the M_R/P_a values back-predicted by the second order polynomial model. The model's performance pertaining to three selected specimens MA-3B, NO-7A, and OS-1B is illustrated in Figures 2, 3, and 4, respectively. The M_R results from these specimens covered the full range of M_R response for the development dataset. Specimen NO-7A shows the best prediction, followed by specimen OS-1B. Specimen MA-3B shows the worst back-prediction. The soil classification results for these specimens indicate lean clay with AASHTO classification of A-6(10), A-6(16) and A-6(21) for OS-1B, NO-7A, and MA-3B, respectively. The U_c results for OS-1B, NO-7A, and MA-3B were 161 kPa (23.3 psi), 272 kPa (39.4 psi), and 310 kPa (45.1 psi), respectively (Ebrahimi 2006). Thus, even though these soils are all classified as A-6 soils, their unconfined compressive strengths were quite different.

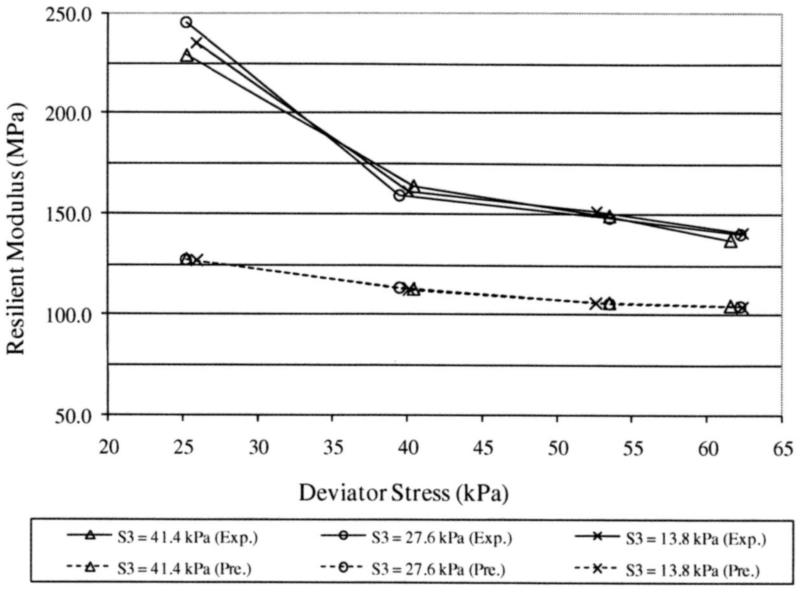

Fig. 2. Resilient modulus from experiment and polynomial model: specimen MA-3B

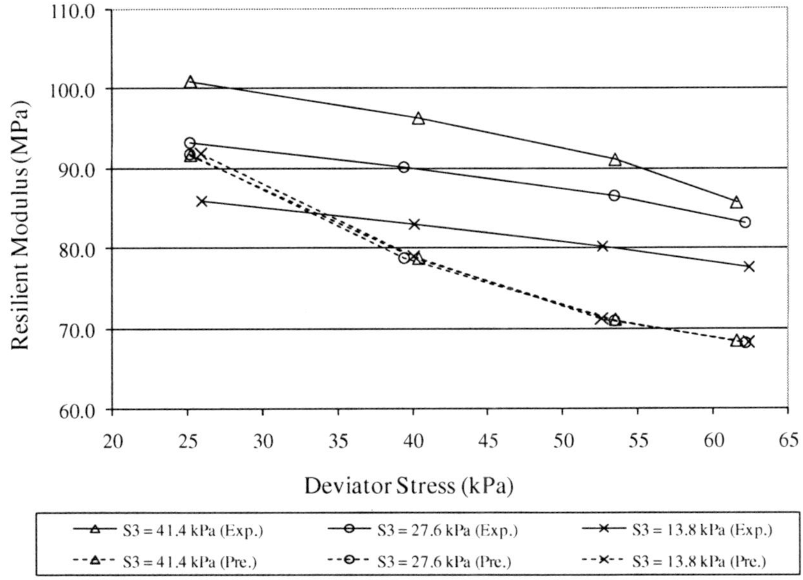

Fig. 3. Resilient modulus from experiment and polynomial model: specimen NO-7A

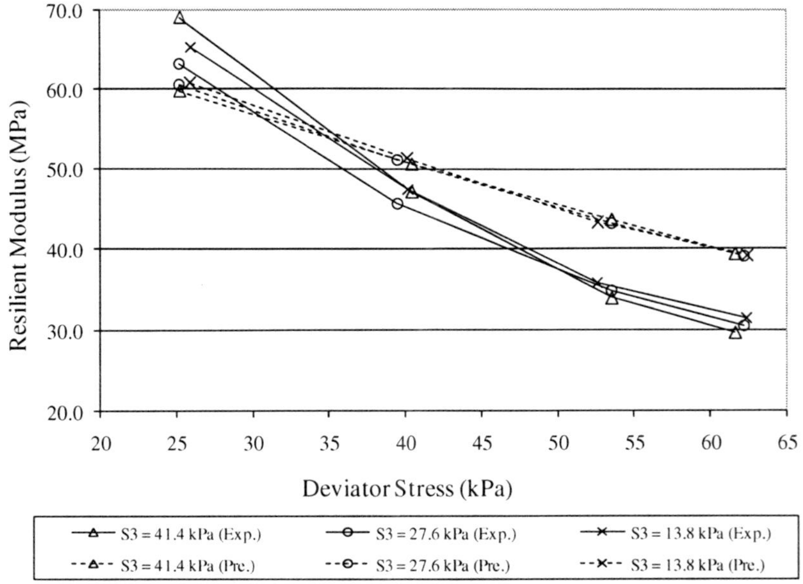

Fig. 4. Resilient modulus from experiment and polynomial model: specimen OS-1B

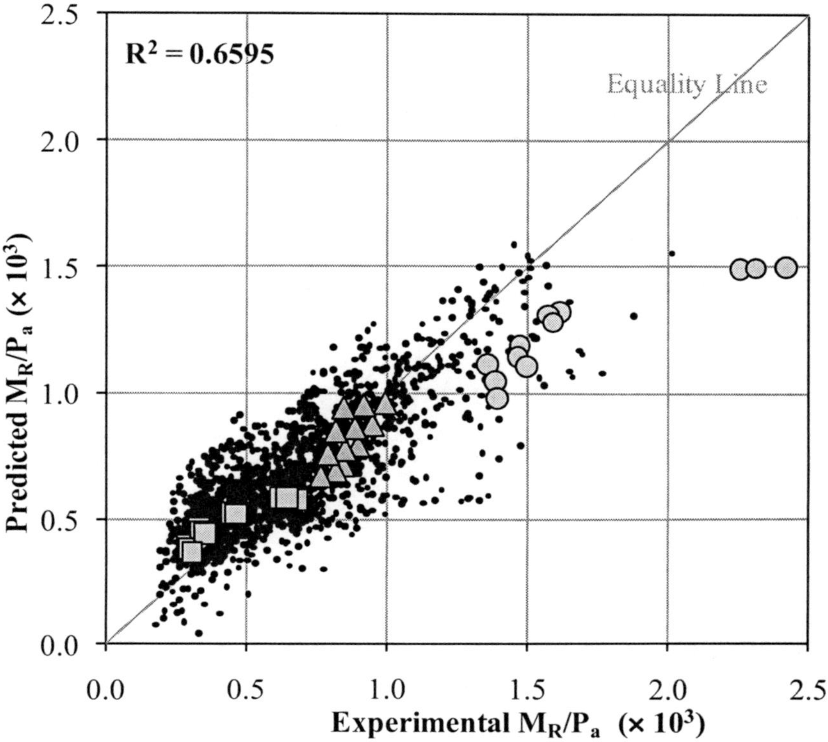

Fig. 5. Comparison of experimental and predicted M_R/P_a for development dataset: factorial model

Overall, it was observed that the M_R values increased with increasing unconfined compressive strength. This may have been a contributing factor for the three specimens exhibiting different levels of correlations between the experimental and predicted M_R.

6.1.2 Factorial Model

Similar to the polynomial model, a factorial model also includes the components of a multiple regression model. However, instead of considering higher order effects of the independent variables, it accounts for interactions among different variables in the model. Different levels of interactions may be incorporated such as interactions between two variables, among three variables, and so on (i.e. $w\times\gamma_d$, $PI\times U_c\times\sigma_d$, $w\times\gamma_d\times PI\times\sigma_d\times\theta$, etc.). A full-factorial regression model consists of all possible products of the independent variables. Moreover, a factorial regression model can be fractional (i.e., fractional exponent) (see e.g., Myers et al. 2001, Montgomery et al. 2006).

A full-factorial model is used in the present study. With seven independent variables and all possible products of the independent variables, the factorial model is a long equation with 128 terms. All the regression constants for this model

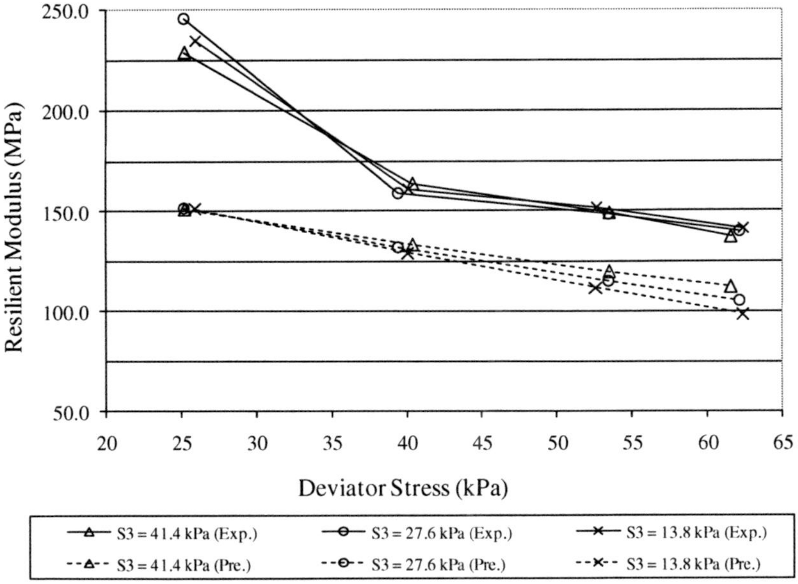

Fig. 6. Resilient modulus from experiment and factorial model: specimen MA-3B

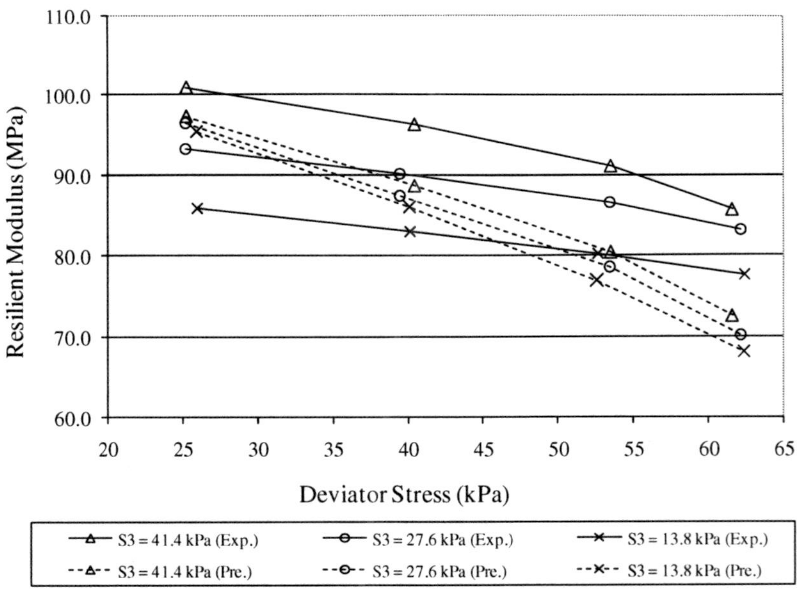

Fig. 7. Resilient modulus from experiment and factorial model: specimen NO-7A

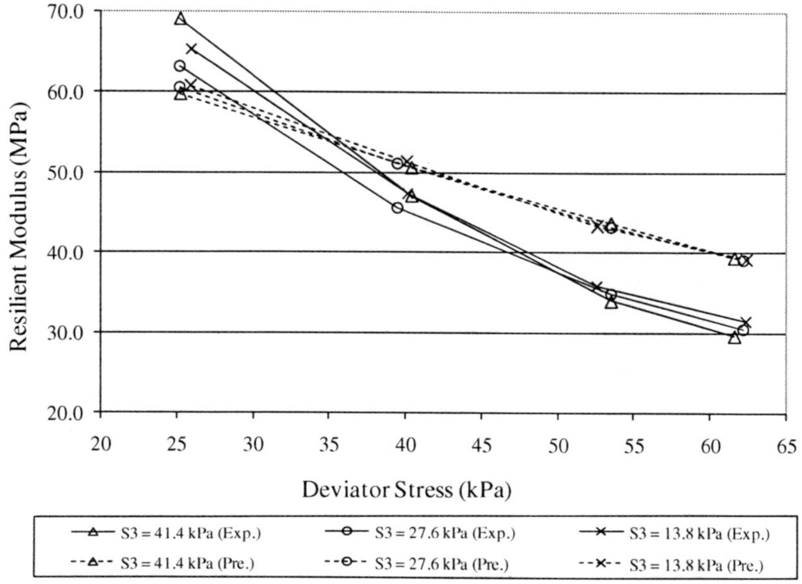

Fig. 8. Resilient modulus from experiment and factorial model: specimen OS-1B

were determined using STATISTICA 7.1. The resulting equation of the FM is presented in Appendix. The R^2 and F values for the FM were 0.6595 and 23.74, respectively. The R^2 is significantly higher than those for the previous models (0.4858). Significant observation was also made by the decrease of the F value from 101.02 for polynomial model to 23.74 for factorial model. Figure 5 shows a plot of experimental versus predicted M_R/P_a values for factorial model. Figures 6, 7, and 8 present a comparison of the back-predicted M_R values against deviatoric stress for specimens MA-3B, NO-7A, and OS-1B. As expected, the factorial model predicted the resilient modulus values of specimen NO-7A very closely, while the prediction for specimen MA-3B is much worse. Furthermore, because of the improvement in R^2 and F values both the goodness of fit of the model and the significance model the observations between the experimental and predicted values are closer. It may therefore be assumed at this point that since the F-value is 23.74 and it is the lowest F value, the factorial model is the most significant regression model for the development dataset.

6.2 Artificial Neural Network Models

In the present study, two feedforward-type ANN models, namely, Radial Basis Function Network (RBFN) and Multi-Layer Perceptrons Network (MLPN), were generated. A commercial software, STATISTICA 7.1, was used to develop these neural network. In the present application, the input layer consists of seven nodes, one node for each of the independent variables, namely w, γ_d, PI, P_{200}, U_c, σ_d, and θ. The output layer consists of one node, representing M_R. For each ANN model developed, a trial and error approach was used to find the number of nodes in the hidden layer(s), in search of the optimum model. After the architecture was set, the development dataset was fed into the model for training. To examine the strengths and weaknesses of the developed models, the predicted M_R values were compared with the experimental values with respect to the R^2 values. Thus, a higher R^2 value was considered a better fit of the development dataset. Previously, several researchers have used R^2 as an indicator of model performance (Tarefder et al., 2005; Rankine and Sivakugan, 2005).

6.2.1 Radial Basis Function Network (RBFN)

The radial basis function network (RBFN) divides the modeling space using hyperspheres. The centers and radii are used to characterize these hyperspheres. The RBFN units respond non-linearly to the distance of points from the center represented by a radial unit. The response surface of a single radial unit is the Gaussian (bell-shaped) function, peaked at the center, and descending outwards (Haykin 1994, Bishop 1995, Statsoft Inc. 2006). Therefore, the RBFN has three layers, namely input, hidden, and output layers. The hidden layer consists of radial units. It models the Gaussian response surface. The two most common methods for assigning the center of the radial units are sub-sampling and K-Means algorithm (Bishop, 1995; Statsoft, Inc., 2006).

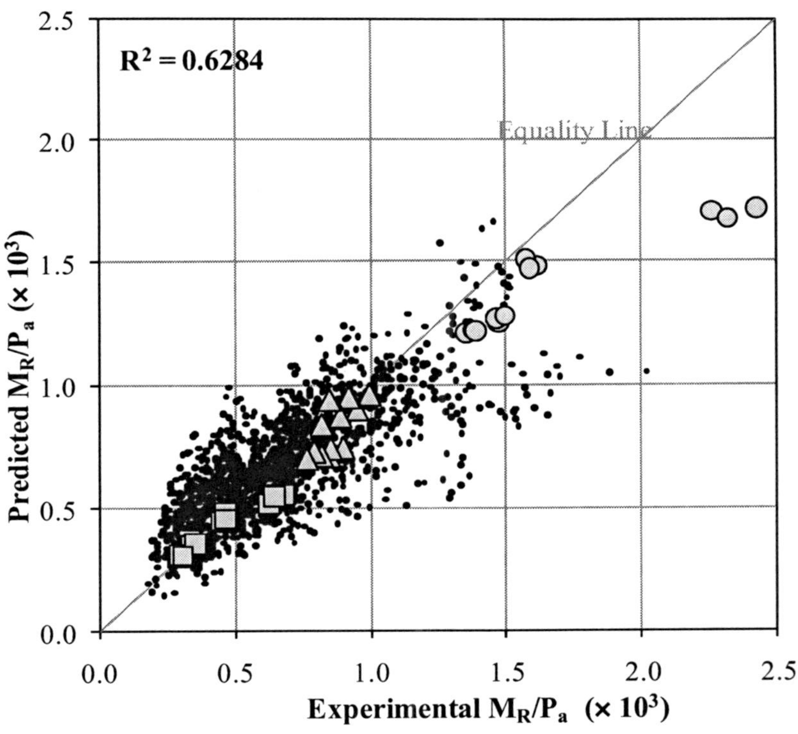

Fig. 9. Comparison of experimental and predicted M_R/P_a for development dataset: radial basis function network (RBFN) model

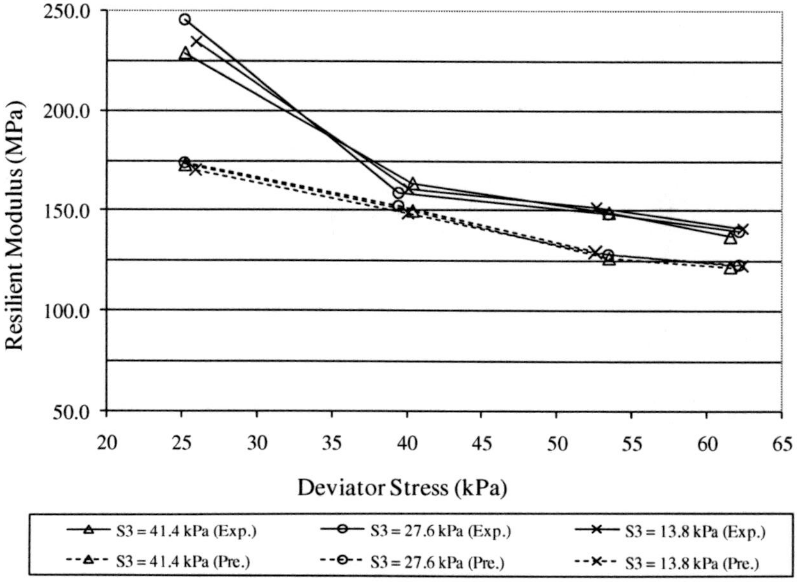

Fig. 10. Resilient modulus from experiment and radial basis function network (RBFN) model: specimen MA-3B

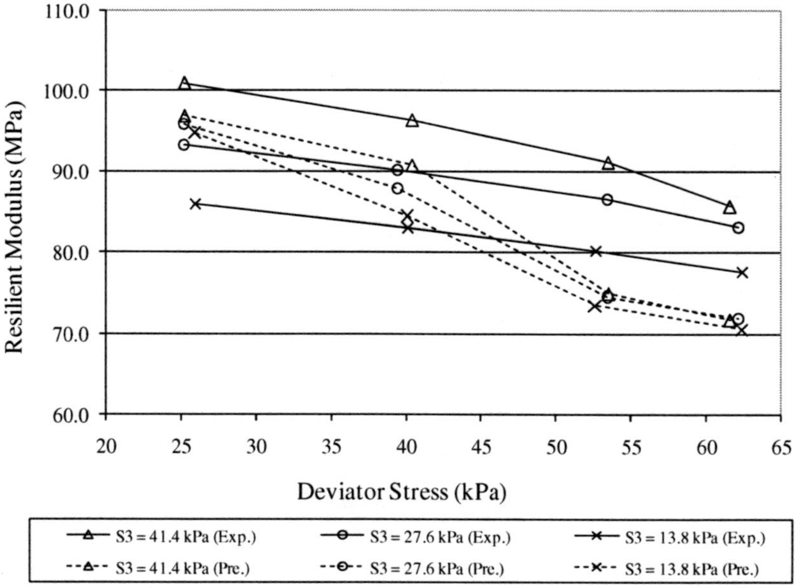

Fig. 11. Resilient modulus from experiment and radial basis function network (RBFN) model: specimen NO-7A

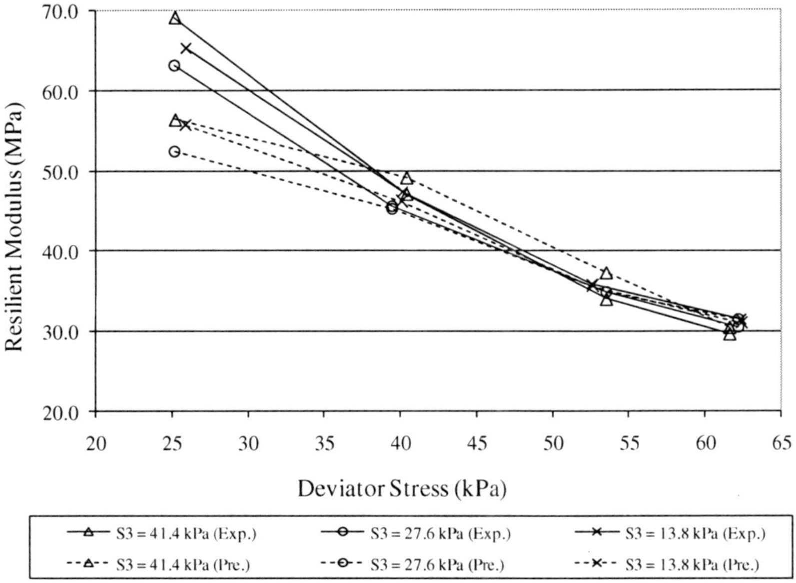

Fig. 12. Resilient modulus from experiment and radial basis function network (RBFN) model: specimen OS-1B

The RBFN model has one hidden layer. A trial and error approach was used to determine the optimum number of node in the hidden layer. Following this approach, the optimum number of node in the hidden layer was found to be 100. The R^2 value of the RBFN model is 0.6284, which is much better than the polynomial model (0.4858) and similar to factorial model (0.6595). Figure 9 shows an overall comparison between experimental and predicted M_R/P_a values for this model. Figures 10, 11, and 12 presents the back-prediction of the experimental M_R values against the deviatoric stress for specimens MA-3B, NO-7A, and OS-1B, respectively. The results are much encouraging for all three specimens and the RBFN model back-predicted the M_R values in both low and high range of M_R values. This improvement in back-prediction may be attributed to the use of hyperspheres in the RBFN model in the form of Gaussian (bell-shaped) function-type response surface (Haykin 1994, Bishop 1995, Bors 2001, Yildirim and Ozyilmaz 2002).

6.2.2 Multi-Layer Perceptrons Network (MLPN)

The MLPN is one of the popular network architecture in use today (Rumelhart and McClelland 1986, Bishop 1995, Narayan, 2002). The MLPN consists of an input layer, a number of hidden layers, and an output layer. In each of the hidden layers, the number of node can be varied. Due to the number of layers and the number of nodes in each layer, the MLPN can adjust the architecture of the network based on the complexity of a problem. In STATISTICA 7.1, the MLPN has up to

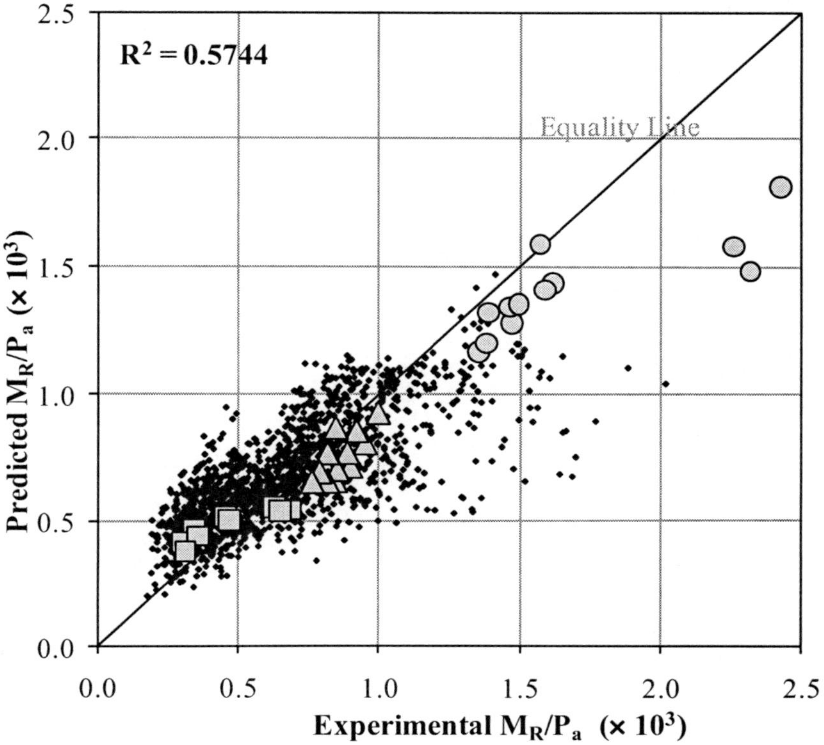

Fig. 13. Comparison of experimental and predicted M_R/P_a for development dataset: multi-layer perceptrons network (MLPN-2) model

three hidden layers available. Each of the nodes in the network performs a biased weighted sum of their inputs and passes this activation level through a transfer function to produce its output. The weights and biases in the network are adjusted using a training algorithm. The training algorithms available in STATISTICA 7.1 are back propagation, conjugate gradient descent, quasi-Newton, and Levenberg-Marquardt (Statsoft Inc. 2006).

In the present study, three MLPN models, henceforth referred to as MLPN-1, MPLN-2, and MLPN-3 models, were developed with different number of hidden layers in each model. The number of nodes in each of the three hidden layers was set at six, based on a trial and error approach. The R^2 values of the MLPN models were found to be 0.5733, 0.5744, and 0.5587 for one, two and three hidden layers, respectively, indicating that the MLPN-3 might have reached over-learning or over-fitting (Bishop 1995). These R^2 values indicate that all three MLPN models are expected to better correlate the M_R/P_a values than the polynomial model (0.4858), but the MLPN models were somewhat worse than the factorial (0.6595) and the RBFN (0.6284) models. Figure 13 shows a comparison between the experimental and predicted values of M_R/P_a values for the MLPN-2 model. Figures 14, 15, and 16 presents the back-prediction of the experimental M_R against the

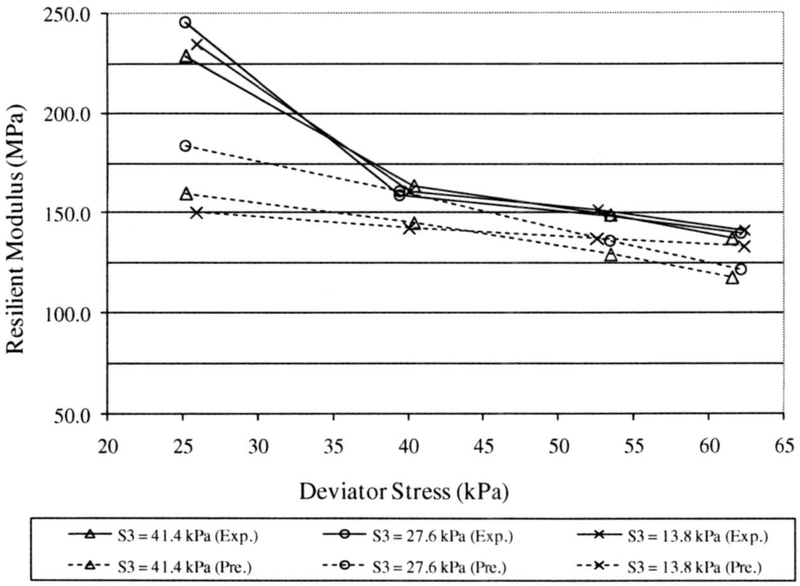

Fig. 14. Resilient modulus from experiment and multilayer perceptrons network (MLPN-2) model: specimen MA-3B

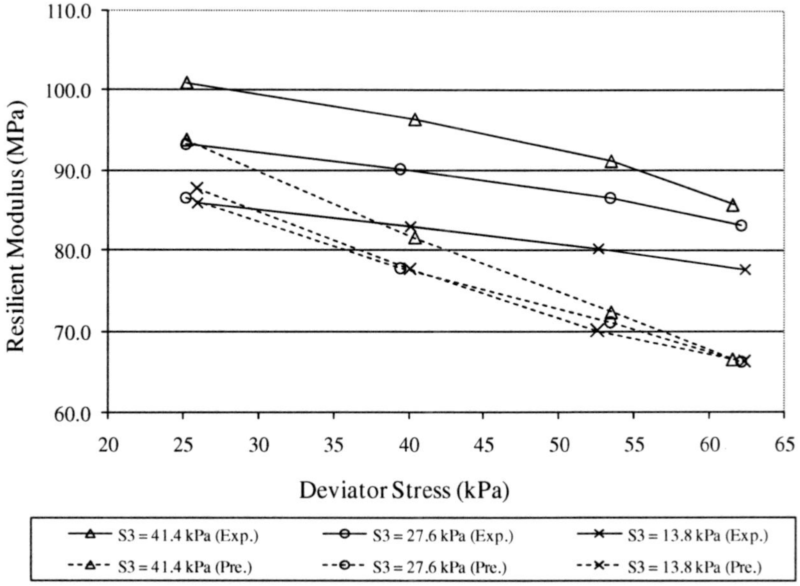

Fig. 15. Resilient modulus from experiment and multilayer perceptrons network (MLPN-2) model: specimen NO-7A

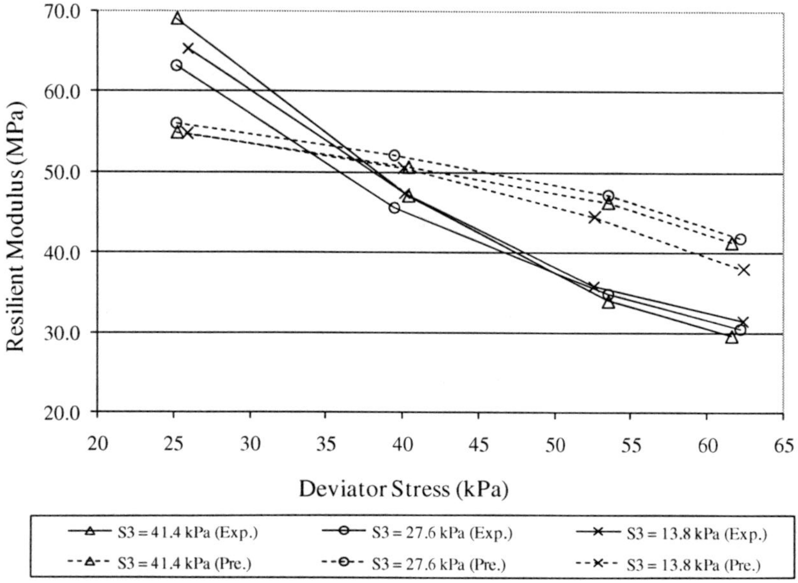

Fig. 16. Resilient modulus from experiment and multilayer perceptrons network (MLPN-2) model: specimen OS-1B

deviatoric stress for the MLPN-2 model for specimens MA-3B, NO-7A, and OS-1B, respectively. The overall predictive quality is slightly worse than the RBFN model, but much better than the polynomial model.

6.3 Comparative Performance of Different Models

Table 4 presents a summary of the R^2 results for all the four models developed here (polynomial, factorial, RBFN, and MLPN-2). The R^2 values ranged from 0.4858 to 0.6595, the polynomial model exhibiting the lowest R^2 value and the factorial model exhibiting the highest. In general, better correlations are observed in the low range of the M_R/P_a values (about 1,000). With increased M_R/P_a, values (beyond 1,000) the differences between the experimental and the predicted values increase. This observation may be due to the distribution of dataset. Only 140 M_R/P_a values out of 1512 M_R/P_a values (approximately 9%) are in the upper range of 1,000. The remaining 91% of the M_R/P_a values for this study are in the lower range of the development dataset. As a result models appear to exhibit difficulty in back-predicting a majority of the resilient modulus values in the dataset that are in the lower range of the M_R/P_a values (Myers et al. 2001, Montgomery 2006, Statsoft Inc. 2006). In the process of lowering the overall error, the back-predicted values in the higher range of M_R/P_a values usually become less than the experimental values (i.e., under-prediction) and the back-prediction in the lower range increase (i.e., over-prediction). The next step in this study is to evaluate the

Table 4. Summary of the regression modeling results

Type of Dataset	Polynomial		Factorial		RBFN		MLPN-2	
	R^2	F	R^2	F	R^2	F	R^2	F
Development	0.4858	101.02	0.6595	23.74	0.6284	NA	0.5744	NA
Evaluation (Overall)	0.5200	NA	0.3634	NA	0.4938	NA	0.5848	NA
Evaluation (WOE)	0.6212	NA	0.0962	NA	0.0251	NA	0.6308	NA
Evaluation (ROE)	0.5523	NA	0.4021	NA	0.5557	NA	0.6026	NA

WOE: Woodward county; ROE: Rogers county; RBFN: Radial basis function network; MLPN: Multi-layer perceptrons network; NA: Not applicable

developed models using a different set of data that were not used in the development phase (Bishop 1995, Hill and Lewicki 2006).

7 Evaluation of Models

The evaluation dataset were separated into soils from Woodward County and Rogers County. Separate comparisons were made for the Woodward County soils (WOE-4B) and the Rogers County soils (ROE-20B), and a comparison was made for both counties together (henceforth called "combined evaluation dataset"). This provides different views on the prediction quality and the importance of datasets on regression analysis (Montgomery et al. 2006, Myers et al. 2001). Additionally, a comparison was made between the differences in the R^2 values of the development dataset and the evaluation dataset.

7.1 Evaluation of Factorial Model

The R^2 value of the combined evaluation dataset was only 0.3634. Figure 17 shows a comparison of the experimental and predicted M_R/P_a values for the combined evaluation dataset. Even though the overall R^2 value for the development dataset was 0.6595, it dropped significantly to 0.3634 for the evaluation dataset (Table 4). The soils from Woodward County (WOE-4B) have the worst predictions with a R^2 value of 0.0962. The full factorial model considered here contains 128 terms in the function, it may be considered a complex function. Therefore, it is possible that the factorial model over-fitted the development dataset and caused a poor prediction in the evaluation dataset (Hill and Lewicki 2006, Montgomery et al. 2006). In the case of present dataset, it appears that the full factorial model has created a condition known as too much wiggle (Sokal and Rohlf 1995, Fernandez-Juricic et al. 2003). Too much wiggle occurs when the equation has too many terms and tries to fit to as many data point as possible. The percent difference in the R^2 between the development dataset and the Woodward County and Rogers County evaluation datasets are 85% and 39%, respectively.

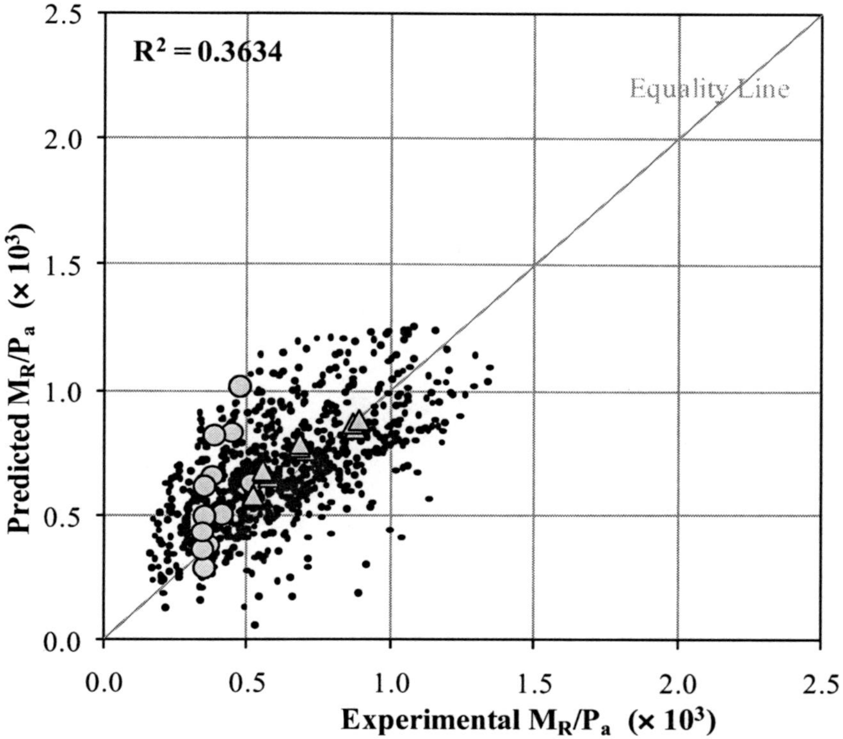

Fig. 17. Comparison of experimental and predicted M_R/P_a for combined evaluation dataset: factorial model

7.2 Evaluation of Second Order Polynomial Model

The second order polynomial model predicted the M_R/P_a values with a R^2 value of 0.5200. A plot of the experimental and predicted M_R/P_a values is given in Figure 18. The results show that the Woodward County (WOE-4B) and the Rogers County soils (ROE-20B) have R^2 values of 0.6212 and 0.5523, respectively. The R^2 values for Woodward and Rogers Counties were approximately 27% and 6.2% higher than the R^2 value for the development dataset. This indicates that the second order polynomial model is capable of predicting the M_R values of the Woodward and Rogers County soils reasonably well, as compared to factorial model.

7.3 Evaluation of RBFN Model

The RBFN model predicted the M_R/P_a values of the combined evaluation dataset with an R^2 value of 0.4938. Figure 19 shows a comparison of the prediction quality of the RBFN model for the combined evaluation dataset. The R^2 for the Rogers County soils was 0.5557, compared to only 0.0251 for the Woodward County soils. Although the R^2 value of the Woodward County soils was very small, it did

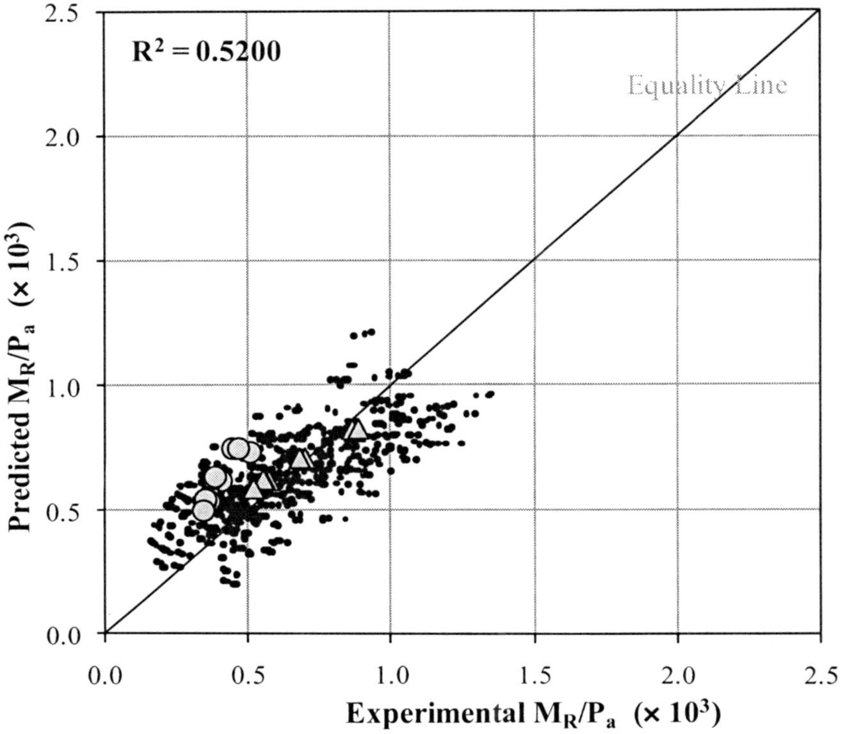

Fig. 18. Comparison of experimental and predicted M_R/P_a for combined evaluation dataset: polynomial model

not have as much influence on the overall R^2 because of the fewer soils involved (5 compared to 29 for Rogers County). Overall, prediction for the Rogers County specimen (ROE-20B) is excellent, while that for the Woodward County specimen (WOE-4B) shows significant difference.

7.4 Evaluation of MLPN-2 Model

The R^2 of the combined evaluation dataset for the MLPN-2 model was found to be 0.5848 (Figure 20). The corresponding R^2 values for the selected Rogers County and the Woodward County specimens were found to be 0.6026 and 0.6308, respectively. Figure 20 also presents the predicted M_R values against the deviatoric stress for specimens WOE-4B (Woodward County) and specimen ROE-20B (Rogers County). The MLPN-2 model predicted the M_R values for specimen (ROE-20B) very well; some difference was observed for specimen WOE-4B. For both specimens, the predicted M_R values were higher than the experimental values. Overall, the MLPN-2 model appears to be the best model for the present (development and evaluation) datasets.

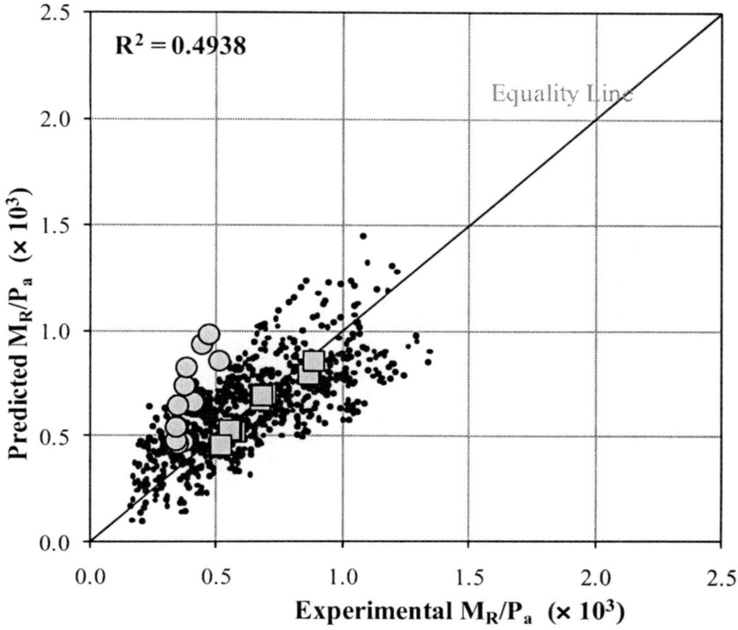

Fig. 19. Comparison of experimental and predicted M_R/P_a for combined evaluation dataset: radial basis function network (RBFN) model

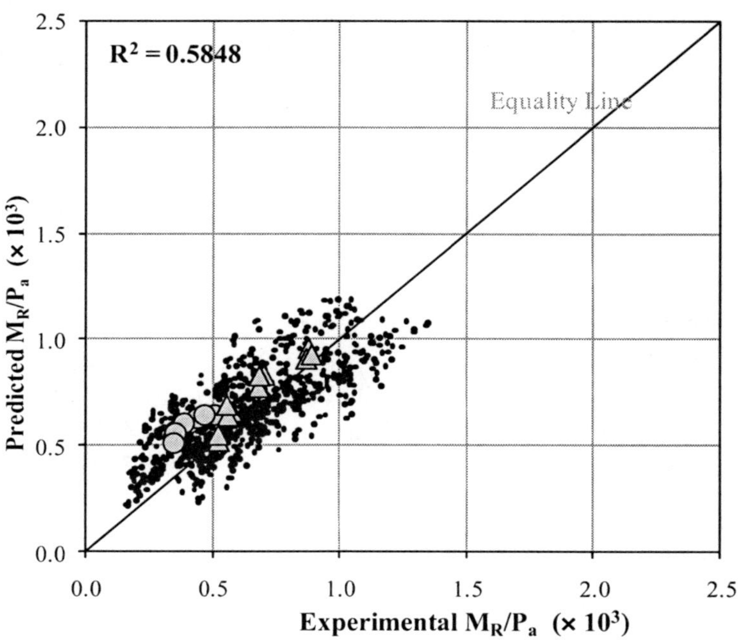

Fig. 20. Comparison of experimental and predicted M_R/P_a for combined evaluation dataset: multilayer perceptrons network (MLPN-2) model

8 Pavement Design

To demonstrate the application of developed model in pavement design, typical pavement sections are designed based on the M_R values obtained experimentally as well as the M_R values predicted by the selected model. The designed sections are then compared and evaluated. The model used in the pavement design is the Multilayer preceptrons network with two hidden layers (MLPN-2), which was determined to be the best prediction model in the present study. In this demonstration exercise, three different subgrade soils are selected – one (WOE-4B, SH 3) from Woodward County and two (ROE-3A, SH 20 and ROE-20B, SH 88) from Rogers County. A design M_R value is predicted using the MLPN-2 model based on the commonly used properties for each selected soil. DARWin 3.1, pavement design software is utilized to determine the design thicknesses. DARWin 3.1, one of the AASHTO softwares, implements the pavement design models presented in the 1993 AASHTO Design Guide. In this software, the structural design (structural number and thickness of each layer) of an asphalt pavement is based on the predicted subgrade M_R values. Finally, the structural numbers and the asphalt pavement thicknesses are calculated based on the experimental M_R values. A comparison of these results is presented in this chapter.

8.1 Resilient Modulus for Pavement Design

The predicted and experimental resilient modulus values for the three selected soils are presented in Table 5. It is seen that the experimental M_R values for soils WOE-4B and ROE-20B are lower than the corresponding predicted values. An opposite trend is seen for the third soil, ROE-3A. The maximum difference between the experimental and predicted M_R values for soil WOE-4B is 19.94 MPa (2,892 psi). The maximum differences for soils ROE-3A and ROE-20B, however, are much lower - 9.86 MPa (1,430 psi) and 7.28 MPa (1,056 psi), respectively.

Table 5. Subgrade resilient moduli for pavement design

Specimen ID and Location	Resilient Modulus	
	Experiment	**Predicted (MLPN-2)**
WOE-4B SH 3, Woodward County	38.34 MPa (5,560 psi)	58.29 MPa (8,452 psi)
ROE-3A SH 20, Rogers County	81.72 MPa (11,850 psi)	71.86 MPa (10,420 psi)
ROE-20B SH 88, Rogers County	64.18 MPa (9,306 psi)	71.46 MPa (10,362 psi)

8.2 Design Parameters

Before one can proceed with the design, there are several design parameters that need to be determined or assumed (AASHTO 2002, Huang 2004). These include

Table 6. Pavement design parameters

Parameter	Value
Design Period (years)	20
Two-Way Traffic (ADT)	11,378
Number of Lanes in Design Direction	2
Percent of All Trucks in Design Lane	80%
Percent Trucks in Design Direction	50%
Percent Heavy Trucks (of ADT) FHWA Class 5 or Greater	3%
Average Initial Truck Factor (ESALs/truck)	2.338
Annual Truck Volume Growth Rate	1.5%
Total Calculated Cumulative ESALs	2,664,208
Reliability Level	80%
Overall Standard Deviation	0.49
Initial Serviceability	4.2
Terminal Serviceability	3.0

design period, traffic data, reliability, and serviceability. The design period for the selected pavements is assumed to be 20 years. The initial two-way traffic for this design is assumed to as 11,378 with 3% of the traffic being heavy trucks (FHWA Class 5 or greater) (Yoder and Witczak 1975, AASHTO 1986, Huang 2004). The equivalent single axle load (ESAL) is calculated from the information presented in Table 6. The ESAL for the present application is found to be 2,664,208. The program determines the ESAL load based on the vehicle type, design period and growth factor. Table 6 presents the reliability (80% with a standard deviation of 0.49) and serviceability values used in this design application. These values are based on the recommendations by the AASHTO Design Guide (AASHTO 1986). The initial and final serviceability values of the pavement are assumed as 4.2 and 3, respectively. These are typical values obtained from the AASHO Road Test (Yoder and Witczak 1975, Huang 2004). These design parameters were kept constant for all three soils. Only the subgrade M_R values are changed and their influence on the structural numbers and the design thickness is examined.

8.3 Pavement Design Results

Based on the design parameters selected in the preceding section, structural numbers and pavement thicknesses are calculated using DARWin 3.1 for each resilient modulus. The design results are summarized in Table 7. As expected, the design structural numbers of soils WOE-4B and ROE-20B from experimental M_R values are higher than the design structural numbers from the predicted M_R values. For the experimental M_R values, the design Structural Numbers (SN) are 4.72, 3.44,

Table 7. Pavement design results

Specimen ID	Design Structural Number (SN)	Required Asphalt Thickness
WOE-4B (Experiment)	4.72	27.18 cm (10.7 in.)
WOE-4B (MLPN-2)	3.98	23.11 cm (9.1 in.)
ROE-3A (Experiment)	3.44	19.81 cm (7.8 in.)
ROE-3A (MLPN-2)	3.64	21.08 cm (8.3 in.)
ROE-20B (Experiment)	3.82	22.10 cm (8.7 in.)
ROE-20B (MLPN-2)	3.65	21.08 cm (8.3 in.)

and 3.82 for soils WOE-4B, ROE-3A, ROE-20B, respectively. For the predicted M_R values, the corresponding design SN are 3.98, 3.64, and 3.65.

In order to convert the design structural number to actual pavement thickness, a flexible pavement section (Asphalt Concrete, AC) is considered. In an AC pavement, the entire SN is converted to AC. Based on a typical value of 0.44 for the asphalt layer coefficient (AASHTO 1986), the required asphalt thickness can be determined by dividing the structural number by the asphalt layer coefficient. Table 7 presents the required AC thickness for the calculated SN pertaining to experimental M_R and predicted M_R.

It is evident that for soil WOE-4B the required AC thickness is under-predicted by 4.07 cm (1.6 in), (approximately 15%). It should be noted that the WOE-4B soil was from western Oklahoma and the MLPN-2 model was developed with no soil from that part of the state. So, this level of difference in pavement thickness may be justifiable. On the other hand, for soils ROE-3A and ROE-20B the differences in the required AC thickness are 1.27 cm (0.5 in), (approximately (-6%) and 1.02 cm (0.4 in), (approximately (+5%), respectively. Both of these soils are from eastern Oklahoma, and the predicted M_R values and the corresponding pavement design thicknesses are closer to the values obtained from the experimental M_R. Overall, the design thicknesses are fairly comparable, indicating that the MLPN-2 model is capable of predicting the design M_R values based on routine soil properties, for pavement design applications.

9 Summary and Conclusions

A total of four, two regression and two artificial neural network (ANN) models, were developed in this study to correlate resilient modulus with routine properties of subgrade soils and state of stress. These models included: polynomial (regression), factorial (regression), RBFN (ANN), and MLPN (ANN) models. A database was developed, containing grain size distribution, Atterberg limits, standard Proctor, unconfined compression and resilient modulus results for 98 soils from 16 different counties in Oklahoma. Of these, 64 soils were used in development, and the remaining 34 soils from two different counties were used in the evaluation of the developed models. The plasticity index, moisture content, dry density, unconfined compressive strength, and resilient modulus results all exhibited normal or near normal distribution, with low skewness and kurtosis values. The distribution of percent passing data, however, were heavily skewed for all sieve sizes, except for percent passing #200 sieve data that exhibited near normal distribution. Overall, the Rogers County soils were characteristically closer to the development dataset than the Woodward County soils. To examine the strengths and weaknesses of the developed models, the predicted M_R values were compared with the experimental values with respect to the R^2 values. Thus, a higher R^2 value was considered a better fit of the development dataset. The following points highlight the assessments and evaluations of these models:

1. A second order polynomial multiple regression model was developed for the development dataset. The R^2 and F values for this model were found to be 0.4858 and 101.02, respectively. Second order polynomial showed the R^2 values for the evaluation dataset (Roger County (0.5523), Woodward County (0.6212), and combined dataset (0.5200)) were relatively high, indicating that this model is comparatively a good model for the combined datasets (development and evaluation).
2. One of the most complicated models considered in this study was a full factorial model. This model had 126 terms, and the R^2 and F values for the development dataset was found to be 0.6595 and 23.74, respectively. Based on these values and not considering the evaluation dataset, this model appeared to be the best regression model. However, the R^2 values for the evaluation dataset (Roger County (0.4021), Woodward County (0.0962), and combined dataset (0.3634)) were relatively low, indicating that even this model was not a good model for the combined datasets (development and evaluation). A high R^2 value (0.4021) for the Rogers County soils is an indicator that these soils are similar to those in the evaluation dataset.
3. For the RBFN model, with one hidden layer, the R^2 value for the development dataset improved further (0.6015). Also, the back predicted response (M_R as a function of deviatoric stress) for three selected specimens (MA-3B, NO-7A, and OS-1B) showed significant improvements. A better R^2 and improvements in back-prediction may be attributed to the use of hyperspheres in the RBFN model in the form of Gaussian (bell-shaped) function-type response surface in this model.
4. The number of hidden layers in the MLPN models in STATISTICA can range from one to three. In the present study, three separate MLPN models (MLPN-1, MPLN-2, and MLPN-3) were developed with different number of hidden layers in each model. The R^2 values of the MLPN models were found to be 0.5733, 0.5744, and 0.5587 for one, two and three hidden layers, respectively, indicating that the MLPN-3 might have reached over-learning or over-fitting. Overall, the MLPN-2 model performed well.
5. Although the RBFN model had higher R^2 than the MLPN-2 model for the development dataset, its predictive capability for the Woodward County soils was poor (R^2 of 0.0251). A lower R^2 value for the Woodward County soils partly results from the differences in the characteristics of these soils, as noted previously. The MLPN-2 model did a much better job in predicting the resilient

modulus of subgrade soils in the evaluation dataset (overall $R^2 = 0.5848$), including the Woodward County soils ($R^2 = 0.6308$).

6. For both selected specimens (WOE-4B and ROE-20B) from the evaluation dataset, the MLPN-2 model under-predicted the M_R values. This may be a potential concern in pavement design applications where lower thicknesses may result from using the predicted M_R values in design.
7. Overall, the MLPN-2 model was found to be the best model for the present development and evaluation datasets. This model as well as the other models could be refined using an enriched database.
8. The MLPN-2 model was used to present the application of developed model in pavement design. Based on a reliability of 80%, the differences in the percent calculated Structural Number for three cases were found to be 15% for Woodward County and 6% and 4% for Rogers county. It is, therefore, possible to predict the M_R for pavement design with a reasonable degree of certainty.

Acknowledgments. The authors are thankful to Burgess Engineering and Testing, Inc. for assistance in laboratory testing of this study. The Oklahoma Department of Transportation provided some of the evaluation data. The authors are also thankful to StatSoft for providing helpful tips that improved the quality of this study. They are also thankful to Dr. Gerald Miller, Dr. Kianoosh Hatami, Dr. Luther White and Dr. Joakim Laguros for their technical comments.

References

AASHTO, Standard specifications for transportation materials and methods of sampling and testing. Transportation Research Board, National Research Council, Washington DC (1986)

American Association of State Highway and Transportation Officials (AASHTO), Guide for mechanistic-empirical design of new and rehabilitated pavement structures. Final report prepared for National Cooperative Highway Research Program. Transportation Research Board, National Research Council, Washington DC (2004)

Bishop, C.: Neural networks for pattern recognition. University Press, Oxford (1995)

Bors, A.G.: Introduction of the Radial Basis Function (RBF) Networks. Online Symp. for Electronics Engineers, DSP Algorithms: Multimedia 1(1), 1–7 (2001)

Carmichael III, R.F., Stuart, E.: Predicting resilient modulus: A study to determine the mechanical properties of subgrade soils. Transportation Research Record 1043, 20–28 (1978)

Dai, S., Zollars, J.: Resilient modulus of Minnesota road research project subgrade soil. Transportation Research Record 1786, 20–28 (2002)

Drumm, E.C., Boateng-Poku, Y., Pierce, T.J.: Estimation of subgrade resilient modulus from standard tests. Journal of Geotechnical Engineering 116(5), 774–789 (1990)

Dunlap, W.S.: A Report on a Mathematical Model Describing the Deformation Characteristics of Granular Materials. Technical Report 1, Project 2-8-62-27, TTI, Texas A&M University, Texas (1963)

Ebrahimi, A.: Regression and neural network modeling of resilient modulus based on routine soil properties and stress states. PhD dissertation, University of Oklahoma, Oklahoma (2006)

Far, M.S.S., Underwood, B.S., Ranjithan, S.R., Kim, Y.R., Jackson, N.: The application of artificial neural networks for estimating the dynamic modulus of asphalt concrete. In: Transportation Research Board 2009 Annual Meeting. CD-ROM Publication, Washington DC (2009)

Farrar, M.J., Turner, J.P.: Resilient modulus of wyoming subgrade soils mountain planins. Consortium Report No. 91-1, The University of Wyoming, Luramie, Wyoming (1991)

Fausett, L.V.: Fundamentals neural networks: Architecture, Algorithms and Applications. Prentice-Hall, Inc., Englewood Cliffs (1994)

Fernandez-Juricic, E.A., Sallent, R.S., Rodriguez-Prieto, I.: Testing the risk-disturbance hypothesis in a fragmented landscape: non-linear responses of house sparrows to humans. The Condor 105, 316–326 (2003)

FHWA (2002) Study of LTPP laboratory resilient modulus test data and response characteristics. Final report, Publication No. FHWA-RD-02-051, Office of Engineering Research and Development, McLean, Virginia (October 2002)

George, K.P.: Resilient Testing of soils using gyratory testing machine. Transportation Research Record 1369, 63–72 (1992)

Gomes, A.C., Gillett, S.: Flexible pavement: Resilient behavior of soils. Balkema, Rotterdam (1996) ISBN 90 54 10 5232

Haykin, W.L.: Neural networks: A comprehensive foundation. Macmillan College Publishing, New York (1994)

Hill, T., Lewicki, P.: STATISTICS methods and applications. StatSoft, Tulsa, Oklahoma (2006)

Hopkins, T.C., Beckham, T.L., Sun, L., Pfalzer, B.: Kentucky geotechnical database. Publication KTC-03-06/SPR-177-98-1F, University of Kentucky Transportation Center, College of Engineering, Lexington, Kentucky (2004)

Huang, Y.H.: Pavement analysis and design, 2nd edn. Prentice Hall, Upper Saddle River (2004)

Khazanovich, L., Celauro, B., Chabourn, B., Zollars, J.: Evaluation of subgrade resilient modulus predictive model for use in mechanistic-empirical pavement design guide. Transportation Research Record 1947, 155–166 (2006)

Kim, D., Kweon, G., Lee, K.: Alternative method of determining resilient modulus of compacted subgrade soils using free-free resonant column test. Transportation Research Record 1577, 62–69 (1997)

Kim, D., Stokoe II, K.H.: Characterization of resilient modulus of compacted subgrade soils using resonant column and torsional shear tests. Transportation Research Record 1369, 83–91 (1992)

Kohonen, T.: Self-organization and associative memory, 3rd edn. Springer, Berlin (1989)

Lee, W., Bohra, N.C., Altschaeffl, A.G., White, T.D.: Resilient modulus of cohesive soils. ASCE Journal of Geotechnical and Geoenvironmental Engineering 123(2), 131–136 (1997)

Li, D., Selig, E.T.: Resilient modulus for fine-grained subgrade soils. ASCE Journal of Geotechnical Engineering 120(6), 939–957 (1994)

Malla, R.B., Joshi, S.: Subgrade resilient modulus prediction models for coarse and fine-grained soils based on long-term pavement performance data. International Journal of Pavement Engineering 9(6), 431–444 (2008)

May, R.W., Witczak, M.W.: Effective granular modulus to model pavement response. Transportation Research Record 810, 1–9 (1981)

Mehrotra, K., Mohan, C.K., Ranka, S.: Elements of artificial neural networks. MIT Press, Cambridge (1996)

Meier, R.W., Alexander, D., Freeman, R.B.: Using artificial neural networks as a forward approach to backcalculation. Transportation Research Record 1570, 126–133 (1996)
Montgomery, D.C., Peck, E.A., Vining, G.G.: Introduction to linear regression analyses. Wiley Series in Probability and Statistics. John Wiley and Sons, Inc., New Jersey (2006)
Moossazadeh, J.M., Witczak, M.W.: Prediction of subgrade moduli for soil that exhibits nonlinear behavior. Transportation Research Record 810, 9–17 (1981)
Myers, R.H., Montgomery, D.C., Vining, G.G.: Generalized linear models: With applications in engineering and the sciences. John Wiley and Sons, Inc., New Jersey (2001)
Najjar, Y.M., Basheer, I.A., Ali, H.E., McReynolds, R.L.: Swelling potential of kansas soils: modeling and validation using artificial neural network reliability approach Transportation. Research Record 1736, 141–147 (2000)
Narayan, S.: Using genetic algorithms to adapt neuron functional forms. In: Proc. Artificial Intelligence and Soft Computing, Banff, Canada (July 2002)
NCHRP, Procedure for resilient modulus of unstabilized aggregate base and subgrade materials. Project 1-28A, Transportation Research Board, Washington D.C (2003)
ODOT, Standards and specifications. Oklahoma Department of Transportation (2000)
Patterson, D.: Artificial neural networks. Prentice Hall, Singapore (1996)
Paute, J.L., Hornych, P.: Flexible pavement: Influence of water content on the cyclic behavior of a silty sand. Balkema, Rotterdam (1996)
Raad, L., Minassian, G.H., Gartin, S.: Characterization of saturated granular bases under repeated loads. Transportation Research Record 1369, 73–82 (1992)
Rada, C., Witczak, W.M.: Comprehensive evaluation of laboratory resilient moduli results for granular material. Transportation Research Record 810, 23–33 (1981)
Rahim, A.M., George, J.P.: Subgrade soil index properties to estimate resilient modulus. In: Transportation Research Board 2004 Annual Meeting. CD-ROM Publication, Transportation Research Board, Washington DC (2004)
Rahim, A.M., George, J.P.: Subgrade soil index properties to estimate resilient modulus. In: Transportation Research Board, Annual Meeting, CD-ROM Publication, Transportation Research Board, Washington DC (2004)
Rankine, R.M., Sivakugan, N.: Prediction of paste backfill performance using artificial neural networks. In: Proc. the 16th ISSMGE (2), Osaka, pp. 1107–1110 (2005)
Ripley, B.D.: Pattern recognition and neural networks. Cambridge University Press, Cambridge (1996)
Rumelhart, D.E., McClelland, J.: Parallel distributed processing, vol. 1. MIT Press, Cambridge (1986)
Seed, H.B., Mitry, F.G., Monosmith, C.L., Chan, C.K.: Prediction of pavement deflection from laboratory repeated load tests. NCHRP report 35, Transportation Research Board, Washington DC (1967)
Shahin, M.A., Jaksa, M.B., Maier, H.R.: Artificial neural network applications in geotechnical engineering. Australian Geomechanics 36(1), 49–62 (2001)
Shahin, M.A., Maier, H.R., Jaksa, M.B.: Data division for developing neural networks applied to geotechnical engineering. Journal of Computing in Civil Engineering 18(2), 105–114 (2004)
Sharma, S., Das, A.: Backcalculation of pavement layer moduli from falling weight deflectometer data using an artificial neural network. Canadian Journal of Civil Engineering 35(1), 57–66 (2008)
Skapura, D.M.: Building neural networks. ACM Press, New York (1996)
Sokal, R.R., Rohlf, F.J.: Biometry: The principles and practice of statistics in biological research. W.H. Freeman and Co., New York (1995)

Speckt, D.F.: A generalized regression neural network. IEEE transactions on neural networks 2(6), 568–576 (1991)

StatSoft Inc.: Electronic statistics textbook. Tulsa, Oklahoma (2006)

Tarefder, R.A., White, L., Zaman, M.: Neural network model for asphalt concrete permeability17(1), 19–27 (2005)

Thompson, M.R., Robnett, Q.L.: Resilient properties of subgrade soils. Final report, FHWA-IL-UI-160, University of Illinois, Urbana, Illinois (1976)

TRB. Use of artificial neural networks in geomechanical and pavement systems. Transportation research circular number E-C012, December 1999, Transportation Research Board, National Research Council, Washington DC (1999)

Uzan, J.: Characterization of granular material. Transportation Research Record 1022, 52–59 (1985)

Yau, A., Von Quintus, H.L.: Study of LTPP laboratory resilient modulus test data and response characteristics. Final report October 2002, FHWA-RD-02-051, USDOT, FHWA (2002)

Yildirim, T., Ozyilmaz, L.: Dimensionality reduction in conic section function neural network. Sadhana 27(6), 675–683 (2002)

Yoder, E.J., Witczak, M.W.: Principles of Pavement Design, 2nd edn. John Wiley & Son, Inc., New York (1975)

Zaman, M.M., Chen, D.H., Laguros, J.G.: Journal of Transportation Engineering 120(6), 967–988 (1994)

Zurada, J.M.: Introduction to artificial neural systems. West St. Paul, Minnesota (1992)

APPENDIX: Equation of Factorial Model

$$
\begin{aligned}
M_R/P_a = {} & 13.2514795 - 438.31923{*}w - 13.311426{*}\gamma_d + 2.41669221{*}PI + 27.2918109{*}P_{200} \\
& - 35.722370{*}U_c - 11.240229{*}\sigma_d + 85.5626222{*}\theta + 278.441637{*}w{*}\gamma_d - \\
& 16.168513{*}w{*}PI - 1.0992907{*}\gamma_d{*}PI + 14.4631546{*}w{*}P_{200} - 9.7399663{*}\gamma_d{*}P_{200} - \\
& 1.7766282{*}PI{*}P_{200} + 332.908859{*}w{*}U_c + 23.0157253{*}\gamma_d{*}U_c - .01410921{*}PI{*}U_c + \\
& 38.7543639{*}P_{200}{*}U_c + 799.099567{*}w{*}\sigma_d + 31.3535709{*}\gamma_d{*}\sigma_d - 3.3563515{*}PI{*}\sigma_d - \\
& 69.037455{*}P_{200}{*}\sigma_d - 47.303113{*}U_c{*}\sigma_d - 279.93578{*}w{*}\theta - 38.643251{*}\gamma_d{*}\theta - \\
& 1.0518368{*}PI{*}\theta - 84.485170{*}P_{200}{*}\theta - 29.795776{*}U_c{*}\theta - 106.57364{*}\sigma_d{*}\theta + \\
& 9.15134484{*}w{*}\gamma_d{*}PI - 28.679440{*}w{*}\gamma_d{*}P_{200} + 21.1200089{*}w{*}PI{*}P_{200} + \\
& .662040611{*}\gamma_d{*}PI{*}P_{200} - 200.60637{*}w{*}\gamma_d{*}U_c + 2.75148494{*}w{*}PI{*}U_c - \\
& .02351306{*}\gamma_d{*}PI{*}U_c - 244.11931{*}w{*}P_{200}{*}U_c - 25.359274{*}\gamma_d{*}P_{200}{*}U_c - \\
& 1.2799791{*}PI{*}P_{200}{*}U_c - 588.87015{*}w{*}\gamma_d{*}\sigma_d + 15.1565883{*}w{*}PI{*}\sigma_d + \\
& .481801078{*}\gamma_d{*}PI{*}\sigma_d - 218.18881{*}w{*}P_{200}{*}\sigma_d + 10.1638379{*}\gamma_d{*}P_{200}{*}\sigma_d + \\
& 2.36460454{*}PI{*}P_{200}{*}\sigma_d - 45.492450{*}w{*}U_c{*}\sigma_d + 12.7119527{*}\gamma_d{*}U_c{*}\sigma_d + \\
& 3.45466718{*}PI{*}U_c{*}\sigma_d + 73.1347282{*}P_{200}{*}U_c{*}\sigma_d + 115.117756{*}w{*}\gamma_d{*}\theta + \\
& 7.15819491{*}w{*}PI{*}\theta - .16922596{*}\gamma_d{*}PI{*}\theta + 319.715817{*}w{*}P_{200}{*}\theta + \\
& 37.2140312{*}\gamma_d{*}P_{200}{*}\theta + .456277589{*}PI{*}P_{200}{*}\theta + 87.1036047{*}w{*}U_c{*}\theta + \\
& 12.8403258{*}\gamma_d{*}U_c{*}\theta + .470678587{*}PI{*}U_c{*}\theta + 17.7135930{*}P_{200}{*}U_c{*}\theta + \\
& 117.516534{*}w{*}\sigma_d{*}\theta + 45.1075781{*}\gamma_d{*}\sigma_d{*}\theta + 4.48108267{*}PI{*}\sigma_d{*}\theta + \\
& 123.422308{*}P_{200}{*}\sigma_d{*}\theta + 69.9906302{*}U_c{*}\sigma_d{*}\theta - 11.825725{*}w{*}\gamma_d{*}PI{*}P_{200} - \\
& 2.1155135{*}w{*}\gamma_d{*}PI{*}U_c + 149.381038{*}w{*}\gamma_d{*}P_{200}{*}U_c + .461733179{*}w{*}PI{*}P_{200}{*}U_c + \\
& .843999848{*}\gamma_d{*}PI{*}P_{200}{*}U_c - 3.1294192{*}w{*}\gamma_d{*}PI{*}\sigma_d + 277.350974{*}w{*}\gamma_d{*}P_{200}{*}\sigma_d - \\
& 14.172917{*}w{*}PI{*}P_{200}{*}\sigma_d + .525633666{*}\gamma_d{*}PI{*}P_{200}{*}\sigma_d + 98.7215822{*}w{*}\gamma_d{*}U_c{*}\sigma_d - \\
& 15.964519{*}w{*}PI{*}U_c{*}\sigma_d - 1.1448209{*}\gamma_d{*}PI{*}U_c{*}\sigma_d - 169.47515{*}w{*}P_{200}{*}U_c{*}\sigma_d - \\
& 22.778234{*}\gamma_d{*}P_{200}{*}U_c{*}\sigma_d - 2.0574402{*}PI{*}P_{200}{*}U_c{*}\sigma_d - 1.1780048{*}w{*}\gamma_d{*}PI{*}\theta - \\
& 136.19640{*}w{*}\gamma_d{*}P_{200}{*}\theta - 5.5904934{*}w{*}PI{*}P_{200}{*}\theta + .582825521{*}\gamma_d{*}PI{*}P_{200}{*}\theta -
\end{aligned}
$$

$$
\begin{aligned}
&32.417984*w*\gamma_d*U_c*\theta - 4.4154766*w*PI*U_c*\theta - .03434428*\gamma_d*PI*U_c*\theta - \\
&35.024894*w*P_{200}*U_c*\theta - 5.1994910*\gamma_d*P_{200}*U_c*\theta + .197336556*PI*P_{200}*U_c*\theta + \\
&12.9337904*w*\gamma_d*\sigma_d*\theta - 14.637167*w*PI*\sigma_d*\theta - 1.4228429*\gamma_d*PI*\sigma_d*\theta - \\
&175.26170*w*P_{200}*\sigma_d*\theta - 51.332213*\gamma_d*P_{200}*\sigma_d*\theta - 4.4923407*PI*P_{200}*\sigma_d*\theta - \\
&150.22381*w*U_c*\sigma_d*\theta - 29.716170*\gamma_d*U_c*\sigma_d*\theta - 2.2982062*PI*U_c*\sigma_d*\theta - \\
&70.820468*P_{200}*U_c*\sigma_d*\theta + 0.00000000*w*\gamma_d*PI*P_{200}*U_c + \\
&0.00000000*w*\gamma_d*PI*P_{200}*\sigma_d + 5.73119283*w*\gamma_d*PI*U_c*\sigma_d + \\
&0.00000000*w*\gamma_d*P_{200}*U_c*\sigma_d + 9.80008993*w*PI*P_{200}*U_c*\sigma_d - \\
&.05009087*\gamma_d*PI*P_{200}*U_c*\sigma_d + 0.00000000*w*\gamma_d*PI*P_{200}*\theta + \\
&1.92867824*w*\gamma_d*PI*U_c*\theta + 0.00000000*w*\gamma_d*P_{200}*U_c*\theta + \\
&1.30268945*w*PI*P_{200}*U_c*\theta - .38745801*\gamma_d*PI*P_{200}*U_c*\theta + \\
&3.36590974*w*\gamma_d*PI*\sigma_d*\theta + 0.00000000*w*\gamma_d*P_{200}*\sigma_d*\theta + \\
&11.4488598*w*PI*P_{200}*\sigma_d*\theta + 1.17782173*\gamma_d*PI*P_{200}*\sigma_d*\theta + \\
&31.7684321*w*\gamma_d*U_c*\sigma_d*\theta + 5.85679607*w*PI*U_c*\sigma_d*\theta + \\
&.607338016*\gamma_d*PI*U_c*\sigma_d*BS + 128.506165*w*P_{200}*U_c*\sigma_d*\theta + \\
&26.8574361*\gamma_d*P_{200}*U_c*\sigma_d*\theta + 1.63623868*PI*P_{200}*U_c*\sigma_d*\theta + \\
&0.00000000*w*\gamma_d*PI*P_{200}*U_c*\sigma_d + 0.00000000*w*\gamma_d*PI*P_{200}*U_c*\theta + \\
&0.00000000*w*\gamma_d*PI*P_{200}*\sigma_d*\theta + 0.00000000*w*\gamma_d*PI*U_c*\sigma_d*\theta + \\
&0.00000000*w*\gamma_d*P_{200}*U_c*\sigma_d*\theta + 0.00000000*w*PI*P_{200}*U_c*\sigma_d*\theta + \\
&0.00000000*\gamma_d*PI*P_{200}*U_c*\sigma_d*\theta - 4.6975464*w*\gamma_d*PI*P_{200}*U_c*\sigma_d*\theta
\end{aligned}
$$

Application of Soft Computing Techniques to Expansive Soil Characterization

Pijush Samui[1], Sarat Kumar Das[2], and T.G. Sitharam[3]

[1] Postdoctoral fellow, Department of civil engineering,
Tampere University of Technology, Tampere, Finland.
pijush.phd@gmail.com
[2] Assistant Professor, Department of Civil Engineering,
National Institute of Technology Rourkela, India-769 008.
saratdas@rediffmail.com
[3] Professor, Department of civil engineering,
Indian Institute of Science, Banglore, India-560012
sitharam@civil.iisc.ernet.in

Abstract. Very often it is difficult to develop mechanistic models for pavement geotechnical engineering problems due to its complex nature and uncertainty in material parameters. The difficulty in mechanistic analysis has forced the engineers to follows certain empirical correlations. The artificial neural network (ANN) is being as an alternate statistical method, mapping in higher-order spaces, such models can go beyond the existing univariate relationships. The applications of ANNs in pavement geotechnical engineering problems is mostly limited to constitutive modeling, with few applications on prediction of soil layer properties using Falling Weight Deflectometer (FWD), prediction of swelling potential and compute the remaining life of flexible pavements. However, ANN is considered as a 'Black box' system being unable to explain interrelation between inputs and output. The ANNs also have inherent drawbacks such as slow convergence speed, less generalizing performance, arriving at local minimum and over-fitting problems. Recently support vector machine (SVM) is being used due to its, better generalization as prediction error and model complexity are simultaneously minimized. SVM is based on statistical learning theory unlike ANNs (biological learning theory). The application of SVM in pavement geotechnical engineering is very much limited and to best of the knowledge such methods have not been applied to pavement geotechnical engineering. However, engineering application of numerical methods is a science as well as an art. This juxtaposition is based on the fact that even though the developed algorithms are based on scientific logic and belong to the special branch of applied mathematics, their successful application to new problems is problem oriented and is an art. As no method can be the panacea to solve all problems to the last details, their application to new areas needs critical evaluation. With above in view, an attempt has been made to develop the art of applying the above artificial intelligence techniques (ANN and SVM) to different pavement engineering problems such as prediction of compaction characteristics, permeability, swelling potential, coefficient of subgrade reaction etc. The parameters associated with the model developments are discussed in terms of guide line for its future

1 Introduction

The pavement geotechnical engineering is a complex problem involving three phase system. The difficulty in mechanistic analysis and uncertainty in soil

K. Gopalakrishnan et al. (Eds.): Intel. & Soft Comp. in Infra. Sys. Eng., SCI 259, pp. 305–323.
springerlink.com

parameters has forced the engineers to follows certain empirical correlations. Potential of artificial neural network (ANN) has been realized as an alternate tool to handle such cases and have been successfully applied in various complex problems. The ANN is being as an alternate statistical method, for solving certain types of problems too complex, too poorly understood, or too resource-intensive to tackle using more-traditional computational methods. The ANN is capable of mapping in higher-order spaces, and such models can go beyond the existing univariate relationships. The applications of ANNs in pavement geotechnical engineering problems is mostly limited to constitutive modeling (Ghaboussi 1992), with a few applications on prediction of soil layer properties using Falling Weight Deflectometer (FWD) (Meier and Rix 1994), prediction of swelling potential (Najjar et al. 1996) and compute the remaining life of flexible pavements(Abdallah et al. 2000). However, ANN is considered as a 'Black box' system being unable to explain interrelation between inputs and output. The ANNs also have inherent drawbacks such as slow convergence speed, less generalizing performance, arriving at local minimum and over-fitting problems.

The biggest challenge in successful application of ANN is when to stop training. If training is insufficient then the network will not be fully trained, where as if training is excessive then it will memorize the training patter or learn noise. When the numbers of data points are scanty the training set is driven to a very small value, but when new data is presented to the network the error is too large which is known as overfitting. The network needs to be equally efficient for new data during testing or validation, which is called as generalization. There are different methods for generalization like early stopping or cross validation (Basheer 2001; Shahin et al. 2002, Das and Basudhar 2006). In case of early stopping criteria the error on the validation/testing set is monitored during the training process and the training is stopped when the error on the testing set begin to rise. In cross validation an independent test set is used to asses the performance of the model at various stages of learning. However, this method is not suitable if data points are scanty.

Recently machine learning algorithms like support vector machine (SVM) and relevance vector machine (RVM) are being used due to its, better generalization as prediction error and model complexity is simultaneously minimized. The SVM and RVM are based on statistical learning theory unlike ANNs (biological learning theory). The application of SVM and RVM in geotechnical engineering is very much limited and to best of the knowledge such methods have not been applied to pavement geotechnical engineering. Engineering application of numerical methods is a science as well as an art. This juxtaposition is based on the fact that even though the developed algorithms are based on scientific logic and belong to the special branch of applied mathematics, their successful application to new problems is problem oriented and is an art. As no method can be the panacea to solve all problems to the last details, their application to new areas needs critical evaluation. There are no fixed rules for developing an ANN model, even though a general framework can be followed based on previous successful applications in such problems. With above in view some of problems related to pavement geotechnical engineering are discussed as follows with introduction to the methodology used.

2 Methodology

2.1 Basic Principles of Artificial Neural Network

A typical structure of ANN consists of a number of processing elements or neurons that are usually arranged in layers; an input layer, an output layer and one or more hidden layers (Figure 1). The input from each processing element in the previous layer is multiplied by an adjustable connection weight (w_{ji}). At each neuron, the weighted input signals are summed and a threshold value (b_j) is added. The combined input (I_j) is then passed through a nonlinear transfer function {f()} to produce the output of processing element. Hence the output (y_k) from the output node can be written as Equation (1).

$$y_k = F(v_k) = F\left[\sum_{j=1}^{N_h} w_{kj}\, f\left[\sum_{l=1}^{N_l} w_{jl} x_l + b_{j0}\right] + b_{k0}\right] \tag{1}$$

The 'learning' or 'training' process in ANN in general, is a nonlinear optimization of an error function. The aim of the training is to minimize the error function to get the optimized weight vectors. This is equivalent to the parameter estimation phase in conventional statistical models. The most commonly used error function is the mean squared error (MSE) function. The error associated with weights and sigmoid function is a highly non-linear optimization with many local minima. Local and global optimization methods are carried out for finding out the weight vectors. As the characteristic of traditional nonlinear programming based optimization method are initial point dependent, the results obtained using back propagation algorithm are sensitive to initial conditions (weight vector) (Shahin et al. 2002). The use of global optimization algorithms like genetic algorithm and simulated annealing though being widely used in other field of engineering (Morshed and Kaluarachchi 1998), in geotechnical engineering use of GA for training ANN is limited (Goh 2002; Goh et al. 2005). In recent past another heuristic global optimization called differential evolution (DE), introduced by Storn and Price (1995) is being used successfully in aerodynamics shape optimization and mechanical design.

The steepest descent algorithm and Levenberg-Marquardt (LM) algorithm which are gradient search algorithms are mostly used in ANNs applied to geotechnical engineering problems (Das 2005). The magnitudes of the weights and biases (parameters) are responsible for the poor generalization of the ANN rather than the number of network parameters.

In the present study, the ANN models are trained with differential evolution and Bayesian regularization method and are defined as DENN and BRNN respectively. The results are compared with that obtained from commonly used Levenberg-Marquardt trained neural networks (LMNN) to discuss the prediction efficiency of the networks. The above neural network models have been developed using MATLAB tool boxes (Math Works 2001). A brief description about the BRNN and DENN is presented here for completeness.

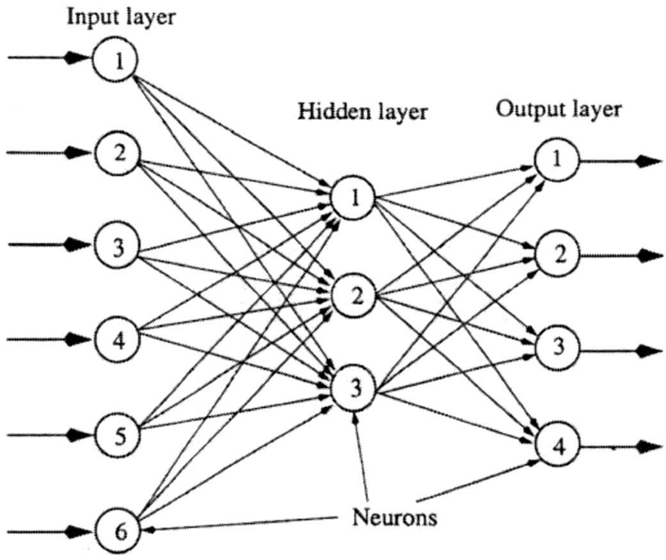

Fig. 1. Typical architecture of a Neural Network

2.1.1 Bayesian Regularization Neural Network (BRNN)

The most commonly used error function is the mean squared error (MSE) function. In LMNN, overfitting is due to unbounded values of weights (parameters) during minimization of the error function, mean square error (MSE). The other method called as regularization, in which the performance function is changed by adding a term that consist of mean square error of weights and biases as given below.

$$MSEREG = \gamma\, MSE + (1 - \gamma)\ MSW \tag{2}$$

Where MSE is the mean square error of the network, γ is the performance ratio and

$$MSW = \frac{1}{n} \sum_{j=1}^{n} w_j^2 \tag{3}$$

This performance function will cause the network to have smaller weights and biases there by forcing networks less likely to be overfit. The optimal regularization parameter λ is determined through Bayesian framework (Demuth and Beale 2000) as the low value of λ will not adequately fit the training data and high value of it may result in over fit. The number of network parameters (weights and biases) are being effectively used by the network can be found out by the above algorithm. The above combination works best when the inputs and targets area scaled in the range [-1, 1] (Demuth and Beale 2000). The above neural network models have been developed using MATLAB tool boxes (Math Works Inc. 2001).

2.1.2 Differential Evolution Neural Network (DENN)

The training of the feed-forward BPNN using DE optimization is known as differential evolution neural network (DENN) (Ilonen *et al.* 2003). The DE optimization is a population based heuristic global optimization method. Unlike other evolutionary optimization, in DE the vectors in current populations are randomly sampled and combined to create vectors for next generation. The real valued cross over factor and mutation factor governs the convergence of the search process. The detail of DENN is available in Ilomen et al. (2003). However, the DENN has not been applied in geotechnical engineering. In the present study, DENN has been implemented using the MATLAB (Math Works Inc. 2001) modeling environment.

In the present study single hidden layer is used and number of hidden layer neuron was obtained by trial and error. In the present study, the generalization was given priority and hence, the model with minimum error for the testing data was considered. The best ANN model was obtained with three neurons in the hidden layers.

2.2 Support Vector Machine

Support Vector Machine (SVM) has originated from the concept of statistical learning theory pioneered by Boser et al. (1992). This study uses the SVM as a regression technique by introducing a ε-insensitive loss function. In this section, a brief introduction on how to construct SVM for regression problem is presented. More details can be found in many publications (Boser et al. 1992; Cortes and Vapnik 1995; Gualtieri et al. 1999; Vapnik 1998). There are three distinct characteristics when SVM is used to estimate the regression function. First of all, SVM estimates the regression using a set of linear functions that are defined in a high dimensional space. Secondly, SVM carries out the regression estimation by risk minimization where the risk is measured using Vapnik's ε-insensitive loss function. Thirdly, SVM uses a risk function consisting of the empirical error and a regularization term which is derived from the structural risk minimization (SRM) principle. Considering a set of training data $\{(x_1, y_1),....,(x_l, y_l)\}, x \in R^n$, $y \in r$. Where x is the input, y is the output, R^N is the N-dimensional vector space and r is the one-dimensional vector space.

The ε-insensitive loss function can be described in the following way

$$L_\varepsilon(y) = 0 \text{ for } |f(x) - y| < \varepsilon \text{ otherwise } L_\varepsilon(y) = |f(x) - y| - \varepsilon \tag{4}$$

This defines an ε tube (Figure 2) so that if the predicted value is within the tube the loss is zero, while if the predicted point is outside the tube, the loss is equal to the absolute value of the deviation minus ε. The main aim in SVM is to find a function $f(x)$ that gives a deviation of ε from the actual output and at the same time is as flat as possible. Let us assume a linear function

$$f(x) = (w.x) + b \quad w \in R^n, \; b \in r \tag{5}$$

Where, w = an adjustable weight vector and b = the scalar threshold. Flatness in the case of (5) means that one seeks a small w. One way of obtaining this is by minimizing the Euclidean norm $\|w\|^2$. This is equivalent to the following convex optimization problem

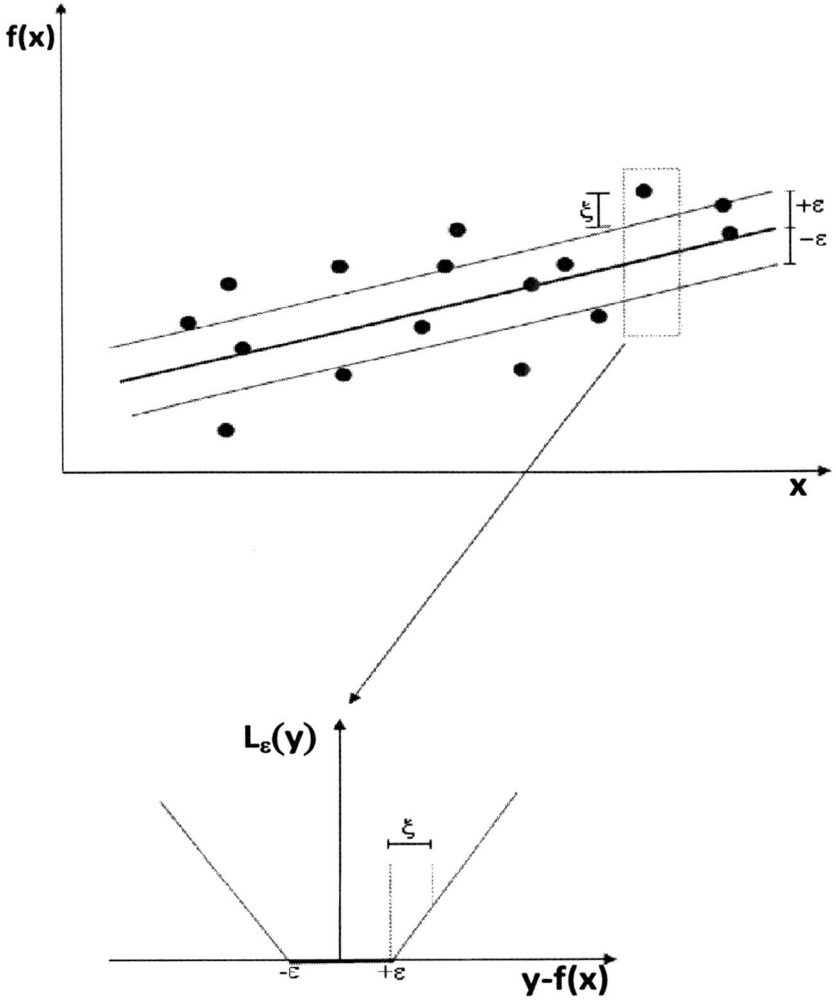

Fig. 2. Prespecified Accuracy ε and Slack Variable ξ in support vector regression [Scholkopf (1997)].

$$\text{Minimize: } \frac{1}{2}\|w\|^2$$

$$\text{Subjected to: } y_i - \left(\langle w.x_i \rangle + b\right) \leq \varepsilon,\ i = 1, 2, \ldots, l$$

$$\left(\langle w.x_i \rangle + b\right) - y_i \leq \varepsilon,\ i = 1, 2, \ldots, l \tag{6}$$

The above convex optimization problem is feasible. Sometimes, however, this may not be the case, or I also may want to allow for some errors. Analogously to the "soft margin" loss function (Bennett and Mangasarian 1992) which was used in SVM by Cortes and Vapnik (1995).As shown in the Figure 2, the parameters ξ_i, ξ_i^* are slack variables that determine the degree to which samples with error more than ε be penalized. In other words, any error smaller than ε does not require ξ_i, ξ_i^* and hence does not enter the objective function because these data points have a value of zero for the loss function. The slack variables (ξ_i, ξ_i^*) has been introduced to avoid infeasible constraints of the optimization problem (6).

$$\text{Minimize: } \frac{1}{2}\|w\|^2 + C\sum_{i=1}^{l}\left(\xi_i + \xi_i^*\right)$$

$$\text{Subjected to: } y_i - \left(\langle w.x_i \rangle + b\right) \leq \varepsilon + \xi_i,\ i = 1, 2, \ldots, l$$

$$\left(\langle w.x_i \rangle + b\right) - y_i \leq \varepsilon + \xi_i^*,\ i = 1, 2, \ldots, l$$

$$\xi_i \geq 0 \text{ and } \xi_i^* \geq 0,\ i = 1, 2, \ldots, l \tag{7}$$

The constant $0<C<\infty$ determines the trade-off between the flatness of f and the amount up to which deviations larger than ε are tolerated (Smola and Scholkopf 2004). This optimization problem (7) is solved by Lagrangian Multipliers (Vapnik 1998), and its solution is given by

$$f(x) = \sum_{i=1}^{nsv}\left(\alpha_i - \alpha_i^*\right)\left(x_i.x\right) + b \tag{8}$$

Where $b = -\left(\frac{1}{2}\right)w.\left[x_r + x_s\right]$, α_i, α_i^* are the Lagrangian Multipliers and nsv is the number of support vectors. An important aspect is that some Lagrange multipliers (α_i, α_i^*) will be zero, implying that these training objects are considered to

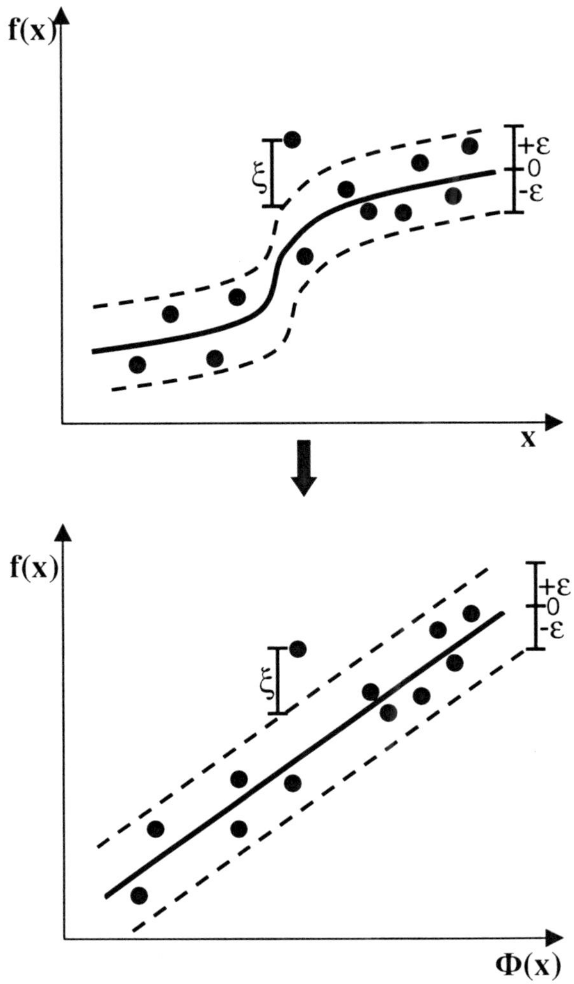

Fig. 3. Concept of nonlinear regression.

be irrelevant for the final solution (sparseness). The training objects with nonzero Lagrange multipliers are called support vectors.

When linear regression is not appropriate, then input data has to be mapped into a high dimensional feature space through some nonlinear mapping (Boser et al. 1992) (see Figure 3). The two steps that are involved are first to make a fixed nonlinear mapping of the data onto the feature space and then carry out a linear regression in the high dimensional space. The input data is mapped onto the feature space by a map Φ(see Figure 3). The dot product given by $\Phi(x_i).\Phi(x_j)$is computed as a linear combination of the training points. The concept of kernel

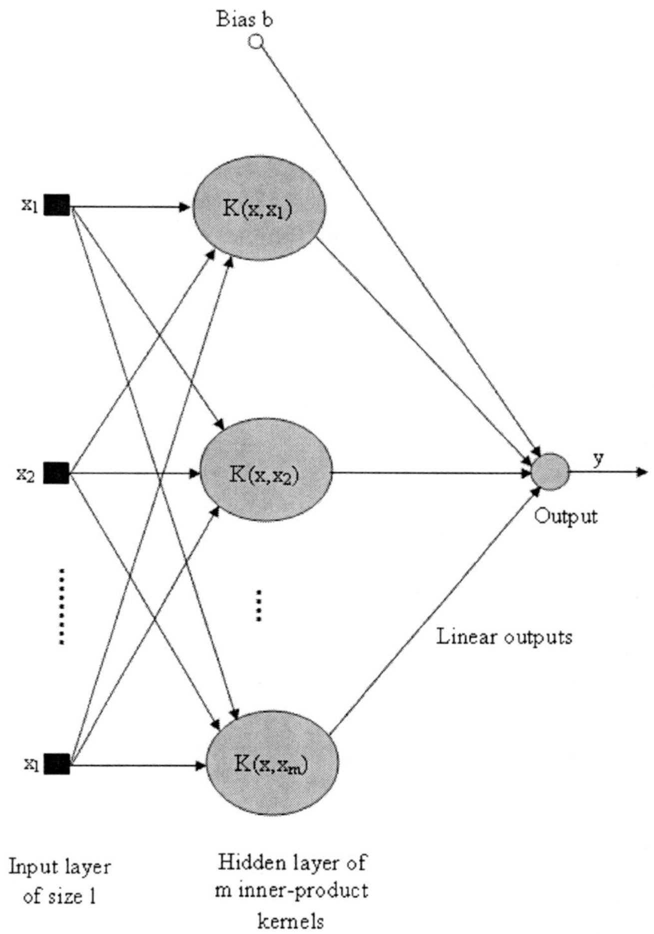

Fig. 4. Architecture of Support Vector Machine (Haykin, 1999).

function [$K(x_i, x_j) = \Phi(x_i).\Phi(x_j)$] has been introduced to reduce the computational demand (Cristianini and Shawe-Taylor 2000, Cortes and Vapnik 1995). So, equitation (5) becomes written as

$$f(x) = \sum_{i=1}^{nsv} \left(\alpha_i - \alpha_i^*\right) K\left(x_i . x_j\right) + b \tag{9}$$

Some common kernels have been used such as polynomial (homogeneous), polynomial (non homogeneous), radial basis function, Gaussian function, sigmoid etc for non-linear cases. Figure 4 shows a typical architecture of SVM.

The successful application of SVM models depends upon suitable parameters like type of kernel function and the parameters C and ε is obtained by trial and error. A large C assigns higher penalties to errors so that the regression is trained to minimize error with lower generalization while a small C assigns fewer penalties to errors; this allows the minimization of margin with errors, thus higher general zation ability. If C goes to infinitely large, SVM would not allow the occurrence of any error and result in a complex model, whereas when C goes to zero, the result would tolerate a large amount of errors and the model would be less complex. With regards to the selection of ε if ε is too large, too few support vectors are selected which leads to a decrease of the final prediction performance. If ε is too small, many support vectors are selected which leads to the risk of overfitting(Thissen et al. 2004). To train the SVM model, three types of kernel function have been used: They are

1. Polynomial
2. Radial basis function
3. Spline

3 Prediction of Swelling Pressure of Expansive Soil

Expansive soil and bedrock underlie more than one third of world's land surface. Each year, damage to buildings, roads, pipelines, and other structures by expansive soils is much higher than damage that are caused by floods, hurricanes, tornadoes, and earthquakes combined Jones and Holtz (1973). The estimated annual cost of damage due to expansive soils is \$1000 million in the USA, £150 million in the UK, and many billions of pounds worldwide Gourley et al. (1993). However, as the hazards due to expansive soils develop gradually and seldom present a threat to life, these have received limited attention, despite their severe effects on the economy. Much of the damage related to expansive soils is not due to a lack of appropriate engineering solutions but to the non recognition of expansive soils and expected magnitude of expansion early in land use and project planning. The damage to foundation on expansive soil can be avoided / minimized by proper identification, classification, quantification of swell pressure and provision of an appropriate design procedure. Swelling potential of clayey soil is a measure of the ability and degree to which such a soil might swell if its environments were changed in a definite way. Hence, the expansive soil is classified based on its potential for swelling. However, there is not a definite expression of swell potential for classification of expansive soils (Nelson and Miller 1992). Holtz (1959) referred to swell potential as the volume change of air-dried undisturbed sample, whereas, Seed et al. (1962) defined it as change in volume of a remoulded sample. Though factors like clay content, Atterberg's limits and mineral types are found to affect the swelling potential, the available literature presents contradicting results. McCormack and Wilding (1975) observed that for soil dominated by illite, clay content to be as reliable in predicting swelling potential, where as Yule and Ritchie (1980) and Gray and Allbrook (2002) reported, there being no relationship between clay percentage and soil swelling. The cation exchange capacity (CEC), saturation moisture and plastic index (PI) are also important indices for estimation of swelling potential Gill and Reaves (1957). Parker et al. (1977) concluded swell index and PI as superior to other indices.

The swelling pressure depends upon various soil parameters such as mineralogy, clay content, Atterberg's limits, dry density, moisture content, initial degree of saturation, etc along with structural and environmental factors. The parameters are interrelated in a complex manner, and it is difficult to model and analyze effectively taking all the above aspects into consideration. However, it can be measured easily with relevant data pertaining to soil, structure and environment. So various statistical/empirical methods have been attempted to predict the swelling pressure based on index properties of soil (Das 2002).

4 Results and Discussion

The data from various sources available in literature (Aciroyd et al. 1988; Savana et al. 1978; Abdujauwad 1994; Abdujauwad et al. 1994) are taken with input parameters, natural moisture content (w_n), dry density (γ_d), LL, PI, clay fraction (CF) and swelling pressure (SP) as output. The total number of data points considered is 230 out of which 167 are taken for training and 63 are taken for testing. The data is normalized between 0 to 1. The maximum, minimum, average and standard deviation for the data used are shown in Table 1 and it can be seen that it covers a wide range of values. The successful application of a method depends upon the identification of suitable input parameters. Table 2 shows the cross correlation between the inputs and output, it can be seen that CF, LL, PI are found to be important input parameters.

The results of different ANN models using the above parameters are shown in Table 3. The correlation coefficient (R) and root means square error (RMSE) are mostly for performance criteria evaluation of ANN models. However, R is a

Table 1. Parameters of the data considered for the present study

	w_n (%)	γ_d (kN/m^3)	LL	PI	Clay Fraction	Swelling pressure (kN/m^2)
Maximum	63.90	15.70	193.00	165.00	97.00	805.00
Minimum	2.70	1.04	26.00	12.00	19.00	3.00
Average	18.31	9.33	90.74	59.06	41.19	122.61
Std Dev.	9.58	5.89	47.77	43.42	14.04	140.90

Table 2. General performance of different neural network models.

ANN models	Training data		Testing data		Overfitting ratio
	R	E	R	E	
DENN	0.95	0.91	0.87	0.75	1.37
BRNN	0.98	0.96	0.90	0.79	2.12
LMNN	0.95	0.90	0.88	0.74	1.43
SVM	0.98	0.96	0.94	0.88	1.40

biased parameter and sometimes, higher values of R may not necessarily indicate better performance of the model because of the tendency of the model to be biased towards higher or lower values (Das and Basudhar 2006), the coefficient of efficiency (E) is also considered. The E is defined as

$$E = \frac{E_1 - E_2}{E_1} \tag{10}$$

Where

$$E_1 = \sum_{t=1}^{N} \left(SP_m - \overline{SP_m}\right)^2$$
$$E_2 = \sum_{t=1}^{N} \left(SP_p - SP_m\right)^2 \tag{11}$$

and SP_m, $\overline{SP_m}$ and SP_p are the measure, average and predicted swelling pressure respectively. The E value compares the modeled and measured values of the variable and evaluates how far the network is able to explain total variance in the data set. The overfitting ratio is defined as the ratio of RMSE for testing and training data and it defines the generalization. It can be seen that comparing the values of R and E values for training and testing data, BRNN is found to better than DENN and LMNN. However, DENN is having good generalization with small overfitting ratio, followed by LMNN and BRNN.

The RMSE value only defined the efficiency of a model as overall; however MAE can reveal the presence of regional areas of poor prediction. Figure 5 and 6 show the value of MAE, AAE and RMSE for different ANN models for training and testing data respectively.

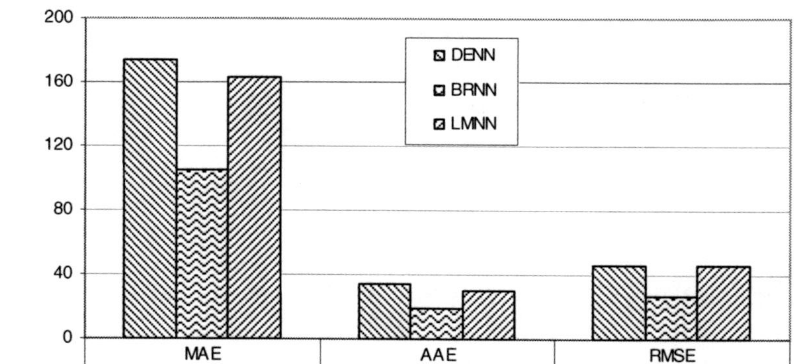

	MAE	AAE	RMSE
DENN	174.44	33.77	46.47
BRNN	104.85	19.31	27.36
LMNN	162.57	30.34	45.64

Fig. 5. Comparison of prediction capabilities of ANN models for training data.

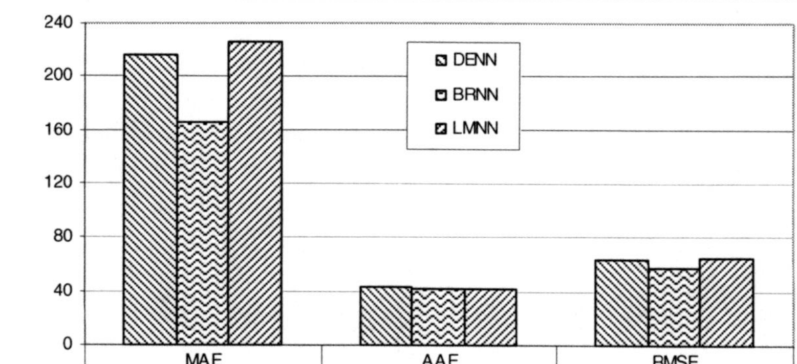

	MAE	AAE	RMSE
DENN	215.55	43.72	63.69
BRNN	166.12	42.53	58.06
LMNN	225.61	41.57	65.15

Fig. 6. Comparison of prediction capabilities of ANN models for testing data.

It can be seen that for training data BRNN is having lowest values of MAE, AAE and RMSE. However, for testing data AAE is comparable for all the methods, but based on MAE and RMSE values BRNN performs better than DENN and LMNN. Hence, based on different statistical performance criteria for the present study it can be concluded that BRNN is better followed by DENN and LMNN.

The ANN is considered as a 'Black box' system due to insufficient explanations to the weight vectors, but methods like Garson's algorithm and connection weight approach have been used utilizing the weight vector to identify the important input vectors (Das and Basudhar 2006). Such a study also made here to compare the above two methods in identifying the important parameters. Table 3 shows the ranking of important input parameters as calculated from Garson's algorithm and connection weight approach with the weights obtained from DENN, BRNN and LMNN. It can be seen from that the ranking of important input parameters as obtained by Garson's algorithm and Connection weight approach are different for BRNN and LMNN, where as for DENN the ranking of 1st and 2nd parameters are same by both the methods.

Table 5 presents the results of SVM models developed and based on R and E values SVM model with radial basis kernel function (SVM-R) found to be more efficient compared to models developed with other kernel functions (SVM-P and SVM-S). From Table 5, it is clear that SVM model employs 65 to 75 % (radial basis function=74.85%, Polynomial kernel=65.26% and spline kernel = 66.46%) of the training patterns as support vectors. So, SVM is remarkable in producing an excellent generalization level while maintaining the sparsest structure. Sparseness means that a significant number of the weights are zero (or effectively zero), which has the consequence of producing compact, computationally efficient models, which in addition are simple and therefore produce smooth functions. In SVM, support vectors represent prototypical examples. The prototypical examples exhibit the essential features of the information content of the data, and thus are able to transform the input data into the specified targets. Figure 7 and 8 show the value of MAE, AAE and RMSE for different SVM models for training and testing

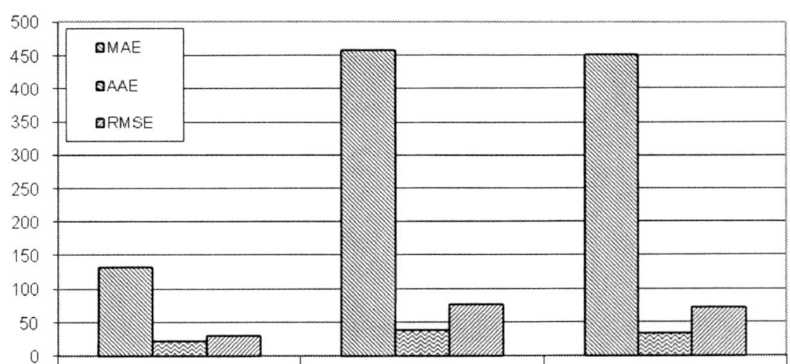

	SVM-R	SVM-P	SVM-S
MAE	131.84	456.62	450.85
AAE	21.60	37.41	33.35
RMSE	29.64	76.58	70.51

Fig. 7. Comparison of prediction capabilities of SVM models for training data.

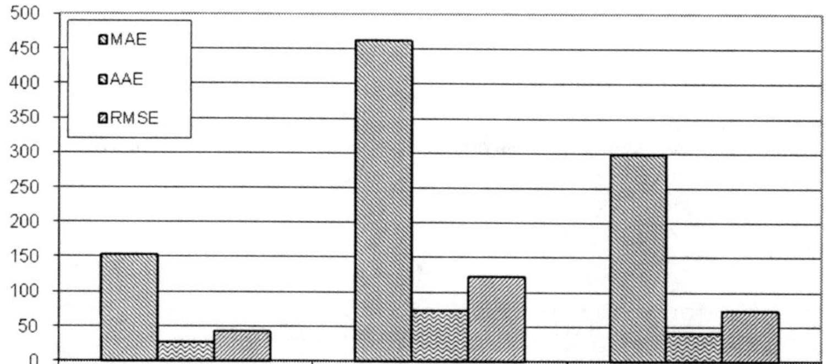

	SVM-R	SVM-P	SVM-S
MAE	152.47	461.06	298.75
AAE	26.89	73.43	40.85
RMSE	41.78	123.16	72.74

Fig. 8. Comparison of prediction capabilities of SVM models and for testing data

data respectively. It can be seen that for training data SVM-R is having lowest values of MAE, AAE and RMSE. In comparison to ANN models SVM-R model is found to better than all the ANN models. The use of the SRM principle in defining cost function provided more generalization capacity with the SVM compared to the ANN, which uses the empirical risk minimization principle. SVM uses only three parameters (radial basis function: σ, C and ε; polynomial kernel: degree of polynomial, C and ε; spline kernel: C and ε). In ANN, there are a larger number of controlling parameters, including the number of hidden layers, number of hidden nodes, learning rate, momentum term, and number of training epochs, transfer functions, and weight initialization methods. Obtaining an optimal combination of these parameters is a difficult task as well. Another major advantage of the SVM is its optimization algorithm, which includes solving a linearly constrained quadratic programming function leading to a unique, optimal, and global solution compared to the ANN. In SVM, the number of support vectors has determined by algorithm rather than by trial-and-error which has been used by ANN for determining the number of hidden nodes.

In this study, a sensitivity analysis has been carried out to extract the cause and effect relationship between the inputs and outputs of the SVM model. The basic idea is that each input of the model is offset slightly and the corresponding change in the output is reported. The procedure has been taken from the work of Liong et al. (2000). According to Liong et al. (2000), the sensitivity (S) of each input parameter has been calculated by the following formula

$$S = \frac{1}{N}\sum_{i=1}^{N}\left(\frac{\%\text{Change in output}}{\%\text{Change in input}}\right)\times 100 \tag{11}$$

Table 4. Relative importance of different input parameters

Input Parameters	Garson's algorithm			Connection weight approach			Ranking of inputs as per relative importance
	Ranking of inputs as per relative importance			Ranking of inputs as per relative importance			
	DENN	BRNN	LMNN	DENN	BRNN	LMNN	SVM-R
w_n	5	5	4	4	4	4	3
γ_d	3	3	3	5	3	3	2
LL	2	1	1	2	2	2	4
PI	1	2	2	1	1	1	1
CF	4	4	5	3	5	5	5

Table 5. General performance of SVM for different kernels

Kernel	C	ε	Training performance		Testing performance		Number of support vector
			Correlation coefficent (R)	Coefficient of determination (E)	Correlation coefficent (R)	Coefficient of determination (E)	
Radial basis function, width(σ) = 2.6	20	0.009	0.979	0.958	0.941	0.887	125
Polynomial, degree = 2	10	0.01	0.865	0.726	0.652	0.018	109
Spline	4	0.01	0.890	0.768	0.859	0.657	112

Where N is the number of data points. The analysis has been carried out on the trained model for radial basis function by varying each of input parameter, one at a time, at a constant rate of 20%. The result of the above analysis is also presented in Table 4. It is observed that similar to ANN analysis using connection weight approach PI is found to be more important parameters followed by γ_d and w_n.

5 Conclusions

The different ANN techniques and SVM model examined here have shown the ability to build accurate models with high predictive capabilities for prediction of swelling pressure of soil from the inputs; natural moisture content ($w_{n)}$, dry density (d), liquid limit (LL), plasticity index (PI) and clay fraction (CF). Based on different statistical performance criteria, the Bayesian regularization neural network (BRNN) model found to be more efficient compared to DENN and LMNN. However, the DENN model found to better in terms of generalization. The performance of the developed SVM model is better than the developed ANN models.

The ranking of important input parameters found to be consistent as per connection weight approach for the ANN model considered here. However, while using Garson's algorithm the ranking found to be different for different ANN models. Developed ANN and SVM models have the advantage that once the model is trained, it can be used as an accurate and quick tool for predicting swelling pressure without a need to perform any manual work such as using tables or charts. Comparison between the ANN and SVM model indicates that SVM model is superior to ANN model for predicting swelling pressure.

References

Abdallah, I., Ferregut, C., Nazarian, S., Melchor Lucero, O.: Prediction of remaining life of flexible pavements with artificial neural network models. In: Tayabji, S.D., Lukanen, E.O. (eds.) Nondestructive testing of Pavements and Backcalulcation of Moduli, ASTM, STP, 1375, vol. 3, pp. 484–498 (2000)

Abdujauwad, S.N.: Swelling behaviour of calcareous clays from the Eastern Provinces of Saudi Arabia. Quarterly Journal of Engg. Geology 27, 333–351 (1994)

Abdujauwad, S.N., Al-Sulaimani, B, Sl-Buraim: Response of structure to heave of expansive soil. Geotechniques (1994)

Aciroyd, L.W., Husain, R.: Residual and Lacustrine black cotton soil of North- East Nigeria. Geotechnique, 113–118 (1986)

Basheer, I.A.: Empirical modeling of the compaction curve of cohesive soil. Canadian Geotechnical Journal 38(1), 29–45 (2001)

Boser, B.E., Guyon, I.M., Vapnik, V.N.: A training algorithm for optimal margin classifier. In: The proceedings of the Fifth Annual ACM Workshop on Computational Learning theory, Pittusburgh, PA, USA, July 27–29 (1992)

Cortes, C., Vpanik, V.N.: Supportvector networks. Machine Learning 20(3), 273–297 (1995)

Cristianini, N., Shawe-Taylor, J.: An introduction to Support vector machine. Cambridge University Press, London (2000)

Dakhsnamurthy, Raman: A simple method of identifying expansive soil. Soils and Foundation 13(1), 97–104 (1973)

Das, B.M.: Principles of Geotechnical Engineering. Brookes- Cole, New York (2002)

Das, S.K.: Application of genetic algorithm and artificial neural network to some geotechnical engineering problem, Ph.D Thesis submitted to Indian Institute of Technology, Kanpur, India (2005)

Das, S.K., Basudhar, P.K.: Undrained lateral load capacity of piles in clay using artificial neural network. Computer and Geotechnics 33(8), 454–459 (2006)

Demuth, H., Beale, M.: Neural Network Toolbox. The Math Works Inc., USA (2000)

Ghaboussi, J.: Potential application of neuro-biological computational models in geotechnical engineering. In: Pande, G.N., Pietruszezak, S. (eds.) Numerical models in geomechanices, Balkemma, Rotterdam, The Netherlands, pp. 543–555 (1992)

Gill, W.R., Reaves, C.A.: Relationships of Atterberg limits and cation-exchange capacity to some physical properties of soil. Soil Sci. Soc. Am. Proc. 21, 491–494 (1957)

Goh, A.T.C.: Probabilistic neural network for evaluating seismic liquefaction potential. Canadian Geotechnical Journal 39, 219–232 (2002)

Goh, A.T.C., Kulhawy, F.H., Chua, C.G.: Bayesian neural network analysis of undrained side resistance of drilled shafts. J. of Geotech. and Geoenv. Engineering, ASCE 131(1), 84–93 (2005)

Gourley, C.S., Newill, D., Shreiner, H.D.: Expansive soils: TRL's research strategy. In: Proc. 1st Int. Symp. on Engineering Characteristics of Arid Soils (1993)

Gray, C.W., Allbrook, R.: Relationships between shrinkage indices and soil properties in some New Zealand soils. Geoderma 108(3-4), 287–299 (2002)

Gualtieri, J.A., Chettri, S.R., Cromp, R.F., Johnson, L.F.: Support vector machine classifiers as applied to AVIRIS data. In: The Summaries of the Eighth JPL Airbrone Earth Science Workshop (1999)

Holtz, W.G.: Expansive clays—properties and problems. Q. Colo. Sch. Mines 54(4), 89–117 (1959)

Ilonen, J., Kamarainen, J.K., Lampinen, J.: Differential Evolution training algorithm for feed-forward neural network. Neural Processing Letters 17, 93–105 (2003)

Jones, D.E., Holtz, W.G.: Expansive soils – The hidden disaster. Civil Engineering ASCE 43(8) (1973)

Liong, S.Y., Lim, W.H., Paudyal, G.N.: River stage forecasting in Bangladesh: neural network approach. Journal of Computing in Civil Engineering 14(1), 1–8 (2000)

MathWork, Inc.: Matlab user's manual. Version 6.5. Natick, The MathWorks, Inc., MA (2001)

McCormack, D.E., Wilding, L.P.: Soil properties influencing swelling in Canfield and Geeburg soils. Soil Sci. Soc. Am. Proc. 39, 496–502 (1975)

Meier, R.W., Rix, G.J.: Backcalculation of flexible pavement moduli using artificial neural networks. In: Transportation research record, 1448, TRB, National Research Council, Washington, DC, pp. 75–82 (1994)

Morshed, J., Kaluarachchi, J.J.: Parameter estimation using artificial neural network and genetic algorithm for free-product migration and recovery. Water resource research AGU 34(5), 1101–1113 (1998)

Najjar, M.Y., Basheer, I.A., McReynold, R.: Neural mod ing of Kansas soil swelling. Transportation Research Record (1526), 14–19 (1996)

Nelsonm, J.D., Miller, D.J.: Expansive Soils: Problem and Practice in Foundation and Pavement Engineering. Wiley, New York (1992)

Okasha, T.M., Abdujauwad, S.N.: Expansive soil in Al-Madinha. Saudi Arabia. Quarterly Journal of Engg. Geology 27, 333–351 (1994)

Park, D., Rilett, L.R.: Forecasting freeway link ravel times with a multi-layer feed forward neural network. Computer Aided Civil and infastructure Engineering 14, 358–367 (1999)

Parker, J.C., Amos, D.F., Kaster, D.L.: An evaluation of several methods of estimating soil volume change. Soil Soc. Am. J. 41, 1059–1064 (1977)

Haykin, S.: Neural Networks. A Comprehensive Foundation. Prentice Hall, Englewood Cliffs (1999)

Samui, P.: Slope stability analysis: a support vector machine approach. Environ. Geol. 56, 255–267 (2008)

Savana, G.S., Rajan, C.R., Srinivasan, B.S., Subrahmanyam, N., Sampathkumar, T.S.: Swelling Characteristics of Black cotton soils of Karnatak. In: Proc. CBIP, 47th Research Session, vol. 3, pp. 109–111 (1978)

Scholkopf, B.: Support vector learning. R. Oldenbourg, Munich (1997)

Seed, H.B., Woodward, J.R., Lundgren, R.J.: Prediction of swelling potential for compacted clays. J. Soil Mech. Found. Div., Am. Soc. Civ. Eng. 88 (SM3), 53–87 (1963)

Shahin, M.A., Maier, H.R., Jaksa, M.B.: Predicting settlement of shallow foundations using neural network. Journal of Geotechnical and Geoenvironmental Engineering ASCE 128(9), 785–793 (2002)

Smola, A.J., Scholkopf, B.A.: A tutorial on support vector regression. Statistics and Computing 14, 199–222 (2004)

Storn, R., Price, K.: Differential Evolution- A simple and efficient adaptive scheme for global optmization over continuous spaces, Technical Report TR-95-012, International Computer Science Institute, Berkeley, CA, USA (1995)

Thissen, U., Pepers, M., Ustuna, B., Melssena, W.J., Buydensa, L.M.C.: Comparing support vector machines to PLS for spectral regression applications. Chemom. Intell. Lab. Syst. 73(3), 169–179 (2004)

Vapnik, V.: Statistical learning theory. Wiley, New York (1998)

Vapnik, V.: The nature of statistical learning theory. Springer, New York (1995)

Yule, D.F., Ritchie, J.T.: Soil shrinkage relationships of Texas vertisols: 1 small cores. Soil Sci. Soc. Am. J. 44, 1285–1291 (1980)

Author Index

Breinigsville, PA USA
19 November 2009
227774BV00005BA/18/P